Ruhul Amin Sarker and Tapabrata Ray (Eds.)

Agent-Based Evolutionary Search

Adaptation, Learning, and Optimization, Volume 5

Series Editor-in-Chief

Meng-Hiot Lim
Nanyang Technological University, Singapore
E-mail: emhlim@ntu.edu.sg

Yew-Soon Ong
Nanyang Technological University, Singapore
E-mail: asysong@ntu.edu.sg

Further volumes of this series can be found on our homepage: springer.com

Vol. 1. Jingqiao Zhang and Arthur C. Sanderson
Adaptive Differential Evolution, 2009
ISBN 978-3-642-01526-7

Vol. 2. Yoel Tenne and Chi-Keong Goh (Eds.)
Computational Intelligence in
Expensive Optimization Problems, 2010
ISBN 978-3-642-10700-9

Vol. 3. Ying-ping Chen (Ed.)
Exploitation of Linkage Learning in Evolutionary Algorithms, 2010
ISBN 978-3-642-12833-2

Vol. 4. Anyong Qing and Ching Kwang Lee
Differential Evolution in Electromagnetics, 2010
ISBN 978-3-642-12868-4

Vol. 5. Ruhul Amin Sarker and Tapabrata Ray (Eds.)
Agent-Based Evolutionary Search, 2010
ISBN 978-3-642-13424-1

Ruhul Amin Sarker and Tapabrata Ray (Eds.)

Agent-Based Evolutionary Search

Springer

Ruhul Amin Sarker
School of Engineering and Information Technology
University of New South Wales
ADFA Campus, Northcott Drive
Canberra, ACT 2600, Australia
Email: r.sarker@adfa.edu.au

Tapabrata Ray
School of Engineering and Information Technology
University of New South Wales
ADFA Campus, Northcott Drive
Canberra, ACT 2600, Australia
Email: t.ray@adfa.edu.au

ISBN 978-3-642-13424-1 e-ISBN 978-3-642-13425-8

DOI 10.1007/978-3-642-13425-8

Adaptation, Learning, and Optimization ISSN 1867-4534

Library of Congress Control Number: 2010929473

Typeset & *Cover Design:* Scientific Publishing Services Pvt. Ltd., Chennai, India.

Printed on acid-free paper

9 8 7 6 5 4 3 2 1

springer.com

Preface

Agent based evolutionary search is an emerging paradigm in computational intelligence offering the potential to conceptualize and solve a variety of complex problems such as currency trading, production planning, disaster response management, business process management etc. There has been a significant growth in the number of publications related to the development and applications of agent based systems in recent years which has prompted special issues of journals and dedicated sessions in premier conferences.

The notion of an *agent* with its ability to sense, learn and act autonomously allows the development of a plethora of efficient algorithms to deal with complex problems. This notion of an agent differs significantly from a restrictive definition of a solution in an evolutionary algorithm and opens up the possibility to model and capture emergent behavior of complex systems through a natural agent-oriented decomposition of the problem space. While this flexibility of representation offered by agent based systems is widely acknowledged, they need to be designed for specific purposes capturing the right level of details and description. This edited volume is aimed to provide the readers with a brief background of agent based evolutionary search, recent developments and studies dealing with various levels of information abstraction and applications of agent based evolutionary systems. There are 12 peer reviewed chapters in this book authored by distinguished researchers who have shared their experience and findings spanning across a wide range of applications.

We hope that this text will serve as a reference for the researchers working in the area of agent based evolutionary search and for practitioners who would be interested to apply concepts outlined here for their own set of problems.

Finally, we would like to thank the contributors, reviewers and the editorial board for their time, effort and support for this timely edition.

January 2010

Ruhul Sarker
Tapabrata Ray

Contents

Agent Based Evolutionary Approach: An Introduction 1
Ruhul A. Sarker, Tapabrata Ray

Multi-Agent Evolutionary Model for Global Numerical Optimization 13
Jing Liu, Weicai Zhong, Licheng Jiao

An Agent Based Evolutionary Approach for Nonlinear Optimization with Equality Constraints 49
Abu S.S.M. Barkat Ullah, Ruhul Sarker, Chris Lokan

Multiagent-Based Approach for Risk Analysis in Mission Capability Planning 77
Lam T. Bui, Axel Bender, Michael Barlow, Hussein A. Abbass

Agent Based Evolutionary Dynamic Optimization 97
Yang Yan, Shengxiang Yang, Dazhi Wang, Dingwei Wang

Divide and Conquer in Coevolution: A Difficult Balancing Act 117
Hemant Kumar Singh, Tapabrata Ray

Complex Emergent Behaviour from Evolutionary Spatial Animat Agents 139
K.A. Hawick, C.J. Scogings

An Agent-Based Parallel Ant Algorithm with an Adaptive Migration Controller 161
Ying Lin, Jun Zhang

An Attempt to Stochastic Modeling of Memetic Systems 179
Aleksander Byrski, Robert Schaefer

Searching for the Effective Bidding Strategy Using Parameter Tuning in Genetic Algorithm 203
Kim Soon Gan, Patricia Anthony, Jason Teo, Kim On Chin

PSO **(Particle Swarm Optimization): One Method, Many Possible Applications** 229
Federico Cecconi, Marco Campenní

***VISPLORE*: Exploring Particle Swarms by Visual Inspection** 255
Namrata Khemka, Christian Jacob

Index 285

List of Contributors

Hussein Abbass
School of SEIT, University College,
University of New South Wales,
Australian Defence Force Academy
(UNSW@ADFA) Canberra ACT,
Australia
h.abbass@adfa.edu.au

Patricia Anthony
Universiti Malaysia Sabah Locked
Bag No. 2073, 88999, Kota Kinabalu,
Sabah, Malaysia
panthony@ums.edu.my

Michael Barlow
School of SEIT, University College,
University of New South Wales,
Canberra ACT, Australia

Axel Bender
Defence Science and Technology
Organization (DSTO), Edinburgh,
South Australia, Australia

Lam T. Bui
School of SEIT, University College,
University of New South Wales,
Canberra ACT, Australia
lam.bui07@gmail.com

Aleksander Byrski
Department of Computer Science,
AGH University of Science and
Technology, Al. Mickiewicza
30, 30-059 Kraków, Poland
olekb@agh.edu.pl

Marco Campenní
LABSS-ISTC, Via Palestro 9,
Rome, Italy
marco.campenni@istc.cnr.it

Federico Cecconi
LABSS-ISTC, Via Palestro 9 Rome,
Italy
federico.cecconi@
istc.cnr.it

K.A. Hawick
Massey University, Albany,
Auckland, New Zealand
K.A.Hawick@massey.ac.nz

Christian Jacob
University of Calgary, Dept. of
Computer Science, Dept. of
Biochemistry and Molecular Biology,
2500 University Drive NW, Calgary,
Alberta T2N 1N4, Canada
cjacob@ucalgary.ca

Licheng Jiao
Institute of Intelligent Information
Processing, Xidian University,
P.O. Box 224, Xi'an 710071, P.R.China
lchjiao@mail.xidian.edu.cn

Namrata Khemka
University of Calgary, Dept. of
Computer Science, 2500 University
Drive NW, Calgary, Alberta
T2N 1N4, Canada.
namrata.khemka@gmail.com

Ying Lin
Department of Computer Science, Sun Yat-Sen University, Guangzhou, 510275, China

Jing Liu
Institute of Intelligent Information Processing, Xidian University, P.O. Box 224, Xi'an 710071, P.R. China
neouma@163.com

Chris Lokan
School of Engineering and Information Technology, UNSW@ADFA, Canberra ACT 2600, Australia

Chin Kim On
Universiti Malaysia Sabah Locked Bag No. 2073, 88999, Kota Kinabalu, Sabah, Malaysia
kimonchin@gmail.com

Tapabrata Ray
School of Engineering and Information Technology, University of New South Wales at Australian Defence Force Academy (UNSW@ADFA), Canberra ACT, Australia
t.ray@adfa.edu.au

Ruhul Sarker
School of Engineering and Information Technology, UNSW@ADFA, Canberra ACT 2600, Australia
r.sarker@adfa.edu.au

Robert Schaefer
Department of Computer Science, AGH University of Science and Technology, Al. Mickiewicza 30, 30-059 Kraków, Poland
schaefer@agh.edu.pl

C.J. Scogings
Massey University, Albany, Auckland, New Zealand
C.Scogings@massey.ac.nz

Hemant Kumar Singh
School of Engineering and Information Technology, University of New South Wales at Australian Defence Force Academy (UNSW@ADFA), Canberra ACT, Australia
h.singh@adfa.edu.au

Gan Kim Soon
Universiti Malaysia Sabah Locked Bag No. 2073, 88999, Kota Kinabalu, Sabah, Malaysia
g_k_s967@yahoo.com

Jason Teo
Universiti Malaysia Sabah Locked Bag No. 2073, 88999, Kota Kinabalu, Sabah, Malaysia
jtwteo@ums.edu.my

Abu S.S.M. Barkat Ullah
School of Engineering and Information Technology, UNSW@ADFA, Canberra ACT 2600, Australia
barkat@adfa.edu.au

Dazhi Wang
School of Information Science and Engineering, Northeastern University, Shenyang 110004, China
wongdz@gmail.com

Dingwei Wang
School of Information Science and Engineering. Northeastern University, Shenyang 110004, China
dwwang@mail.neu.edu.au

Yang Yan
School of Information Science and Engineering, Northeastern University, Shenyang 110004, China
yanyangmail@163.com

Shengxiang Yang
Department of Computer Science, University of Leicester, University Road, Leicester LE1 7RH, United Kingdom
s.yang@mcs.le.ac.uk

Jun Zhang
Department of Computer Science, Sun Yat-Sen University, Guangzhou, 510275, China
junzhang@ieee.org.

Weicai Zhong
Institute of Intelligent Information Processing, Xidian University, P.O. Box 224, Xi'an 710071, P.R. China
wakenov@gmail.com

Agent Based Evolutionary Approach: An Introduction

Ruhul A. Sarker and Tapabrata Ray

Abstract. Agent based evolutionary approach is a new paradigm to efficiently solve a range of complex problems. The approach can be considered as a hybrid scheme which combines an agent system with an evolutionary algorithm. In this chapter, we provide an introduction to an evolutionary algorithm and an agent based system which leads to the foundation of the agent based evolutionary algorithm. The strengths and weaknesses of these algorithms are analyzed. In addition, the contributions in this book are also discussed.

1 Introduction

Over the last two decades, Evolutionary Computation (EC), which is based on the principle of evolution (survival of the fittest), has shown tremendous success in solving complex real-world problems [1, 2]. Engineers and scientists with quite different backgrounds have worked together to tackle some of the most difficult problems using a promising set of stochastic search algorithms - Evolutionary Algorithms (EAs). Evolutionary computation is still one of the hot topics in the area of computer science and optimization. Within the last decade, solving optimization and optimization based decision problems using EAs has become an important application area. Due to its parallelism and some intelligent properties such as self-organization, adaptation and self-learning, EAs have been applied successfully to many complex problems where classical approaches are either unavailable or generally lead to unsatisfactory results. In recent years, the interest in EAs has been growing dramatically specifically for solving structured and semi-structured optimization problems [3]. The structured problems are one that can be fully defined with its decision variables, parameters, objective and constraints. The structured problem can usually be expressed mathematically. The un-structured problems cannot be defined explicitly. The semi-structured problems are in between the two.

Ruhul Sarker · Tapabrata Ray
School of Engineering and Information Technology (SEIT),
University of New South Wales, Australian Defence Force Academy (UNSW@ADFA),
Canberra ACT, Australia
e-mail: r.sarker@adfa.edu.au, t.ray@adfa.edu.au

R.A. Sarker and T. Ray (Eds.): Agent-Based Evolutionary Search, ALO 5, pp. 1–11.
springerlink.com

As an approach, an agent or multi-agent system is quite different from an evolutionary system or computation. The agents are usually recognized as intelligent agents. A multi-agent system (MAS) is composed of multiple interacting intelligent agents. Typically multi-agent systems refer to software agents. However, the agents in a multi-agent system could equally well be robots, humans, human teams or their combination. An interesting aspect of the multi-agent systems is that the systems can manifest self-organization and complex behaviors even though the individual strategies of all their agents are simple. The multi-agent systems can be used to solve problems which are either difficult or impossible for an individual agent or a monolithic system (also known as a single approach or an integrated approach) to solve. Examples of problems which are appropriate to multi-agent systems include online trading, disaster response, and modelling social structures. They can also be used to solve structured problems such as scheduling and transportation.

The great success of EC in solving extremely complex structured and semi-structured optimization problems from various disciplines, and the success of MAS in solving semi- and un-structured problems, has been reflected through the growing number of publications and a corresponding increase in specialized conferences and journals. Despite these conferences and journals, a large amount of application-specific work is widely scattered throughout the publications of many diverse disciplines, including Operations Research, Optimization, Computer Science, Business, and Engineering, and presented at their related conferences, thus reflecting the general applicability and success of EAs and MAS.

The framework of EAs is well-defined and well-accepted by the research community. However, there is no restriction to allow some variations within its steps. To enhance the performance of EAs, hybridization with another method or methods is very common in practice. For example, a memetic algorithm is simply an evolutionary algorithm integrated with a local search method. Note that, in this chapter, agents will be considered as software agents and MAS will be discussed in the context of optimization and optimization based decision problems. MASs do not have a well defined framework like EAs. Sycara [4] stated that there is a lack of a proven methodology enabling designers to clearly structure applications as multi-agent systems. This cannot be seen as a negative point for MAS because this gives the flexibility in designing MAS for a wide range of problems. They are suitable for solving problems where there are different entities, people or organizations with different goals (possibly conflicting) and proprietary information. They are also useful where the fitness and constraint functions cannot be expressed explicitly [5].

EAs deal with a population of individuals. These individuals have a common goal though all of them may not achieve their goals. However, it can be argued that, due to the process of selection, crossover and mutation, the individuals work cooperatively and collectively to achieve their goals. These individuals may be defined as agents under certain assumptions. As required by multi-agent systems, the agent properties, such as autonomy, communication, proactiveness, learning and reactivity, can potentially be used to enhance the performance of EAs. To satisfy these properties and achieve the goals, the agents engage in interactions involving a wide range of hard and soft methodologies. Most of these methodologies are not used within the current framework of EAs although there is no restriction of doing so without involving MAS formally.

In this chapter, we have briefly discuss evolutionary algorithms, multiagent systems, and their integration for solving decision making problems. The agent based evolutionary algorithms are briefly reviewed and analyzed. The contributions in this book are also briefly described.

2 Evolutionary Algorithms

Evolutionary computation is the study of computational systems which uses ideas from natural evolution and adaptation. Evolutionary algorithms (EAs) are known as stochastic search procedures. EAs have two prominent features which distinguish themselves from other search algorithms [3]. First, they are all population-based (consider a set of possible solution points in contrast to single solution in conventional optimization). Second, there is communications and information exchange among individuals in a population.

EAs are widely used for optimization problem solving. EAs can be regarded as a population-based version of generate-and-test search. They use search operators like crossover and mutation to *generate* new solutions, and use selection to *test* and identify better solutions. It can be noted here that there is no strict rule to use crossover and mutation to *generate* new solutions in evolutionary computation. In principle, one may use any search procedure to generate new solutions that will increase the probability of finding a global optimum. This is also true for selection schemes.

3 Agent and Multi-Agent System

An agent (or intelligent agent) is defined as an autonomous entity which observes and acts upon an environment and directs its activity towards achieving goals. Agents may also learn or use their knowledge to achieve the defined goals. In computer science, the term *intelligent agent* may be used to refer to a software agent that has some intelligence, regardless of whether it is a rational agent or not. For example, autonomous programs used for operator assistance or data mining are also referred as 'intelligent agents'.

Agent Characteristics

The agents in a multi-agent system have several important characteristics as follows.

- *Autonomy*: Agents are at least partially autonomous. Agents operate without the direct intervention of humans or others.
- *Reactivity:* Agents perceive their environment and respond in a timely fashion to the observed changes.

- *Pro-activeness*: Agents do not simply act in response to their environment, they are able to exhibit goal-oriented behavior by taking the initiative.
- *Social ability or communicative:* Agents interact with other agents using some kind of agent-communication language.
- *Learning or adaptive:* Agents change their behavior based on previous experience.
- *Local views*: No agent has a full global view of the system, or the system is too complex for an agent to make practical use of such knowledge.
- *Decentralization*: There is no designated controlling agent.

Depending on the tasks on hand and agent environment, an agent may or may not satisfy all the characteristics listed above. Further details on agent properties can be found in Wooldridge and Jannings [6] and Franklin and Graesser [7].

Agent Classification

Russell and Norvig [8] grouped agents into five classes based on their degree of perceived intelligence and capability as follows.

- *Simple reflex agents*: These agents act only on the basis of the current perception. The agent function is based on the *condition-action rule*: if condition then action. This agent function only succeeds when the environment is fully observable.
- *Model-based reflex agents*: These agents can handle partially observable environments. These agents store abstract structures which describes the un-observable environment. However, they choose an action in the same way as the reflex agent.
- *Goal-based agents*: These agents are model-based agents and store information of desirable situations. The agents then select one from multiple possibilities that reaches a goal state.
- *Utility-based agents*: These agents distinguish between goal states and non-goal states. It is possible to define a measure of how desirable a particular state is. This measure can be obtained through the use of a *utility function* which maps a state to a measure of the utility of the state.
- *Learning agents*: Learning allows the agents to initially operate in unknown environments and to become more competent than what its initial knowledge alone might have allowed.

Based on the applications and purposes, agents are also named as decision agents, input agents, processing agents, spatial agents, world agents, believable agents, physical agents, and temporal agents.

Agent Architecture

For problem solving, the agent architecture is the heart of an agent system. The architecture of an agent is the computational structure (/methodology) that, along with the dynamic knowledge represented within it, generates the agent's behavior in the environment. The architecture must contain structures that enable representing knowledge, representing and achieving goals, interacting with the environment, and coping with unexpected occurrences. Depending on the nature of the environment, other agents (either human or virtual) in the environment, and the kinds of task the agent should perform in the environment, other capabilities may also need to be supported in the agent's architecture; for example, coordination and collaboration, language use, learning, and humanlike behavior and affect [8].

An agent architecture is necessary to structure the programming task. The agent architectures can broadly be divided into three categories as follows.

- *Deliberative Architectures*: Also known as classical approach. A deliberative agent or agent architecture contains an explicitly represented, symbolic model of the world, and makes the decisions (/performs actions) via logical reasoning, based on pattern matching and symbolic manipulation [6].
- *Reactive Architectures*: A reactive agent determines the next executable behavior based on current perceptions. A reactive architecture does neither include any kind of central symbolic world model nor use any complex symbolic reasoning [9].
- *Hybrid Architectures*: A reactive agent may not yield the most efficient behavior in achieving the overall goal because it makes the decision quickly by performing either little or no formal search. In contrast, a deliberative agent takes a rather long time to make a decision because it collects and evaluates all possible behaviors in selecting the best behavior that achieves the goal. A combination of these two architectures can achieve the goal efficiently which is known as a hybrid system [10].

A good number of agent architectures are reported in the literature. The subsumption architecture and the Belief Desire Intentions (BDI) architecture are among the most popular ones. The sub-sumption architecture is a reactive architecture that does not maintain any state [11]. It supports the use of many behaviors and provides a way to specify how behaviors inhibit each other. These inhibition relationships are used to determine which behavior is executed at each time. The BDI architecture provides a more complex control mechanism [12]. A BDI agent has a set of desires and a set of plans associated with each desire. The agent also has a set of beliefs which include everything it knows about the state of the environment and the agents in it, as well as the agent's internal state. The BDI control mechanism then chooses which of the plans to invoke by first finding the current desires, then the set of plans that satisfy these desires, and then choosing one of them.

Agent Language

Agent language or programming is very important for implementation of an agent system. By an agent language, we mean a system that allows one to program hardware or software computer systems in terms of some of the concepts represented by agent systems. At the very least, we expect such a language to include some structure correspondence to an agent. However, we might also expect to see some other attributes, such as beliefs, goals, or other mentalistic notions, used to program agents.

4 Integration of MAS and EAs

In this section, agent-based evolutionary algorithms, evolutionary algorithm based agent systems, sequential use of MAS and EAs, and their variants are discussed and analyzed. There are three types of algorithms or frameworks that associate MAS and EAs which are as follows.

- Agent is responsible for actions and behaviors,
- Agent represents a candidate solution and
- MAS, EAs and their variants are used sequentially.

In the first type, the agent represents the type of actions and behaviors that guide the system in solving the problem. Here, the agents may use EAs for learning or improving certain functionality. Vacher *et al.* [13] proposed a multiagent system where an agent is composed of functions of action, knowledge and behaviours. In their system, a genetic algorithm is used to determine a set of functions for each agent that guides the system in solving the problem efficiently. Nunes and Oliveira [14] and Iantovics and Enachescu [15] used EAs as learning algorithms within MAS. Milano and Roli [16] proposed a multiagent metaheuristic architecture where an agent as a system is able to build a solution, to move over a landscape, to communicate with other agents, to be active (i.e. goal oriented) and possibly, to be adaptive. This way, a new algorithm (that combines several metaheuristics) can easily be designed by choosing the agents to be involved and by defining their interactions.

In the second type, an agent represents a candidate solution. In EAs, an individual can be defined as an agent under some assumptions. So the population of individuals can be considered as a population of agents [17]. As defined by Barkat Ullah *et al.* [18], an agent contains not only a candidate solution but also agent specific additional information such as agent learning techniques and learning rate. These agents live in a 2-D cellular / lattice-like environment [17, 18, 19], graph structure [20], chain-like or ring-like environment [21]. Each agent aims at increasing its own energy by cooperating or competing with its neighbours. Four evolutionary operations may be performed on an agent: competition, cooperation, mutation and self-learning. Competition uses a kind of tournament to identify whether the agent is a loser or winner. If it is a loser, it is replaced by a new agent generated by the system. Cooperation is basically an information exchange mechanism through the crossover operator. Self-learning could be based on local search techniques or rule based analysis. No global selection method is required in this approach. In some lattice-like environment, each agent is fixed on a lattice-point and can only interact with its four neighbours [17]. However, eight neighbours are also used in [18]. Hippolyte *et al.* [22] considered that agents may move and meet other agents in the entire environment.

In the third type, MAS and EAs can be applied either in sequence or iteratively in order to solve certain classes of problems. In solving a dynamic job-shop scheduling problem, Li and Du [23] proposed multiagent consultations for initial task allocations and then applied hybrid genetic algorithms for generating optimal rescheduling. For a multiple travelling salesman problem, Giardini and Kalmar-Nagy [24] implemented a genetic algorithm for finding near-optimal solutions for each vehicle (vehicles are defined as agents). These solutions are then used to create an initial multi-vehicle plan through negotiation and sharing between the agents. This plan is further optimized by using an evolutionary algorithm. Liu and Tang [25] proposed a multiagent system where a genetic algorithm is used as a continuous novelty generator (/generation of alternatives), not as an optimizer. Such a generator stimulates the imagination of agents and extends their thinking spaces for ultimate decision making. In the system proposed by Guo *et al.* [5], different agents have different functions. One of five kinds of agents uses an interactive genetic algorithm as part of the overall decision process.

5 Agent-Based Evolutionary Algorithms

If compared EAs with MAS, the individuals in EAs do not have the autonomy which is a prime condition in MAS. However, EAs inherently hold some MAS properties such as a population of individuals, self-organization, adaptation and self-learning. The individuals in EAs work collectively (not individually) to attain a common but global goal. In most cases, the environment is given and expressed by complex functions and conditions. There are competitions, in some form, to be selected for crossover through the raking and selection processes, and to be selected for the next generation (i.e., for survival). In the crossover operation, the selected individuals cooperate with each other and exchange information (/genetic materials). This exchange of information is considered as communication between the individuals. The individuals may further communicate among themselves for to check relative locations on the fitness landscape. The ranking process may also be considered as a way of indirectly communicating for certain information. Memetic algorithm uses additional local search algorithms which can be recognized as local learning i.e. learning about the locality.

As discussed in the last section, Type-1 system is basically a multiagent system, and Type-3 is based on both multiagent and evolutionary schemes. However, Type-2 is a system which incorporates some aspects of agents and multiagent systems within an evolutionary framework. As evident in the literature, the use of this system is growing in solving different practical problems such as constraint satisfaction [26], job-shop scheduling [27], electrical machine design [22], combinatorial optimization [17], and numerical optimization [18, 28].

As discussed earlier, in Type-2 system, the individuals in EAs are defined as agents with or without further features and assumptions. As an example of further assumptions, Barkat Ullah *et al.* [18] stated that "The individuals in the population of EAs are not agents rather, based on the individual's belief and learning experiences, each agent stands (or supports) for one of the individuals of the population". Such an assumption would help to incorporate some features within agent

/individual representation in EAs. In addition, the individuals can then apply some agent properties within EAs, such as communication, cooperation, competition, and learning for enhancing the performance of EAs. Davidsson *et al.* [29] indicated that we must capitalize the strength of two approaches in a new hybrid method as they complement each other.

The agent environment includes the agent's social network structure, and the size of the neighborhood for interactions. As discussed in the previous section, in Type-2 system, agent environment (such as lattice-like and graph) is defined in a way that each agent can be compared with its neighbours locally, and agents can cooperate and compete with their neighbours [17, 18, 19, 28]. However, the lattice-like environment is somewhat similar to 2-D cellular genetic algorithm structure.

Barkat Ullah *et al.* [30] stated that "In reality, beside reproduction, an individual learns and gains experiences in different ways during its life time which improves the individual's genetic materials". De Jong [31] stated that the agent behavior is basically a combination of "nature and nurture", which are both inherited and learned components. De Jong [31] also indicated that evolution operates at the population level while "lifetime learning" occurs at the individual level. The agents learn throughout their life span which improves their quality (fitness value). This learning process can be chosen by the individual agents independently. For example, in an optimization problem solving, the local search techniques could be learning processes for an agent. To give a flavor of individual's decision making ability and agent learning, Barkat Ullah [18] proposed multiple local search algorithms from which an agent will choose a learning method based on its learning experience. It is worth noting here that the combination of agents' local view and the global search ability of EAs are used to balance between exploitation and exploration.

In every move of an agent, it may apply quantitative and qualitative judgment, or reasoning, based on its own belief, social interaction, knowledge and intelligence. If introduced, a wider reasoning capability of agents, which is currently limited, will make a clear difference between the agent-based EAs and EAs alone. In addition, the agent's learning experiences can be systematically archived and retrieved when needed. The use of such knowledge will make the agent-based EAs more attractive for practical problem solving. A computational tool, based on agent-based EAs, for automated problem solving will be extremely useful for complex decision processes.

6 A Brief Description on the Content of This Book

This volume consisting of 12 chapters present the key ideas surrounding recent developments in the area of agent based evolutionary search and their applications for complex problem solving. The first chapter provides an introduction to agent based evolutionary search, their features and key enhancements introduced in such models. It clearly lays the foundation of agent based evolutionary search and highlights the key differences between an evolutionary algorithm, memetic algorithm and an agent based evolutionary search. The second chapter introduces a model for global optimization using multiagent systems, where the agents engage in co-operation and competition among them to solve large scale optimization problems involving decomposable functions. Multiagent systems are particularly attractive in solving problems that are amenable to *divide and conquer* strategies as illustrated in this chapter. The third chapter addresses an important aspect of modelling and solving equality constrained optimization problems using an agent based approach. Handling equality constraints through evolutionary models is known to be extremely difficult and inefficient and this chapter introduces the concepts of learning that can be used within a multiagent system to efficiently solve such classes of problems. The notion of lifespan learning and various models of agent based memetic algorithms are presented to solve a class of equality constrained optimization problems. The fourth chapter presents a multiagent based approach to the solution of a complex task of risk analysis and mission capability planning. The next chapter investigates the performance of an agent based system to solve dynamic optimization problems and in particular tracking a dynamic optimum. Dynamic optimization is an area where the performance of traditional evolutionary algorithms is far from satisfactory and the above study highlights some of the benefits that can be achieved through an agent based system. The sixth chapter provides a re-look into coevolution and the issues surrounding the *divide and conquer* strategies for a range of optimization problems. The next chapter on emergent behaviour of evolutionary spatial agents illustrates the richness and the flexibility offered by agent based systems to model and gain useful insights of complex behaviours such as evolution of spiral battlefront patterns, pack formation, tribal segregation etc. The next chapter on agent based parallel ant algorithm with an adaptive migration controller highlights the benefits of hybrid agent based systems in solving large scale optimization problems. Chapter nine presents two multiagent system models, namely the evolutionary multiagent system and immunological multiagent system where the concepts of life time learning of an agent is utilised for computationally expensive classification problems. Self adaptation and tuning of parameters of a genetic and an evolutionary algorithm has long been of interest to the researchers and is discussed in the next chapter in the quest to identify the most effective bidding strategy. The next chapter highlights the benefits of particle swarm optimization models and their similarities with agent based systems while the final chapter presents a useful framework for visualizing solutions of population based stochastic optimization models such as particle swarms to uncover underlying dynamics of the system.

References

1. Fischer, M., Leung, Y.: Geocomputational modelling techniques and applications. Springer, Berlin (2001)
2. Barnett, W., Chiarella, C., Keen, S., Marks, R., Schnabl, H.: Complexity and Evolution. Cambridge University Press, Cambridge (2000)
3. Sarker, R., Kamruzzaman, J., Newton, C.: Evolutionary optimization (EvOpt): A brief review and analysis. International Journal of Computational Intelligence and Applications 3(4), 311–330 (2003)

4. Sycara, K.: Multiagent systems. AI Magazine 19(2), 79–92 (1998)
5. Guo, Y.-N., Cheng, J., Gong, D.-W., Yang, D.-Q.: Knowledge-inducing interactive genetic algorithms based on multi-agent. In: Jiao, L., Wang, L., Gao, X.-b., Liu, J., Wu, F. (eds.) ICNC 2006. LNCS, vol. 4221, pp. 759–768. Springer, Heidelberg (2006)
6. Wooldridge, M., Jannings, N.: Intelligent agents: theory and practice. Knowledge Engineering Review 10(2), 115–162 (1995)
7. Franklin, S., Graesser, A.: Is it an agent, or just a program? A taxonomy for autonomous agents. In: Jennings, N.R., Wooldridge, M.J., Müller, J.P. (eds.) ECAI-WS 1996 and ATAL 1996. LNCS, vol. 1193, pp. 21–35. Springer, Heidelberg (1997)
8. Russell, S., Norvig, P.: Artificial intelligence: a modern approach. Prentice Hall, Upper Saddle River (2003)
9. Scheutz, M., Schermerhorn, P.: Steps towards a systematic investigation of possible evolutionary trajectories from reactive to deliberable control systems. In: Proc. of Artificial Life VIII, pp. 283–292. MIT Press, Cambridge (2002)
10. Wu, M., Cao, W.-H., Peng, J., She, J.-H., Chen, X.: Balanced reactive-deliberative architecture for multi-agent system for simulation league of robocup. International Journal of Control, Automation, and Systems 7(6), 945–955 (2009)
11. Brooks, R.A.: Intelligence without representation. Artificial Intelligence 47, 349–355 (1991)
12. Bratman, M., Isreal, D., Pollack, M.: Plans and resource-bound practical reasoning. Computational Intelligence 4, 349–355 (1988)
13. Vacher, J.-P., Galinho, T., Lesage, F., Cardon, A.: Genetic algorithms in a multi-agent system. In: IEEE International Joint Symposia on Intelligence and Systems, pp. 17–26 (1998)
14. Nunes, L., Oliveira, E.: Learning from multiple sources. In: Proceedings of the Third International Joint Conference on Autonomous Agents and Multiagent Systems, vol. 3, pp. 1106–1113 (2004)
15. Iantovics, B., Enăchescu, C.: Intelligent complex evolutionary agent-based systems. In: Proceedings of the 1st International Conference on Bio-Inspired Computational Methods used for Difficult Problems Solving, AIP, pp. 116–124 (2008)
16. Milano, M., Roli, A.: MAGMA: a multiagent architecture for metaheuristics. IEEE Transactions on Systems, Man, and Cybernetics, Part B 34(2), 925–941 (2004)
17. Liu, J., Zhong, W., Jiao, L.: A multiagent evolutionary algorithm for combinatorial optimization problems. IEEE Transactions on Systems, Man and Cybernetics, Part B: Cybernetics 40(1), 229–240 (2010)
18. Barkat Ullah, A.S.S.M., Sarker, R., Cornforth, D., Lokan, C.: AMA: A new approach for solving constrained real-valued optimization problems. Soft Computing 13(8-9), 741–762 (2009)
19. Zhang, J., Liang, C., Huang, Y., Wu, J., Yang, S.: An effective multiagent evolutionary algorithm integrating a novel roulette inversion operator for engineering optimization. Applied Mathematics and Computation 211, 392–416 (2009)
20. Drezewski, R., Siwik, L.: Agent-based co-operative co-evolutionary algorithm for multi-objective optimization. In: Rutkowski, L., Tadeusiewicz, R., Zadeh, L.A., Zurada, J.M. (eds.) ICAISC 2008. LNCS (LNAI), vol. 5097, pp. 388–397. Springer, Heidelberg (2008)
21. Liu, B., Duan, T., Li, Y.: One improved agent genetic algorithm - ring-like agent genetic algorithm for global numerical optimization. Asia-Pacific Journal of Operational Research 26(4), 479–502 (2009)

22. Hippolyte, J.-L., Bloch, C., Chatonnay, P., Espanet, C., Chamagne, D.: A self-adaptive multiagent evolutionary algorithm for electrical machine design. In: Proceedings of the 9th annual conference on Genetic and evolutionary computation, pp. 1250–1255 (2007)
23. Li, Q., Du, L.: Research on hybrid-genetic algorithm for mas based job-shop dynamic scheduling. In: 2009 Second International Conference on Intelligent Computation Technology and Automation, pp. 404–407. IEEE Press, Los Alamitos (2009)
24. Giardini, G., Kalmar-Nagy, T.: Genetic algorithm for multi-agent space exploration. In: 2007 AIAA InfoTech at Aerospace Conference, vol. 2, pp. 1146–1160 (2007)
25. Liu, H., Tang, M.: Evolutionary design in a multi-agent design environment. Applied Soft Computing 6(2), 207–220 (2005)
26. Zhong, W., Liu, J., Jiao, L.: An agent model for binary constraint satisfaction problems. In: Raidl, G.R., Gottlieb, J. (eds.) EvoCOP 2005. LNCS, vol. 3448, pp. 260–269. Springer, Heidelberg (2005)
27. Zhong, W., Liu, J., Jiao, L.: Job-shop scheduling based on multiagent evolutionary algorithm. In: Wang, L., Chen, K., S. Ong, Y. (eds.) ICNC 2005. LNCS, vol. 3612, pp. 925–933. Springer, Heidelberg (2005)
28. Zhong, W., Liu, J., Xue, M., Jiao, L.: A multiagent genetic algorithm for global numerical optimization. IEEE Transactions on Systems, Man and Cybernetics, Part B 34, 1128–1141 (2004)
29. Davidsson, P., Persson, J., Holmgren, J.: On the integration of agent-based and mathematical optimization techniques. Agent and Multiagent Systems: Technologies and Applications, 1–10 (2007)
30. Barkat Ullah, A.S.S.M., Sarker, R., Cornforth, D., Lokan, C.: An agent-based memetic algorithm (ama) for solving constrained optimization problems. In: IEEE Congress on Evolutionary Computation (CEC 2007), pp. 999–1006 (2007)
31. De Jong, K.A.: Evolving intelligent agents: A 50 year quest. IEEE Computational Intelligence Magazine 3, 12–17 (2008)

Multi-Agent Evolutionary Model for Global Numerical Optimization

Jing Liu, Weicai Zhong, and Licheng Jiao

Abstract. Agent-based computation has been studied for several years in the field of distributed artificial intelligence and has been widely used in other branches of computer science. In this chapter, first, multi-agent systems and genetic algorithms are integrated to form a new algorithm, namely Multi-Agent Genetic Algorithm (MAGA), for solving the global numerical optimization problem. An agent in MAGA represents a candidate solution to the optimization problem in hand. All agents live in a latticelike environment, with each agent fixed on a lattice-point. In order to increase energies, they compete or cooperate with their neighbors, and they can also use knowledge. Making use of these agent-agent interactions, MAGA realizes the purpose of minimizing the objective function value. Theoretical analyses show that MAGA converges to the global optimum. Second, Macro-Agent Evolutionary Model (MacroAEM) is proposed with the intrinsic properties of decomposable functions in mind. In this model, a sub-function forms a macro-agent, and 3 new behaviors, namely competition, cooperation, and selfishness, are developed for macro-agents to optimizing objective functions. Finally, the MacroAEM model is integrated with MAGA, which results a new algorithm, namely Hierarchical Multi-Agent Genetic Algorithm (HMAGA), especially for optimizing decomposable functions. The convergence of HMAGA is also analyzed theoretically and the results show that HMAGA also converges to the global

Jing Liu
Institute of Intelligent Information Processing, Xidian University, P.O. Box 224,
Xi'an 710071, P.R. China
Tel.: +86-29-88202661
e-mail: neouma@163.com

Weicai Zhong
Institute of Intelligent Information Processing, Xidian University, P.O. Box 224,
Xi'an 710071, P.R. China
Tel.: +86-29-88202661
e-mail: wakenov@gmail.com

Licheng Jiao
Institute of Intelligent Information Processing, Xidian University, P.O. Box 224,
Xi'an 710071, P.R. China
Tel.: +86-29-88201023
e-mail: lchjiao@mail.xidian.edu.cn

R.A. Sarker and T. Ray (Eds.): Agent-Based Evolutionary Search, ALO 5, pp. 13–48.
springerlink.com

optima. To validate the performance of MAGA, MacroAEM, and HMAGA, benchmark functions are used. The scalability of MAGA along the problem dimension is studied with great care. The results show that MAGA achieves a good performance when the dimensions are increased from 20 to 10,000. Moreover, even when the dimensions are increased to as high as 10,000, MAGA still can find high quality solutions at a low computational cost. Therefore, MAGA has a good scalability and is a competent algorithm for solving high dimensional optimization problems. The experimental results about HMAGA show that HMAGA achieves a good performance, too, especially for high-dimensional functions. Apart from this, the analyses on time complexity demonstrate that HMAGA has a good scalability.

Keywords: Evolutionary algorithms, genetic algorithms, multi-agent systems, numerical optimization, decomposable functions.

1 Introduction

Agent-based computation has been studied for several years in the field of distributed artificial intelligence [1]-[4] and has been widely used in other branches of computer science [5], [6]. Problem solving is an area that many multi-agent-based applications are concerned with. It includes the following subareas: distributed solutions to problems, solving distributed problems, and distributed techniques for problem solving [3], [4]. Reference [6] introduced an application of distributed techniques for solving constraint satisfaction problems. They solved the 7000-queen problem by an energy-based multi-agent model. Enlightened by them, we integrates multi-agent systems with GAs to form a new algorithm, Multi-Agent Genetic Algorithm (MAGA), for solving the global numerical optimization problem. In MAGA, all agents live in a latticelike environment. MAGA realizes the ability of agents to sense and act on the environment that they live in. During the process of interacting with the environment and other agents, each agent increases its energy as much as possible, so that MAGA can achieve the ultimate purpose of minimizing the objective function value.

Being similar to MAGA, cellular genetic algorithms [7]-[9] also use a lattice-based population. In cellular GAs, each individual is located in a cell of the lattice. All operations of cellular GAs and traditional GAs are identical except that there is a neighborhood structure in the former while there is no neighborhood structure in the latter. In essence, cellular GAs are greedy techniques and can present the same problem of premature convergence of traditional GAs [9]. That is, cellular GAs are only a kind of techniques for enabling a fine-grained parallel implementation of GAs. But in MAGA, each individual is considered as an agent, which has its own purpose and behaviors. The experimental results show that MAGA achieves a good performance even for the functions with 10,000 dimensions, which illustrate that MAGA overcomes the problem of premature convergence of traditional GAs in some degree.

From another viewpoint, multi-agent systems are computational systems in which several agents interact or work together in order to achieve goals. Agents in such systems may be homogeneous or heterogeneous, and may have common goals or distinct goals [10], [11]. We can consider a complex function optimization problem as a multi-agent system, if it can be divided into several subproblems. A subproblem forms an agent, purposes of agents form the environment where all agents live, and computation resource is considered as the environmental resource. Each agent has own purpose and behavior. On one hand, to achieve their purposes, agents will adopt selfish behaviors to obtain the resource as much as possible in the environment; on the other hand, since the purposes of the agents are in conflict with the global purpose, agents have to adopt cooperative behaviors under the pressure of the environment.

What are advantages from the viewpoint of a multi-agent system? As we known, the complexity of function optimization problem can be measured by its dimensionality and the intensity of nonlinear epistasis between its variables. If the dimensionality is higher and the nonlinear interactions between variables are more intensive, then the problem becomes more complex and the search algorithm is easier to be trapped in the local optima. Apparently, "the divide and conquer" strategy could be effective if the complex problem can be decomposed into a number of easier subproblems. Accordingly, transforming the function optimization problem into a multi-agent system right embodies the thought of "the divide and conquer". Therefore, for the decomposable functions, we integrate multi-agent systems with evolutionary algorithms and propose a Macro-Agent Evolutionary Model (MacroAEM), which provides a new way of solving large-scale complex function optimization problems.

Although MAGA achieves a good performance in solving functions with 10,000 dimensions, MAGA can only solve the problems with up to 1,000 dimensions for a kind of Rosenbrock-like function which has the strong nonlinear interaction between variables. Thus, we incorporate MAGA into MacroAEM model and propose a Hierarchical Multi-Agent Genetic Algorithm (HMAGA). In HMAGA, macro-agents are organized into a hierarchical structure, and MAGA is considered as one of macro-agents' behaviors. Literature [12] also proposed a scalable macro-evolutionary algorithm based on function decomposition. Their algorithm consisted of a local and global evolution process. In evolutionary iterations, the former first optimizes each sub-function independently, then the latter optimizes the original function which is synthesized by all sub-functions. The two processes are iterated in turn until the terminal condition is satisfied. Differently, MacroAEM uses the bottom-up strategy to synthesize the original function step by step. Because MacroAEM does not re-divide the high-level optimized sub-functions into low-level sub-functions, solutions would not degenerate repeatedly.

2 Multi-Agent Genetic Algorithm

According to [4], [6], an agent is a physical or virtual entity that essentially has the following properties: (a) it is able to live and act in the environment; (b) it is able to sense its local environment; (c) it is driven by certain purposes and (d) it has some reactive behaviors. Multi-agent systems are computational systems in which several agents interact or work together in order to achieve goals. As can be seen, the meaning of an agent is very comprehensive, and what an agent represents is different for different problems. In general, four elements should be defined when multi-agent systems are used to solve problems. The first is the meaning and the purpose of each agent. The second is the environment where all agents live. Since each agent has only local perceptivity, so the third is the definition of the local environment. The last is the behaviors that each agent can take to achieve its purpose. In what follows, the definitions of these elements for global numerical optimization problems are described.

Agent for Numerical Optimization

A global numerical optimization can be formulated as solving the following objective function

$$\text{minimize } f(\boldsymbol{x}),\ \ \boldsymbol{x}=(x_1,\cdots,x_n)\in\mathcal{S} \tag{1}$$

where $\mathcal{S}\subseteq\mathcal{R}^n$ defines the search space which is an n-dimensional space bounded by the parametric constraints $\underline{x}_i\le x_i\le\overline{x}_i$, i=1,…,n. Thus, $\mathcal{S}=[\underline{\boldsymbol{x}},\ \overline{\boldsymbol{x}}]$, where $\underline{\boldsymbol{x}}=(\underline{x}_1,\ \underline{x}_2,\ ...,\ \underline{x}_n)$ and $\overline{\boldsymbol{x}}=(\overline{x}_1,\ \overline{x}_2,\ ...,\ \overline{x}_n)$. Because many 'x' notations have been used throughout this chapter, they are explained explicitly to prevent confusion. The '$\boldsymbol{x}$' in boldface represents a real-valued vector in the search space, and the 'x_i' with a subscript represents a component in the vector '$\boldsymbol{x}$'. The boldfaced '$\underline{\boldsymbol{x}}$' with an underline represents the vector of the lower bound of the search space, and the '$\underline{x}_i$' with a subscript and an underline represents a component in the vector '$\underline{\boldsymbol{x}}$'. The boldfaced '$\overline{\boldsymbol{x}}$' with an overline represents the vector of the upper bound of the search space, and the '$\overline{x}_i$' with a subscript and an overline represents a component in the vector '$\overline{\boldsymbol{x}}$'. An agent for numerical optimization problems is defined as follows:

Definition 1: An agent, $\boldsymbol{a}$, represents a candidate solution to the optimization problem in hand. The value of its energy is equal to the negative value of the objective function,

$$\boldsymbol{a}\in\mathcal{S}\quad\text{and}\quad Energy(\boldsymbol{a})=-f(\boldsymbol{a}) \tag{2}$$

The purpose of $\boldsymbol{a}$ is to increase its energy as much as possible.

As can be seen, each agent carries all variables of the objective function to be optimized. In order to realize the local perceptivity of agents, the environment is organized as a latticelike structure, which is defined as follows:

Definition 2: All agents live in a latticelike environment, L, which is called an agent lattice. The size of L is $L_{size}\times L_{size}$, where L_{size} is an integer. Each agent is fixed on a lattice-point and it can only interact with its neighbors. Suppose that the agent located at (i, j) is represented as $L_{i,j}$, i, j=1,2,…,L_{size}, then the neighbors of $L_{i,j}$, $Neighbors_{i,j}$, are defined as follows:

$$Neighbors_{i,j}=\{L_{i',j},\ L_{i,j'},\ L_{i'',j},\ L_{i,j''}\} \tag{3}$$

where $i' = \begin{cases} i-1 & i \neq 1 \\ L_{size} & i = 1 \end{cases}$, $j' = \begin{cases} j-1 & j \neq 1 \\ L_{size} & j = 1 \end{cases}$, $i'' = \begin{cases} i+1 & i \neq L_{size} \\ 1 & i = L_{size} \end{cases}$, $j'' = \begin{cases} j+1 & j \neq L_{size} \\ 1 & j = L_{size} \end{cases}$.

Therefore, the agent lattice can be represented as the one in Fig.1. Each circle represents an agent, the data in a circle represents its position in the lattice, and two agents can interact with each other if and only if there is a line connecting them.

In traditional GAs, those individuals that will generate offspring are usually selected from all individuals according to their fitness. Therefore, the global fitness distribution of a population must be determined. But in nature, a global selection does not exist, and the global fitness distribution cannot be determined either. In fact, the real natural selection only occurs in a local environment, and each individual can only interact with those around it. That is, in some phase, the natural evolution is just a kind of local phenomenon. The information can be shared globally only after a process of diffusion.

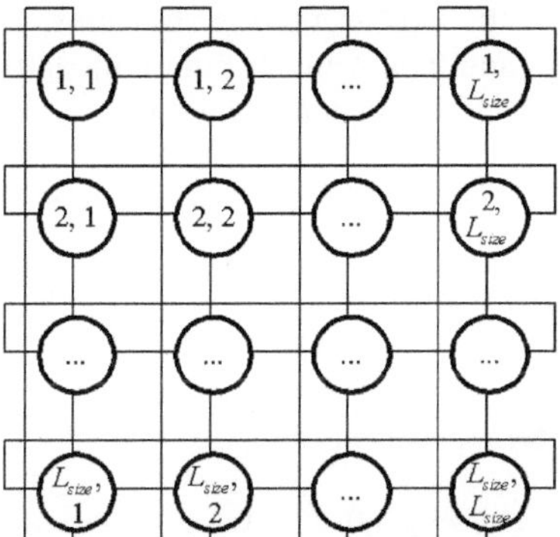

Fig. 1 The model of the agent lattice

In the aforementioned agent lattice, to achieve their purposes, agents will compete or cooperate with others so that they can gain more resources. Since each agent can only sense its local environment, its behaviors of competition and cooperation can only take place between the agent and its neighbors. There is no global selection at all, so the global fitness distribution is not required. An agent interacts with its neighbors so that information is transferred to them. In such a manner, the information is diffused to the whole agent lattice. As can be seen, the model of the agent lattice is more close to the real evolutionary mechanism in nature than the model of the population in traditional GAs.

Four Evolutionary Operators for Agents

To achieve its purposes, each agent has some behaviors. In addition to the aforementioned behaviors of competition and cooperation, each agent can also increase its energy by using its knowledge. On the basis of such behaviors, four evolutionary operators are designed for the agents. The neighborhood competition operator and the neighborhood orthogonal crossover operator realize the behaviors of competition and cooperation, respectively. The mutation operator and the self-learning operator realize the behaviors of making use of knowledge. Suppose that the four operators are performed on the agent located at (i, j), $L_{i,j}=(l_1,l_2,\ldots,l_n)$, and $Max_{i,j}=(m_1,m_2,\ldots,m_n)$ is the agent with maximum energy among the neighbors of $L_{i,j}$, namely, $Max_{i,j}\in Neighbors_{i,j}$ and $\forall \boldsymbol{a}\in Neighbors_{i,j}$, then $Energy(\boldsymbol{a})\leq Energy(Max_{i,j})$.

Neighborhood competition operator: If $L_{i,j}$ satisfies (4), it is a winner; otherwise it is a loser.

$$Energy(L_{i,j})>Energy(Max_{i,j}) \tag{4}$$

If $L_{i,j}$ is a winner, it can still live in the agent lattice. If $L_{i,j}$ is a loser, it must die, and its lattice-point will be occupied by $Max_{i,j}$. $Max_{i,j}$ has two strategies to occupy the lattice-point, and it selects them with probability P_o. If $U(0,1)<P_o$, occupying strategy 1 is selected; otherwise occupying strategy 2 is selected, where $U(\cdot,\cdot)$ is a uniform random number generator. In the two occupying strategies, $Max_{i,j}$ first generates a new agent, $New_{i,j}=(e_1,e_2,\ldots,e_n)$, and then $New_{i,j}$ is put on the lattice-point.

In occupying strategy 1, $New_{i,j}$ is determined by,

$$e_k=\begin{cases}\underline{x}_k & (m_k+U(-1,1)\times(m_k-l_k))<\underline{x}_k\\ \overline{x}_k & ((m_k+U(-1,1)\times(m_k-l_k))>\overline{x}_k\\ m_k+U(-1,1)\times(m_k-l_k) & \text{otherwise}\end{cases},\quad k=1,\ldots,n \tag{5}$$

In occupying strategy 2, $Max_{i,j}$ is first mapped on [0, 1] according to,

$$m_k'=(m_k-\underline{x}_k)/(\overline{x}_k-\underline{x}_k),\qquad k=1,\ldots,n \tag{6}$$

Then, $New_{i,j}'=(e_1',e_2',\cdots,e_n')$ is determined by,

$$\begin{aligned}New_{i,j}'=(m_1',\ m_2',\ \cdots,\ m_{i_1-1}',\ m_{i_2}',\ m_{i_2-1}',\ \cdots,\\ m_{i_1+1}',\ m_{i_1}',\ m_{i_2+1}',\ m_{i_2+2}',\ \cdots,\ m_n')\end{aligned} \tag{7}$$

where $1<i_1<n$, $1<i_2<n$, and $i_1<i_2$.

Finally, $New_{i,j}$ is obtained by mapping $New_{i,j}'$ back to $[\underline{x}_k,\ \overline{x}_k]$ according to,

$$e_k=\underline{x}_k+e_k'\cdot(\overline{x}_k-\underline{x}_k),\qquad k=1,\ldots,n \tag{8}$$

In this operator, two strategies play different roles. When $L_{i,j}$ is a loser, it perhaps still has some useful information, so occupying strategy 1, a kind of heuristic crossover, is in favor of reserving some information of a loser. Occupying strategy 2 is similar to the inversion operation in AEA [13]. It has the function of random searching, but is better than random searching in that it makes use of the information of a winner. Therefore, occupying strategy 1 puts emphasis on exploitation while occupying strategy 2 on exploration.

Neighborhood orthogonal crossover operator: The orthogonal crossover operator is a new operator proposed by [14]. It generates new individuals by means of the orthogonal design. Because we usually have no information about the location of the global optimum, and an orthogonal array can specify a small number of individuals that are scattered uniformly over the search space, the orthogonal crossover operator can generate a small, but representative sample of potential individuals. In MAGA, this operator is performed on $L_{i,j}$ and $Max_{i,j}$ to achieve the purpose of cooperation. In what follows, its basic concept is introduced, and for more details, see [14].

Suppose that the search space defined by $L_{i,j}$ and $Max_{i,j}$ is $[\underline{\boldsymbol{x}}_{LM}, \overline{\boldsymbol{x}}_{LM}]$ where

$$\begin{cases} \underline{\boldsymbol{x}}_{LM} = \left(\min(l_1, m_1),\ \min(l_2, m_2),\ \ldots,\ \min(l_n, m_n)\right) \\ \overline{\boldsymbol{x}}_{LM} = \left(\max(l_1, m_1),\ \max(l_2, m_2),\ \ldots,\ \max(l_n, m_n)\right) \end{cases} \tag{9}$$

The domain of the ith dimension is quantized into $\beta_{i,1}, \beta_{i,2}, \ldots, \beta_{i,Q_2}$ where

$$\beta_{i,j} = \begin{cases} \min(l_i, m_i) & j=1 \\ \min(l_i, m_i)+(j-1)\cdot\left(\frac{|l_i-m_i|}{Q_2-1}\right) & 2\le j\le Q_2-1 \\ \max(l_i, m_i) & j=Q_2 \end{cases} \tag{10}$$

(F-1) integers, $(k_1, k_2, \ldots, k_{F-1})$, are generated randomly such that $1<k_1<k_2< \ldots <k_{F-1}<n$, and then create the following F factors for any agent $\boldsymbol{a}=(x_1, x_2, \ldots, x_n)$,

$$\boldsymbol{f}_1 = \left(x_1, \ldots, x_{k_1}\right), \boldsymbol{f}_2 = \left(x_{k_1+1}, \ldots, x_{k_2}\right), \ldots, \boldsymbol{f}_F = \left(x_{k_{F-1}+1}, \ldots, x_n\right) \tag{11}$$

Q_2 levels for the ith factor $\boldsymbol{f}_i$ are defined as follows:

$$\begin{cases} \boldsymbol{f}_i(1) = \left(\beta_{k_{i-1}+1,1}, \beta_{k_{i-1}+2,1}, \ldots, \beta_{k_i,1}\right) \\ \boldsymbol{f}_i(2) = \left(\beta_{k_{i-1}+1,2}, \beta_{k_{i-1}+2,2}, \ldots, \beta_{k_i,2}\right) \\ \qquad \cdots\ \cdots \\ \boldsymbol{f}_i(Q_2) = \left(\beta_{k_{i-1}+1,Q_2}, \beta_{k_{i-1}+2,Q_2}, \ldots, \beta_{k_i,Q_2}\right) \end{cases} \tag{12}$$

The orthogonal array, $L_{M_2}\left(Q_2^F\right) = \left[b_{i,j}\right]_{M_2\times F}$, is applied to generate the following M_2 agents:

$$\begin{cases} \left(\boldsymbol{f}_1\left(b_{1,1}\right), \boldsymbol{f}_2\left(b_{1,2}\right), \ldots, \boldsymbol{f}_F\left(b_{1,F}\right)\right) \\ \left(\boldsymbol{f}_1\left(b_{2,1}\right), \boldsymbol{f}_2\left(b_{2,2}\right), \ldots, \boldsymbol{f}_F\left(b_{2,F}\right)\right) \\ \qquad \cdots\ \cdots \\ \left(\boldsymbol{f}_1\left(b_{M_2,1}\right), \boldsymbol{f}_2\left(b_{M_2,2}\right), \ldots, \boldsymbol{f}_F\left(b_{M_2,F}\right)\right) \end{cases} \tag{13}$$

Finally, the agent with maximum energy among the M_2 agents is selected to replace $L_{i,j}$. The details about the construction of the orthogonal array were given in [14]. For the sake of simplicity, in MAGA, Q_2, F and M_2 are fixed at 3, 4 and 9, respectively, which are same as those of [14]. Therefore, no parameter needs to be tuned for this operator, and the orthogonal array is $L_9(3^4)$, which is shown in (14).

$$L_9(3^4)=\begin{bmatrix} 1 & 1 & 1 & 1 \\ 1 & 2 & 2 & 2 \\ 1 & 3 & 3 & 3 \\ 2 & 1 & 2 & 3 \\ 2 & 2 & 3 & 1 \\ 2 & 3 & 1 & 2 \\ 3 & 1 & 3 & 2 \\ 3 & 2 & 1 & 3 \\ 3 & 3 & 2 & 1 \end{bmatrix} \tag{14}$$

Mutation operator: A new agent, $New_{i,j}=(e_1,e_2,\ldots,e_n)$, is generated as,

$$e_k=\begin{cases} l_k & U(0,1)<\frac{1}{n} \\ l_k+G(0,\frac{1}{t}) & \text{otherwise} \end{cases}, \quad k=1,\ldots,n \tag{15}$$

where $G(0,\frac{1}{t})$ is a Gaussian random number generator, and t is the evolution generation.

Then, $L_{i,j}$ is replaced by $New_{i,j}$. This operator is similar to the Gaussian mutation operator used in the evolutionary programming and the evolution strategy, but it only performs a small perturbation on some variables of $L_{i,j}$.

Self-learning operator: Agents have knowledge which is related to the problems that they are designed to solve. According to our experiences, for numerical optimization problems, integrating local searches with GAs can improve the performance. There are several ways to realize the local searches. [15] uses a small scale GA as the local searcher and obtains a good performance. Enlightened by their idea, we propose the self-learning operator which uses a small scale MAGA to realize the behavior of using knowledge. In order to be distinguished from the other parameters in MAGA, all symbols of the parameters in this operator begin with an '*s*'.

In the self-learning operator, first of all, an agent lattice, sL, is generated. The size of sL is $sL_{size}\times sL_{size}$, and all agents, $sL_{i',j'}$, $i',j'=1,2,\cdots,sL_{size}$, are generated according to,

$$sL_{i',j'}=\begin{cases} L_{i,j} & i'=1 \text{ and } j'=1 \\ New_{i',j'} & \text{otherwise} \end{cases} \tag{16}$$

where $New_{i',j'}=(e_{i',j',1},e_{i',j',2},\cdots,e_{i',j',n})$ is determined by,

$$e_{i',j',k}=\begin{cases} \underline{x}_k & l_k\cdot U(1-sRadius,\ 1+sRadius)<\underline{x}_k \\ \overline{x}_k & l_k\cdot U(1-sRadius,\ 1+sRadius)>\overline{x}_k \\ l_k\cdot U(1-sRadius,\ 1+sRadius) & \text{otherwise} \end{cases}, \quad k=1,\ldots,n. \tag{17}$$

where $sRadius\in[0, 1]$ represents the search radius.

Next, the neighborhood competition operator and the mutation operator are iteratively performed on *sL*. Finally, $L_{i,j}$ is replaced by the agent with maximum energy found during the above process. For more clarity, ALGORITHM 1 describes the details of this operator.

ALGORITHM 1. Self-learning operator

sL^r represents the agent lattice in the *r*th generation, and $sL^{r+1/2}$ is the mid-lattice between sL^r and sL^{r+1}. $sBest^r$ is the best agent among sL^0, sL^1, ..., sL^r, and $sCBest^r$ is the best agent in sL^r. sP_m is the probability to perform the mutation operator, and *sGen* is the number of generations.

Step 1: Generate sL^0 according to (16) and (17), update $sBest^0$, and $r \leftarrow 0$;

Step 2: Perform the neighborhood competition operator on each agent in sL^r, obtaining $sL^{r+1/2}$;

Step 3: For each agent in $sL^{r+1/2}$, if $U(0,1)<sP_m$, perform the mutation operator on it, obtaining sL^{r+1};

Step 4: Find $sCBest^{r+1}$ in sL^{r+1}, if $Energy(sCBest^{r+1})>Energy(sBest^r)$, then $sBest^{r+1} \leftarrow sCBest^{r+1}$; otherwise $sBest^{r+1} \leftarrow sBest^r$, $sCBest^{r+1} \leftarrow sBest^r$;

Step 5: If $r<sGen$, then $r \leftarrow r+1$, go to Step 2;

Step 6: $L_{i,j} \leftarrow sBest^r$.

Implementation of MAGA

In MAGA, the neighborhood competition operator is performed on each agent. As a result, the agents with low energy are cleaned out from the agent lattice so that there is more developing space for the agents with high energy. The neighborhood orthogonal crossover operator and the mutation operator are performed on each agent with probabilities P_c and P_m, respectively. In order to reduce the computational cost, the self-learning operator is only performed on the best agent in each generation, but it has an important effect on the performance of MAGA. In general, the four operators utilize different methods to simulate the behaviors of agents, and play different roles in MAGA. ALGORITHM 2 describes the details of MAGA.

ALGORITHM 2. Multi-Agent Genetic Algorithm

L^t represents the agent lattice in the *t*th generation, and $L^{t+1/3}$ and $L^{t+2/3}$ are the mid-lattices between L^t and L^{t+1}. $Best^t$ is the best agent among L^0, L^1, ..., L^t, and $CBest^t$ is the best agent in L^t. P_c and P_m are the probabilities to perform the neighborhood orthogonal crossover operator and the mutation operator.

Step 1: Initialize L^0, update $Best^0$, and $t \leftarrow 0$;
Step 2: Perform the neighborhood competition operator on each agent in L^t, obtaining $L^{t+1/3}$;
Step 3: For each agent in $L^{t+1/3}$, if $U(0,1)<P_c$, perform the neighborhood orthogonal crossover operator on it, obtaining $L^{t+2/3}$;
Step 4: For each agent in $L^{t+2/3}$, if $U(0,1)<P_m$, perform the mutation operator on it, obtaining L^{t+1};
Step 5: Find $CBest^{t+1}$ in L^{t+1}, and then perform the self-learning operator on $CBest^{t+1}$;
Step 6: If $Energy(CBest^{t+1})>Energy(Best^t)$, then $Best^{t+1} \leftarrow CBest^{t+1}$; otherwise $Best^{t+1} \leftarrow Best^t$, $CBest^{t+1} \leftarrow Best^t$;
Step 7: If termination criteria are reached, output $Best^t$ and stop; otherwise $t \leftarrow t+1$, go to Step 2.

Convergence of MAGA

Obviously, encoding a variable, x_i, of the objective function with an M-bit sequence is equivalent to quantizing its search space, $[\underline{x}_i, \overline{x}_i]$, into 2^M values, and the precision ε is equal to $(\overline{x}_i - \underline{x}_i)/2^M$. Therefore, the convergence of MAGA is analyzed by real coding directly. Suppose that the required precision is ε. Thus, the search space $\mathcal{S}$ can be changed to a discrete space. Its number of elements, $|\mathcal{S}|$, is equal to $\prod_{i=1}^{n} \lceil (\overline{x}_i - \underline{x}_i)/\varepsilon \rceil$, and each element is a candidate solution, namely an agent. Let $\mathbb{E} = \{Energy(\boldsymbol{a}) \mid \boldsymbol{a} \in \mathcal{S}\}$. It is obvious that $|\mathbb{E}| \le |\mathcal{S}|$, and $\mathbb{E}$ can be represented as $\mathbb{E} = \{E^1, E^2, ..., E^{|\mathbb{E}|}\}$, where $E^1 > E^2 > ... > E^{|\mathbb{E}|}$.

This immediately gives us the opportunity to partition $\mathcal{S}$ into a collection of nonempty subsets, $\{\mathcal{S}^i \mid i = 1, 2, ..., |\mathbb{E}|\}$, where

$$\mathcal{S}^i = \{\boldsymbol{a} \mid \boldsymbol{a} \in \mathcal{S} \text{ and } Energy(\boldsymbol{a}) = E^i\} \tag{18}$$

$$\begin{gathered} \sum_{i=1}^{|\mathbb{E}|} |\mathcal{S}^i| = |\mathcal{S}|; \quad \mathcal{S}^i \neq \varnothing, \ \forall i \in \{1, 2, \cdots, |\mathbb{E}|\}; \\ \mathcal{S}^i \cap \mathcal{S}^j = \varnothing, \ \forall i \neq j; \quad \bigcup_{i=1}^{|\mathbb{E}|} \mathcal{S}^i = \mathcal{S} \end{gathered} \tag{19}$$

Obviously, E^1 is the global optimum, and $\mathcal{S}^1$ consists of all agents whose energies are E^1.

In MAGA, the number of agents remains constant throughout the evolutionary process. Let $\mathcal{L}$ stand for the set of all agent lattices. Because an agent lattice may contain multiple copies of one or more agents, the number of elements in $\mathcal{L}$ is $|\mathcal{L}| = \binom{|\mathcal{S}| + L_{size} \times L_{size} - 1}{L_{size} \times L_{size}}$.

Suppose that the energy of an agent lattice, L, is equal to $Energy(L)$, which is determined by,

$$Energy(L) = \max\left\{Energy(L_{i,j}) \mid i, j = 1, \cdots, L_{size}\right\} \tag{20}$$

Thus $\forall L \in \mathcal{L},\ E^{|\mathbb{E}|} \leq Energy(L) \leq E^1$. Therefore, $\mathcal{L}$ can be partitioned into a collection of nonempty subsets $\left\{\mathcal{L}^i \mid i = 1, 2, \cdots, |\mathbb{E}|\right\}$, where

$$\mathcal{L}^i = \left\{L \mid L \in \mathcal{L} \text{ and } Energy(L) = E^i\right\} \tag{21}$$

$$\begin{gathered}\sum_{i=1}^{|\mathbb{E}|} |\mathcal{L}^i| = |\mathcal{L}|; \quad \mathcal{L}^i \neq \varnothing,\ \forall i \in \{1, 2, \cdots, |\mathbb{E}|\}; \\ \mathcal{L}^i \cap \mathcal{L}^j = \varnothing,\ \forall i \neq j; \quad \bigcup_{i=1}^{|\mathbb{E}|} \mathcal{L}^i = \mathcal{L}\end{gathered} \tag{22}$$

$\mathcal{L}^1$ consists of all the agent lattices whose energies are E^1.

Let L^{ij}, $i = 1, 2, \cdots, |\mathbb{E}|$, $j = 1, 2, \cdots, |\mathcal{L}^i|$, stand for the jth agent lattice in $\mathcal{L}^i$. In any generation, the four evolutionary operators transform the agent lattice, L^{ij}, to another one, L^{kl}, and this process can be viewed as a transition from L^{ij} to L^{kl}. Let $p_{ij.kl}$ be the probability of transition from L^{ij} to L^{kl}, $p_{ij.k}$ be the probability of transition from L^{ij} to any agent lattice in $\mathcal{L}^k$, and $p_{i.k}$ be the probability of transition from any agent lattice in $\mathcal{L}^i$ to any agent lattice in $\mathcal{L}^k$. It is obvious that

$$p_{ij.k} = \sum_{l=1}^{|\mathcal{L}^k|} p_{ij.kl}, \quad \sum_{k=1}^{|\mathbb{E}|} p_{ij.k} = 1, \quad p_{i.k} \geq p_{ij.k} \tag{23}$$

Based on the concepts above and [16], [17], the convergence of MAGA is proved as follows:

Theorem 1[18]: Let $\boldsymbol{P}$: $n' \times n'$ be a reducible stochastic matrix which means that by applying the same permutations to rows and columns, $\boldsymbol{P}$ can be brought into the form $\begin{pmatrix} \boldsymbol{C} & \boldsymbol{0} \\ \boldsymbol{R} & \boldsymbol{T} \end{pmatrix}$, where $\boldsymbol{C}$: $m \times m$ is a primitive stochastic matrix and $\boldsymbol{R}, \boldsymbol{T} \neq \boldsymbol{0}$. Then

$$\boldsymbol{P}^\infty = \lim_{k\to\infty} \boldsymbol{P}^k = \lim_{k\to\infty} \begin{pmatrix} \boldsymbol{C}^k & \boldsymbol{0} \\ \sum\limits_{i=0}^{k-1} \boldsymbol{T}^i \boldsymbol{R} \boldsymbol{C}^{k-i} & \boldsymbol{T}^k \end{pmatrix} = \begin{pmatrix} \boldsymbol{C}^\infty & \boldsymbol{0} \\ \boldsymbol{R}^\infty & \boldsymbol{0} \end{pmatrix} \tag{24}$$

is a stable stochastic matrix with $\boldsymbol{P}^\infty = \boldsymbol{1}'\boldsymbol{p}^\infty$, where $\boldsymbol{p}^\infty = \boldsymbol{p}^0 \boldsymbol{P}^\infty$ is unique regardless of the initial distribution, and $\boldsymbol{p}^\infty$ satisfies: $p_i^\infty > 0$ for $1 \leq i \leq m$ and $p_i^\infty = 0$ for $m < i \leq n'$.

Theorem 2: In multi-agent genetic algorithm,

$$\forall i, k \in \{1, 2, \cdots, |\mathbb{E}|\},\ p_{i,k} = \begin{cases} > 0, & k \leq i \\ = 0, & k > i \end{cases}.$$

Proof: $\forall L^{ij} \in \mathcal{L}^i$, $i = 1, 2, \cdots, |\mathbb{E}|$, $j = 1, 2, \cdots, |\mathcal{L}^i|$, $\exists \boldsymbol{a}^* = (x_1, \cdots, x_n) \in L^{ij}$, $Energy(\boldsymbol{a}^*) = E^i$. Suppose that L^{ij} changes to L^{kl} after the four evolutionary operators are performed. If L^{ij} is the agent lattice in the *t*th generation, then L^{kl} is the one in the (*t*+1)th generation. Therefore, L^{ij} and L^{kl} are labeled as L^t and L^{t+1}, respectively.

Firstly, (25) can be obtained according to Step 6 in ALGORITHM 2,

$$\begin{aligned} Energy(CBest^{t+1}) &= Energy(Best^{t+1}) \\ &\geq Energy(Best^t) = Energy(\boldsymbol{a}^*) \end{aligned} \tag{25}$$

Therefore,

$$\begin{aligned} Energy(L^{t+1}) \geq Energy(L^t) &\Rightarrow k \leq i \\ &\Rightarrow \forall k > i, p_{ij.kl} = 0 \\ &\Rightarrow \forall k > i, p_{ij.k} = \sum\nolimits_{l=1}^{|\mathcal{L}^k|} p_{ij.kl} = 0 \\ &\Rightarrow \forall k > i, p_{i.k} = 0 \end{aligned} \tag{26}$$

Secondly, in the neighborhood competition operator, $\boldsymbol{a}^*$ must be a winner because its energy is greater than that of any other agents in L^t, so $\boldsymbol{a}^* \in L^{t+1/3}$. The probability of $\boldsymbol{a}^* \in L^{t+2/3}$ is (1-P_c) because the probability to perform the neighborhood orthogonal crossover operator on $\boldsymbol{a}^*$ is P_c. Therefore, the probability to perform the mutation operator on $\boldsymbol{a}^*$ is:

$$Pr_1 = (1 - P_c) \cdot P_m > 0 \tag{27}$$

$\exists \boldsymbol{a}' \in L^{t+1}$, $Energy(\boldsymbol{a}') = E^k$. Suppose that there are n_1 variables, $x_1', \cdots, x_{n_1}'$, in $\boldsymbol{a}'$ which are different from the corresponding ones in $\boldsymbol{a}^*$. Then the probability to generate $\boldsymbol{a}'$ from $\boldsymbol{a}^*$ by the mutation operator is:

$$Pr_2 = \left(1 - \tfrac{1}{n}\right)^{(n-n_1)} \cdot \prod_{i=1}^{n_1} \tfrac{\sqrt{t}}{\sqrt{2\pi}} e^{-\frac{t(x_i - x_i')^2}{2}} > 0 \tag{28}$$

Thus, the probability of transition from L^{ij} to any agent lattice in $\mathcal{L}^k$ by the four evolutionary operators is:

$$p_{ij.k} > Pr_1 \times Pr_2 > 0 \tag{29}$$

Therefore, $\forall k \leq i$, $p_{i.k} \geq p_{ij.k} > 0$. □

It follows from this theorem that there is always a positive probability to transit from an agent lattice to the one with identical or higher energy and a zero probability to the one with lower energy. Thus, once MAGA enters $\mathcal{L}^1$, it will never go out.

Theorem 3: Multi-agent genetic algorithm converges to the global optimum.

Proof: It is clear that one can consider each $\mathcal{L}^i$, $i = 1, 2, \cdots, |\mathbb{E}|$, as a state in a homogeneous finite Markov chain. According to theorem 2, the transition matrix of the Markov chain can be written as follows:

$$\boldsymbol{P} = \begin{pmatrix} p_{1.1} & 0 & \cdots & 0 \\ p_{2.1} & p_{2.2} & \cdots & 0 \\ \vdots & \vdots & \ddots & \vdots \\ p_{|\mathbb{E}|.1} & p_{|\mathbb{E}|.2} & \cdots & p_{|\mathbb{E}|.|\mathbb{E}|} \end{pmatrix} = \begin{pmatrix} \boldsymbol{C} & \boldsymbol{0} \\ \boldsymbol{R} & \boldsymbol{T} \end{pmatrix} \tag{30}$$

Obviously, $\boldsymbol{R} = \left(p_{2.1},\ \ p_{3.1},\ \ \cdots,\ \ p_{|\mathbb{E}|.1}\right)^T > 0$, $\boldsymbol{T} \neq \boldsymbol{0}$, $\boldsymbol{C} = \left(p_{1.1}\right) = (1) \neq \boldsymbol{0}$.

According to theorem 1, $\boldsymbol{P}^\infty$ is given by,

$$\boldsymbol{P}^\infty = \lim_{k\to\infty} \boldsymbol{P}^k = \lim_{k\to\infty} \begin{pmatrix} \boldsymbol{C}^k & \boldsymbol{0} \\ \sum_{i=0}^{k\text{-}1} \boldsymbol{T}^i \boldsymbol{R} \boldsymbol{C}^{k-i} & \boldsymbol{T}^k \end{pmatrix} = \begin{pmatrix} \boldsymbol{C}^\infty & \boldsymbol{0} \\ \boldsymbol{R}^\infty & \boldsymbol{0} \end{pmatrix} \tag{31}$$

where $\boldsymbol{C}^\infty = 1$, $\boldsymbol{R}^\infty = (1,1,\cdots,1)^T$. Thus, $\boldsymbol{P}^\infty$ is a stable stochastic matrix, and

$$\boldsymbol{P}^\infty = \begin{pmatrix} 1 & 0 & \cdots & 0 \\ 1 & 0 & \cdots & 0 \\ \vdots & \vdots & \ddots & \vdots \\ 1 & 0 & \cdots & 0 \end{pmatrix} \tag{32}$$

Therefore,

$$\lim_{t\to\infty} Pr\{Energy(L^t) = E^1\} = 1 \tag{33}$$

where Pr stands for the probability. This implies that multi-agent genetic algorithm converges to the global optimum. □

Experimental Studies

In order to test the performance of MAGA, 10 benchmark functions have been used:

Generalized Schwefel's Problem 2.26:

$f_1(\boldsymbol{x}) = \sum_{i=1}^n (-x_i \sin\sqrt{|x_i|})$, $\mathcal{S} = [-500,\ 500]^n$;

Generalized Rastrigin's Function:

$f_2(\boldsymbol{x}) = \sum_{i=1}^n [x_i^2 - 10\cos(2\pi x_i) + 10]$, $\mathcal{S} = [-5.12,\ 5.12]^n$;

Ackley's Function:

$$f_3(\boldsymbol{x}) = -20\exp\left(-0.2\sqrt{\tfrac{1}{n}\sum_{i=1}^n x_i^2}\right) - \exp\left(\tfrac{1}{n}\sum_{i=1}^n \cos(2\pi x_i)\right) + 20 + e,\ \mathcal{S} = [-32,\ 32]^n;$$

Generalized Griewank Function:

$f_4(\boldsymbol{x}) = \frac{1}{4000}\sum_{i=1}^n x_i^2 - \prod_{i=1}^n \cos(x_i/\sqrt{i}) + 1$, $\mathcal{S} = [-600,\ 600]^n$;

Generalized Penalized Function 1:

$$f_5(\boldsymbol{x}) = \tfrac{\pi}{n}\left\{10\sin^2(\pi y_1) + \sum_{i=1}^{n-1}(y_i-1)^2\left[1+10\sin^2(\pi y_{i+1})\right] + (y_n-1)^2\right\} + \sum_{i=1}^{n} u(x_i,10,100,4),$$

$$u(x_i,a,k,m) = \begin{cases} k(x_i-a)^m & x_i > a \\ 0 & -a \le x_i \le a, \\ k(-x_i-a)^m & x_i < -a \end{cases}$$

$$y_i = 1 + \tfrac{1}{4}(x_i+1),\ \mathcal{S} = [-50,50]^n;$$

Generalized Penalized Function 2:

$$f_6(\boldsymbol{x}) = \tfrac{1}{10}\left\{\sin^2(3\pi x_1) + \sum_{i=1}^{n-1}(x_i-1)^2\left[1+\sin^2(3\pi x_{i+1})\right] + (x_n-1)^2\left[1+\sin^2(2\pi x_n)\right]\right\} + \sum_{i=1}^{n} u(x_i,5,100,4),$$

$$\mathcal{S} = [-50,50]^n;$$

Sphere Model: $f_7(\boldsymbol{x}) = \sum_{i=1}^{n} x_i^2$, $\mathcal{S} = [-100,100]^n$;

Schwefel's Problem 2.22:

$$f_8(\boldsymbol{x}) = \sum_{i=1}^{n}|x_i| + \prod_{i=1}^{n}|x_i|,\ \mathcal{S} = [-10,10]^n;$$

Schwefel's Problem 1.2:

$$f_9(\boldsymbol{x}) = \sum_{i=1}^{n}\left(\sum_{j=1}^{i} x_j\right)^2,\ \mathcal{S} = [-100,100]^n;$$

Schwefel's Problem 2.21:

$$f_{10}(\boldsymbol{x}) = \max_i\left\{|x_i|,\ 1 \le i \le n\right\},\ \mathcal{S} = [-100,100]^n;$$

f_1-f_6 are multimodal functions where the number of local minima increases with the problem dimension. For example, the number of local minima of f_2 is about $10n$ in the given search space. f_7-f_{10} are unimodal functions. Some parameters must be assigned to before MAGA is used to solve problems. $L_{size}\times L_{size}$ is equivalent to the population size in traditional GAs, so L_{size} can be chosen from 5 to 10. P_o determines whether MAGA puts emphasis on exploitation or on exploration. That is, when P_o<0.5, MAGA puts emphasis on searching in the new space, while when P_o>0.5, on making use of available information. It is better to let P_c be small than 0.5, otherwise it will greatly increase the computational cost. P_m is similar to the mutation probability. The self-learning operator is a small scale MAGA, so its four parameters can be chosen easily. On account of the computational cost, it is better to let sL_{size} be small than 5 and choose *sGen* from 5 to 10. *sRadius* controls the size of the local search area, so it is better to assign a small value to *sRadius*. sP_m is similar to P_m. In the following experiments, the parameter settings are: L_{size}=5, P_o=0.2, P_c=0.1, P_m=0.1, sL_{size}=3, *sRadius*=0.2, sP_m=0.05, *sGen*=10.

A. Descriptions of the Compared Algorithms

Since MAGA is compared with FEP [19], OGA/Q [14], BGA [20], and AEA [13] in the following experiments, we first give a brief description of the four algorithms.

1) FEP [19]: This is a modified version of the classical evolutionary programming (CEP). It is different from CEP in generating new individuals. Suppose that the selected individual is $\boldsymbol{x}=(x_1,\cdots,x_n)$. In CEP, the new individual, $\boldsymbol{x}'=(x_1',\cdots,x_n')$, is generated as follows: $x_i'=x_i+\eta_i N_i(0,1)$, $i=1,2,...,n$, where η_i's are standard deviations for Gaussian mutations, and $N(0,1)$ denotes a normally distributed one-dimensional random number with a mean zero and 1of standard deviation. In FEP, $\boldsymbol{x}'$ is generated as follows: $x_i'=x_i+\eta_i\delta_i$, $i=1,2,...,n$, where δ_i is a Cauchy random variable with the scale parameter 1.

2) OGA/Q [14]: This is a modified version of the classical genetic algorithm (CGA). It is the same as CGA, except that it uses the orthogonal design to generate the initial population and the offspring of the crossover operator.

3) BGA [20]: It is based on artificial selection similar to that used by human breeders, and is a recombination of evolution strategies (ES) and GAs. BGA uses truncation selection as performed by breeders. This selection scheme is similar to the $(\mu,\ \lambda)$-strategy in ES. The search process of BGA is mainly driven by recombination, making BGA a genetic algorithm. Thus, BGA can be described by $\left(P_g^0,\ N,\ T,\ \Gamma,\ \Delta,\ HC,\ F,\ term\right)$. P_g^0 is the initial population, N the size of the population, T the truncation threshold, Γ the recombination operator, Δ the mutation operator, HC the hill-climbing method, F the fitness function and *term* the termination criterion.

4) AEA [13]: This is a modified version of BGA. Besides the new recombination operator and the mutation operator, each individual of AEA is coded as a vector with components all in the unit interval, and inversion is applied with some probability to the parents before recombination is performed.

B. Comparison between FEP, OGA/Q, and MAGA on Functions with 30 Dimensions

FEP [19] and OGA/Q [14] are two methods proposed recently and obtain good performances on numerical optimization problems. [14] compared OGA/Q with traditional GAs and five existing algorithms, and the results showed that OGA/Q outperforms all the other methods. In [19], the termination criterion of FEP was to run 1500 generations for f_3 and f_5-f_7, 2000 generations for f_4 and f_8, 5000 generations for f_2 and f_9-f_{10}, and 9000 generations for f_1. In [14], the termination criterion of OGA/Q was the quality of the solution cannot be further improved in successive 50 generations after 1000 generations. Since the termination criteria of FEP and OGA/Q are different, to make a fair comparison, we let the computational cost of MAGA be less than those of FEP and OGA/Q, and compare the qualities of their final solutions at the given computational cost. Therefore, the termination criterion of MAGA is to run 150 generations for each function. The results averaged over 50 trials are shown in Table 1, where n=30.

As can be seen, MAGA finds the exact global optimum, 0, in all trials for six out of ten functions. For all the ten functions, both the mean function value and the mean number of function evaluations of MAGA are much better than those of FEP. For f_1, f_5 and f_6, the solutions of MAGA are better than those of OGA/Q, and for the other functions, the solutions of MAGA are as good as those of OGA/Q. Moreover, the mean number of function evaluations of MAGA is about 10,000 for all functions, while that of OAGA/Q ranges from 100,000 to 300,000. Therefore, the computational cost of MAGA is much lower than those of OGA/Q. To summarize, the results show that MAGA outperforms FEP and OGA/Q, and is competent for the numerical optimization problems.

Table 1 The comparison between FEP, OGA/Q, and MAGA on functions with 30 dimensions

f	f_{min}	Mean function value (standard deviation)			Mean number of function evaluations		
		FEP	OGA/Q	MAGA	FEP	OGA/Q	MAGA
f_1	-12569.5	-12554.5 (52.6)	-12569.4537 (6.447×10^{-4})	**-12569.4866 (7.121×10^{-12})**	900,000	302,166	**10,862**
f_2	0	4.6×10^{-2} (1.2×10^{-2})	0 (0)	0 (0)	500,000	224,710	**11,427**
f_3	0	1.8×10^{-2} (2.1×10^{-3})	4.440×10^{-16} (3.989×10^{-17})	**4.440×10^{-16} (0)**	150,000	112,421	**9,656**
f_4	0	1.6×10^{-2} (2.2×10^{-2})	0 (0)	0 (0)	200,000	134,000	**9,777**
f_5	0	9.2×10^{-6} (3.6×10^{-6})	6.019×10^{-6} (1.159×10^{-6})	**1.142×10^{-18} (4.390×10^{-18})**	150,000	134,556	**10,545**
f_6	0	1.6×10^{-4} (7.3×10^{-5})	1.869×10^{-4} (2.615×10^{-5})	**1.039×10^{-17} (4.196×10^{-17})**	150,000	134,143	**11,269**
f_7	0	5.7×10^{-4} (1.3×10^{-4})	0 (0)	0 (0)	150,000	112,559	**9,502**
f_8	0	8.1×10^{-3} (7.7×10^{-4})	0 (0)	0 (0)	200,000	112,612	**9,591**
f_9	0	1.6×10^{-2} (1.4×10^{-2})	0 (0)	0 (0)	500,000	112,576	**9,479**
f_{10}	0	0.3 0.5	0 (0)	0 (0)	500,000	112,893	**9,603**

C. Performance of MAGA on Functions with 20~1000 Dimensions

Because the size of the search space and the number of local minima increase with the problem dimension, the higher the dimension is, the more difficult the problem is. Therefore, this experiment studies the performance of MAGA on functions with 20~1000 dimensions. The termination criterion of MAGA is one of the objectives, $|f_{best} - f_{min}| < \varepsilon \cdot |f_{min}|$ or $|f_{best}| < \varepsilon$ if $f_{min} = 0$, is achieved, where f_{best} and f_{min} represent the best solution found until the current generation and the global optimum, respectively. To be consistent, $\varepsilon = 10^{-4}$ is used for all functions. Table 2 gives the mean number and the standard deviations of function evaluations of MAGA averaged over 50 trials.

As can be seen, for f_1 and f_2, when the dimension increases from 20 to 1000, the number of evaluations increases to about 20,000. For f_5 and f_6, when the dimension increases to 1000, the number of evaluations only increases to 11,214 and 17,829. For the six other functions, MAGA obtains high quality solutions ($\varepsilon = 10^{-4}$) only with thousands of evaluations at all selected dimensions. In addition, because ε is assigned to 10^{-4} in the termination criterion, small standard deviations are obtained for all functions at all selected dimensions.

BGA [20] and AEA [13] are also tested on f_1-f_4 with 20, 100, 200, 400, 1000 dimensions. In [20] and [13], the termination criteria of BGA and AEA were the same as that of MAGA, but they used $\varepsilon = 10^{-4}$ for f_1, $\varepsilon = 10^{-1}$ for f_2, and $\varepsilon = 10^{-3}$ for f_3 and f_4. Therefore, a comparison is made between BGA, AEA, and MAGA, which is shown in Table 3.

Table 2 The mean number of function evaluations (the standard deviations) of MAGA on functions with 20~1000 dimensions

n	f_1	f_2	f_3	f_4	f_5	f_6	f_7	f_8	f_9	f_{10}
20	2,483	4,301	3,583	2,566	2,827	3,745	2,420	2,956	4,151	6,823
	(0.2)	(1.9×10^{-5})	(9.3×10^{-6})	(2.3×10^{-5})	(1.9×10^{-5})	(1.9×10^{-5})	(1.6×10^{-5})	(1.0×10^{-5})	(1.7×10^{-5})	(6.5×10^{-6})
100	5,443	10,265	5,410	4,447	4,907	7,929	4,199	5,638	6,351	8,920
	(0.9)	(1.6×10^{-5})	(8.8×10^{-6})	(2.1×10^{-5})	(2.8×10^{-5})	(2.1×10^{-5})	(1.3×10^{-5})	(7.9×10^{-6})	(1.3×10^{-5})	(5.9×10^{-6})
200	7,284	14,867	6,051	5,483	6,870	9,732	4,966	6,757	6,949	9,307
	(2.7)	(2.9×10^{-5})	(6.9×10^{-6})	(2.1×10^{-5})	(3.2×10^{-5})	(2.2×10^{-5})	(1.6×10^{-5})	(8.7×10^{-6})	(1.1×10^{-5})	(6.6×10^{-6})
400	12,368	17,939	6,615	6,249	9,305	12,820	5,576	7,753	7,474	9,662
	(6.1)	(2.7×10^{-5})	(7.1×10^{-6})	(1.8×10^{-5})	(3.7×10^{-5})	(3.3×10^{-5})	(1.2×10^{-5})	(8.3×10^{-6})	(1.0×10^{-5})	(5.4×10^{-6})
800	19,992	20,306	7,069	6,883	10,572	16,070	6,079	8,692	7,902	9,823
	(11)	(3.6×10^{-5})	(7.5×10^{-6})	(1.7×10^{-5})	(3.0×10^{-5})	(3.2×10^{-5})	(1.4×10^{-5})	(8.3×10^{-6})	(1.3×10^{-5})	(5.2×10^{-6})
1000	22,827	20,083	7,288	7,358	11,214	17,829	6,273	9,465	8,024	9,945
	(14)	(3.7×10^{-5})	(8.6×10^{-6})	(1.8×10^{-5})	(3.6×10^{-5})	(3.5×10^{-5})	(1.4×10^{-5})	(7.2×10^{-6})	(1.1×10^{-5})	(5.2×10^{-6})

Table 3 The mean number of function evaluations of BGA, AEA, and MAGA on $f_1 \sim f_4$ with 20~1000 dimensions

n	f_1			f_2		
	BGA	AEA	MAGA	BGA	AEA	MAGA
20	16,100	**1,603**	2,483	3,608	**1,247**	4,301
100	92,000	**5,106**	5,443	25,040	**4,789**	10,265
200	248,000	8,158	**7,284**	52,948	**10,370**	14,867
400	700,000	13,822	**12,368**	112,634	23,588	**17,939**
1000	-	23,867	**22,827**	337,570	46,024	**20,083**
n	f_3			f_4		
	BGA	AEA	MAGA	BGA	AEA	MAGA
20	19,420	7,040	**3,583**	66,000	3,581	**2,566**
100	53,860	22,710	**5,410**	361,722	17,228	**4,447**
200	107,800	43,527	**6,051**	748,300	36,760	**5,483**
400	220,820	78,216	**6,615**	1,630,000	61,975	**6,249**
1000	548,306	160,940	**7,288**	-	97,660	**7,358**

As can be seen, the number of evaluations of MAGA is much smaller than that of BGA for all the four functions. For f_1, when $n \leq 100$, the number of evaluations of MAGA is slightly greater than that of AEA, while when $200 \leq n \leq 1000$, it is slightly smaller than that of AEA. For f_2, when $n\leq 200$, the number of evaluations of MAGA is greater than that of AEA, while when $400\leq n\leq 1000$, it is smaller than that of AEA. For both f_3 and f_4, the number of evaluations of MAGA is much smaller than that of AEA at all dimensions. In general, MAGA obtains better solutions ($\varepsilon = 10^{-4}$) at a lower computational cost than BGA and AEA, and displays a good performance in solving high dimensional problems.

D. Performance of MAGA on Functions with 1000~10,000 Dimensions

In order to study the scalability of MAGA along the problem dimension further, in this experiment, MAGA is used to optimize f_1-f_{10} with higher dimensions. The problem dimension is increased from 1000 to 10,000 in steps of 500. To the best of our knowledge, no researchers have ever optimized the functions with such high dimensions by means of evolution. Therefore, no existing results can be used for a direct comparison. In the previous two subsections, MAGA was compared with four algorithms, and the results show that MAGA is much better than FEP and BGA. Although the performance of OGA/Q is also good, the size of the memory to store the sampling points of the orthogonal design increases with the problem dimension dramatically. For example, when the dimension is 200, a 604.3M memory is needed. So OGA/Q cannot be used to optimize high-dimensional functions in its current form.

We implement AEA and run it with the high dimensions in which MAGA is tested. Both MAGA and AEA used the termination criterion in the above and $\varepsilon = 10^{-4}$. Fig.2 shows the mean number of function evaluations of MAGA and AEA, where the results of MAGA are averaged over 50 trials and the results of AEA are averaged over 10 trials since the running time of AEA is much longer than that of MAGA. Because the number of evaluations of AEA is much greater than that of MAGA, the results of AEA and MAGA for each function are depicted in two figures so that the effect of dimensions on the performance of MAGA can be shown more expressively. The figures in the same row represent the results of the same function, where the left one is the result of AEA and the right one is that of MAGA. In order to study the complexity of MAGA further, in Fig.2, the number of function evaluations is approximated by $O(n^a)$ ($0<a<2$). For more clarity, the comparisons between AEA and MAGA in the mean number of function evaluations for functions with 10,000 dimensions and the derived $O(n^a)$ are shown in Table 4.

Table 4 The comparisons between AEA and MAGA in the mean number of function evaluations for functions with 10,000 dimensions and the derived $O(n^a)$

		f_1	f_2	f_3	f_4	f_5
Mean FE (n=10,000)	AEA	1,239,098	1,458,428	841,827	1,416,717	821,632
	MAGA	**195,292**	**14,315**	**7,961**	**28,815**	**27,645**
$O(n^a)$	AEA	$O(n^{1.03})$	$O(n^{1.62})$	$O(n^{0.78})$	$O(n^{1.35})$	$O(n^{1.01})$
	MAGA	$\mathbf{O(n^{0.78})}$	-	$\mathbf{O(n^{0.06})}$	$\mathbf{O(n^{0.41})}$	$\mathbf{O(n^{0.39})}$
		f_6	f_7	f_8	f_9	f_{10}
Mean FE (n=10,000)	AEA	3,530,502	3,198,123	5,512,354	509,812	3,098,611
	MAGA	**121,370**	**7,784**	**12,233**	**9,672**	**10,219**
$O(n^a)$	AEA	$O(n^{1.04})$	$O(n^{1.08})$	$O(n^{1.12})$	$O(n^{1.24})$	$O(n^{1.29})$
	MAGA	$\mathbf{O(n^{0.80})}$	$\mathbf{O(n^{0.11})}$	$\mathbf{O(n^{0.15})}$	$\mathbf{O(n^{0.08})}$	$\mathbf{O(n^{0.02})}$

As can be seen, the number of evaluations of AEA dramatically increases with the dimensions, and much greater than that of MAGA. In MAGA, for f_1 and f_6, although the number of evaluations obviously increases with the dimension, they are only 195,292 and 121,370, respectively, even when dimensions increase to 10,000. For f_2, the dimension has no obvious effect on the number of evaluations, and especially when n>1000, it still lies between 10,000 and 20,000. For f_4, f_5, the number of evaluations slowly increases with the dimension, and are only 28,815 and 27,645 even when n increases to 10,000. For f_3, f_7, f_8, f_9, and f_{10}, MAGA only uses about 10,000 evaluations even when n increases to 10,000.

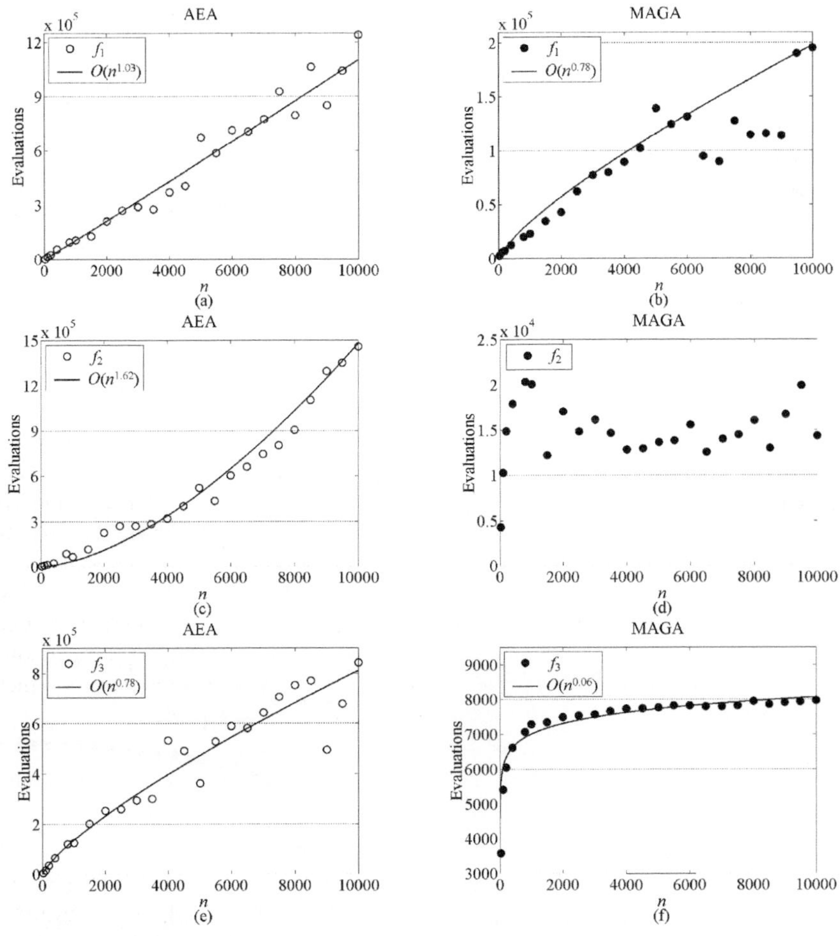

Fig. 2 The comparison between AEA and MAGA on f_1-f_{10} with 20~10,000 dimensions

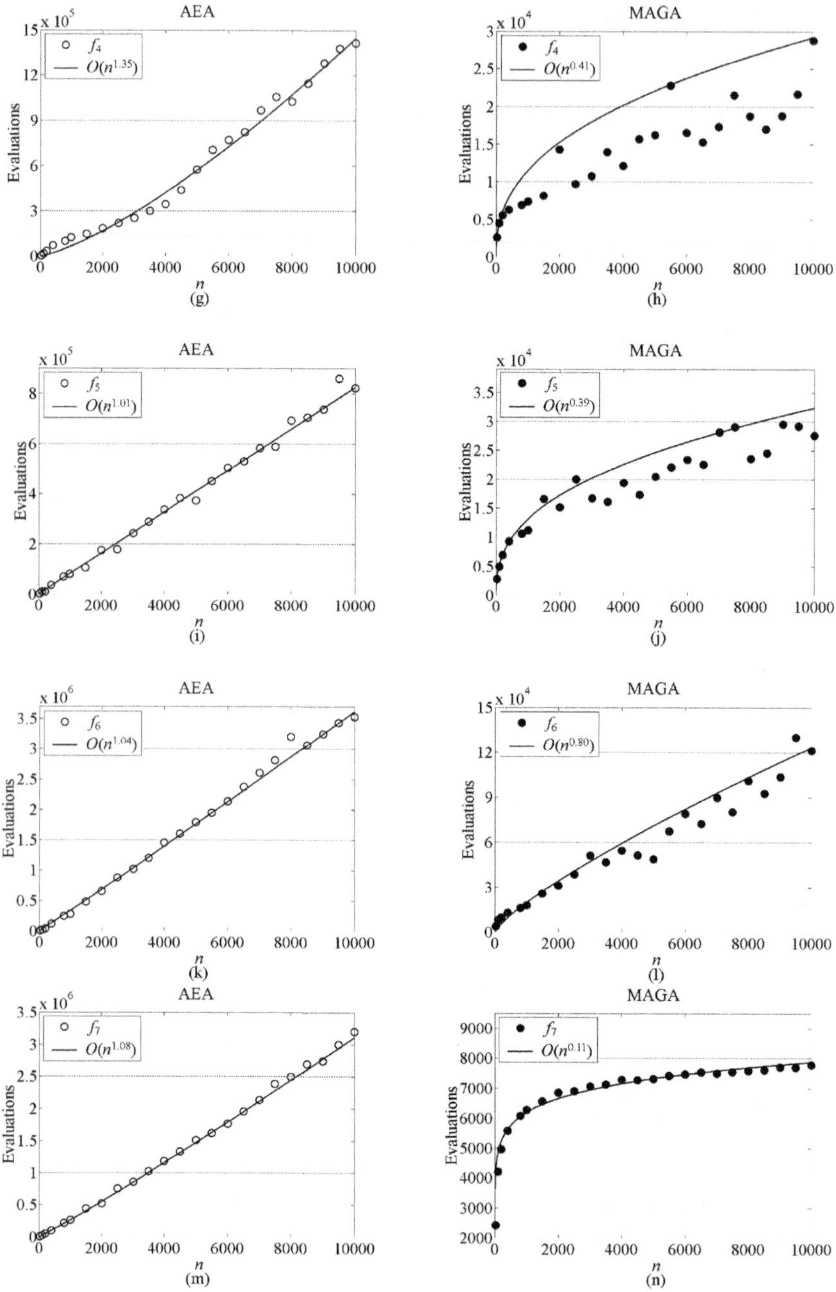

Fig. 2 *(continued)*

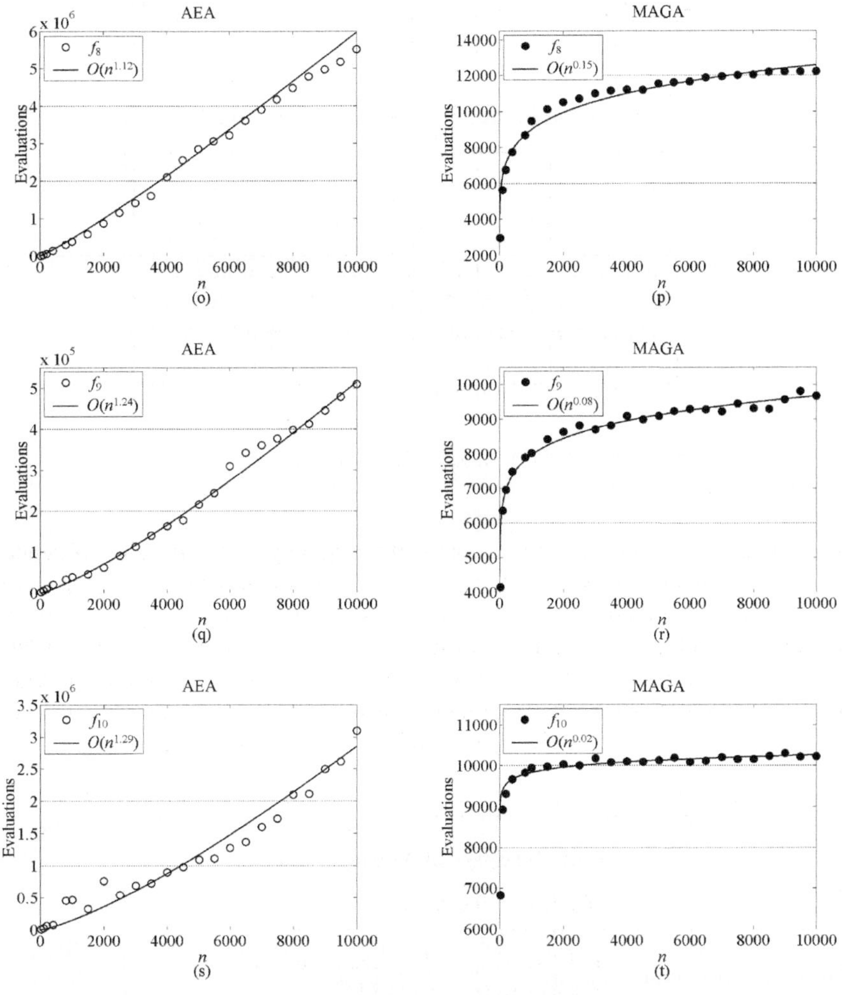

Fig. 2 (*continued*)

From the results approximated by $O(n^a)$ we can see that the complexities of AEA for 9 out of the 10 functions are worse than $O(n)$, and only the complexity of f_3 is $O(n^{0.78})$. At the same time, we can see that the complexities of MAGA for all the 10 functions are very low, which are better than $O(n)$ for the 9 functions other than f_2. Although the number of evaluations of f_2 does not change with the dimension obviously, it is smaller than 25,000 at all dimensions. The worst complexities of MAGA among the 10 functions are those of f_1 and f_6, which are $O(n^{0.78})$ and $O(n^{0.80})$, respectively. The complexities for f_4 and f_5 are $O(n^{0.41})$ and $O(n^{0.39})$, respectively, and Fig.2(h) and (j) shows that most results of the two functions are better than these complexities. The complexities for f_3 and f_7-f_8 are only about $O(n^{0.1})$. f_7-f_8 are unimodal functions and MAGA displays similar performances for these functions. Therefore, the complexity of MAGA for unimodal functions is very low.

3 Multi-Agent Evolutionary Model for Decomposable Function Optimization

Decomposable Functions

There are a number of ways to decompose the complex functions. In order to obtain the minimization of the original function, the sub-functions should be consistent to the original function to some extent, that is, solutions of sub-functions should provide some useful information for the global solutions, so we need to add some restrictions on the ways of decomposing the original function. Definition 3 gives one definition of decomposable function.

Definition 3: If $f(\boldsymbol{x})$ can be decomposed into m sub-functions $f_i^s(\boldsymbol{x}_i^s)$, $i=1, 2, ..., m$, and $f(\boldsymbol{x})=\mathrm{OP}_{i=1}^{m} f_i^s(\boldsymbol{x}_i^s)$, then $f(\boldsymbol{x})$ is called a decomposable function. Herein, $\boldsymbol{x}\in\boldsymbol{\mathcal{S}}\subseteq\boldsymbol{\mathcal{R}}^n$ and $\boldsymbol{x}_i^s\subseteq\boldsymbol{x}$, $\bigcup_i \boldsymbol{x}_i^s=\boldsymbol{x}$, OP=$\{\Sigma, \Pi\}$. If OP=$\Sigma$, then $f(\boldsymbol{x})$ is called an additive decomposable function; if OP=Π, then $f(\boldsymbol{x})$ is called a multiplicative decomposable function. Furthermore, if $\forall\, \boldsymbol{x}_i^s, \boldsymbol{x}_j^s$, $i, j=1, 2, \ldots, m$ and $i\neq j$, we have $\boldsymbol{x}_i^s\cap\boldsymbol{x}_j^s=\varnothing$, then $f(\boldsymbol{x})$ is called a separable function. Similarly, if OP=Σ, then $f(\boldsymbol{x})$ is called an additive separable function; if OP=Π, then $f(\boldsymbol{x})$ is called a multiplicative separable function.

Because the multiplicative function can be transformed into the additive function through logarithmic operation, in this chapter, we only consider the additive function. Apparently, for the separable additive function, if $\exists\boldsymbol{x}^*\in\boldsymbol{\mathcal{S}},\ \boldsymbol{x}^*=\{\boldsymbol{x}_1^{s*}, \boldsymbol{x}_2^{s*}, ..., \boldsymbol{x}_m^{s*}\}$, $\forall\boldsymbol{x}\in\boldsymbol{\mathcal{S}},\ \boldsymbol{x}=\{\boldsymbol{x}_1^s, \boldsymbol{x}_2^s, ..., \boldsymbol{x}_m^s\}$, and $f_1^s(\boldsymbol{x}_1^{s*})\le f_1^s(\boldsymbol{x}_1^s)$, $f_2^s(\boldsymbol{x}_2^{s*})\le f_2^s(\boldsymbol{x}_2^s)$, ..., $f_m^s(\boldsymbol{x}_m^{s*})\le f_m^s(\boldsymbol{x}_m^s)$, then we have $f(\boldsymbol{x}^*)\le f(\boldsymbol{x})$. Thus it can be seen that the purpose of each sub-function in the separable function is consistent to the global purpose, namely, optima of sub-functions are a part of global optima of the original function, so we only need to optimize the each sub-function independently. However, if $\boldsymbol{x}_i^s\cap\boldsymbol{x}_j^s\neq\varnothing$, then it cannot guarantee that the purpose of each sub-function is consistent to the global purpose, that is, optima of sub-functions may not be a part of global optima of the original function. Sometimes, minimization of a sub-function will lead to, to some extent, degrade the current optima of the original function. On the other hand, the solutions of each sub-function contain some useful information for minimizing the original function minimize because some potential search space has been exploited by them. Therefore, it is necessary to make use of the information contained in the solutions of sub-functions. In the MacroAEM model, we use the cooperation behavior among macro-agents to compromise the variables in conflict. On the other hand, since variables in conflict are $\boldsymbol{x}_i^s\cap\boldsymbol{x}_j^s\,(\neq\varnothing)$, so we will minimize $|\,\boldsymbol{x}_i^s\cap\boldsymbol{x}_j^s\,|$ in the process of function decomposition.

Macroagent

Definition 4: If $f(\boldsymbol{x})$ is an additive decomposable function, namely $f(\boldsymbol{x})=\sum_{i=1}^{m} f_i^s(\boldsymbol{x}_i^s)$, then k sub-functions $f_{i_j}^s(\boldsymbol{x}_{i_j}^s)$, j=1,2,…, k form a macro-agent, $\mathcal{MA}$, where 1≤k≤m, 1≤i_j≤m. The energy of $\mathcal{MA}$ is $\mathbb{E}(\mathcal{MA})=-\sum_{j=1}^{k} f_{i_j}^s(\boldsymbol{x}_{i_j}^s)$, and its purpose is to find a vector $\bigcup_{j=1}^{k}\boldsymbol{x}_{i_j}^s$, which maximize its energy $\mathbb{E}(\mathcal{MA})$. To achieve its purpose, $\mathcal{MA}$ will adopt some behaviors.

An agent represents a vector in MAGA. To distinguish, herein we use a macro-agent to represent a sub-function. For function optimization problems, we always expect to obtain the high quality solutions with as less as possible computation cost. Therefore, computation resource can be considered as the environment resource. Due to the resource is limited, macro-agents will compete with others for obtaining more resources. Obviously, macro-agents with the high energy will survive, which reflects the rule of "survival of the fittest". Accordingly, macro-agents will adopt some behaviors (e.g selfishness) to increase their own energy. Since $\boldsymbol{x}_i^s \cap \boldsymbol{x}_j^s = \boldsymbol{x}_{ij}^s \neq \varnothing$, minimizing $f_i^s(\boldsymbol{x}_i^s)$ may not minimize the $f_j^s(\boldsymbol{x}_j^s)$. Thus if a macro-agent increases its energy blindly, the environment would certainly generate the negative reaction since the final goal is maximizing the original function, so that it need to cooperate other macro-agents to increase their energy further. Therefore, it can be seen that the behaviors of the macro-agents can be classified into three types: competition behavior, selfish behavior, and cooperation behavior. Accordingly, a macro-agent can be represented as the following structure:

Macro-Agent $\mathcal{MA} \triangleq$ {
  Candidate vector set $\boldsymbol{x}^s$: $\boldsymbol{x}^s = \bigcup_{i=1}^{k}\boldsymbol{x}_i^s$;
  sub-function set $f^s(\boldsymbol{x}^s)$: $f^s(\boldsymbol{x}^s)=\{f_1^s(\boldsymbol{x}_1^s), f_2^s(\boldsymbol{x}_2^s), ..., f_k^s(\boldsymbol{x}_k^s)\}$;
  Energy $\mathbb{E}$: $\mathbb{E}(\mathcal{MA})=-\sum_{i=1}^{k} f_i^s(\boldsymbol{x}_i^s)$;
  Behavior set $\mathbb{A}$: $\mathbb{A}$={competition behavior, selfishness behavior, cooperation behavior};
}

Macroagent Evolutionary Model

In the initialization step, each macro-agent contains a subset of the sub-functions set. Due to the global purpose is minimization of the original function, we expect to obtain a macro-agent which contains all sub-functions. This process is similar to the process of constructing building blocks in genetic algorithms. Therefore, we consider a macro-agent as an individual and evolve it in a manner of evolution. In the evolutionary process, we always expect that the population could contain all building blocks which can construct the final solution of the problem, so if the original function is decomposed as $f(\boldsymbol{x})=\sum_{i=1}^{m} f_i^s(\boldsymbol{x}_i^s)$ and each sub-function forms a macro-agent, then we should guarantee that the copy number of each macro-agent is not less than 1 in the initial population, namely the size of the initial population is not less than m. Furthermore, in the each evolutionary iteration, we should also guarantee that the union of sub-function sets of several macro-agents in population contains all sub-functions. Thus it is necessary to add some constrains on the behaviors of the macro-agents.

Definition 5: $\forall \mathcal{MA}_1$, $\mathcal{MA}_2$, if $\mathcal{MA}_1\left(f^s(\boldsymbol{x}^s)\right)\subseteq \mathcal{MA}_2\left(f^s(\boldsymbol{x}^s)\right)$ or $\mathcal{MA}_2\left(f^s(\boldsymbol{x}^s)\right)\subseteq \mathcal{MA}_1\left(f^s(\boldsymbol{x}^s)\right)$, then $\mathcal{MA}_1$ and $\mathcal{MA}_2$ are homogeneous macro-agents, otherwise are heterogeneous macro-agents. If $\mathcal{MA}_1$ and $\mathcal{MA}_2$ are heterogeneous macro-agents and $\mathcal{MA}_1(\boldsymbol{x}^s)\bigcap \mathcal{MA}_2(\boldsymbol{x}^s)=\varnothing$, then they are complete heterogenous macro-agents, otherwise are incomplete heterogeneous macro-agents. Herein $\mathcal{MA}_i(\cdot)$ operator is defined to extract the part which the macro-agent $\mathcal{MA}_i$ corresponds.

From the definition 5, two homogeneous macro-agents contain the same information w.r.t the global minimization of the original function. In order to reduce the redundant and save the environmental resource, we only need to keep one of them. Consequently, the relationship between them is competition. Because there are no conflicts between two complete heterogenous macro-agents, the relationship between them is independent. However there exist variables in conflict between two incomplete heterogenous macro-agents, so they must compromise each other to construct solutions of union of two sub-functions. Therefore the relationship between them certainly is cooperation. Accordingly, 3 behaviors of the macro-agents can be described as follows.

Selfish behavior: the behavior of the macro-agent maximizing its energy.

Competition behavior: For two homogeneous macro-agents $\mathcal{MA}_1$ and $\mathcal{MA}_2$, without loss of generality, suppose $\mathcal{MA}_1\left(f^s(\boldsymbol{x}^s)\right)\subseteq \mathcal{MA}_2\left(f^s(\boldsymbol{x}^s)\right)$ exists. If $\mathcal{MA}_1\left(f^s(\boldsymbol{x}^s)\right)\subset \mathcal{MA}_2\left(f^s(\boldsymbol{x}^s)\right)$, then $\mathcal{MA}_1$ will lose and die in the competition; if $\mathcal{MA}_1\left(f^s(\boldsymbol{x}^s)\right)= \mathcal{MA}_2\left(f^s(\boldsymbol{x}^s)\right)$ and $\mathbb{E}(\mathcal{MA}_1)>\mathbb{E}(\mathcal{MA}_2)$, then $\mathcal{MA}_2$ will lost and die in competition, otherwise $\mathcal{MA}_1$ will die. If the lost macro-agents are directly and simply cleaned out from the population, then the useful information they obtained in the evolutionary iterations will be wasted, so it is need to transfer the useful information into the wined macro-agents. Aiming at the practical problems, we can design a number of ways to transfer the information. Generally, the transferring ways are related to the selfishness behavior adopted since the selfishness behavior actually represent specific optimization methods used. For instance, if the selfish behavior is maximization of the macro-agent's energy by an evolutionary way and each macro-agent has a corresponding population, then we can design such a transferring way that new individuals are generated by crossover among the individuals with high energy in the population the lost macro-agent corresponds and the individuals in the population the wined macro-agent corresponds.

Cooperation behavior: For the two incomplete heterogeneous macro-agents $\mathcal{MA}_1$ and $\mathcal{MA}_2$, the variables in conflict are $\boldsymbol{x}_{12}^s = \mathcal{MA}_1(\boldsymbol{x}^s) \cap \mathcal{MA}_2(\boldsymbol{x}^s)$. Similarly, aiming at the practical problems, we also can design a number of ways to compromise two macro-agents $\mathcal{MA}_1$ and $\mathcal{MA}_2$. For instance, we can average the values of $\boldsymbol{x}_{12}^s$ which $\mathcal{MA}_1(\boldsymbol{x}^s)$ and $\mathcal{MA}_2(\boldsymbol{x}^s)$ correspond. In addition, the compromising ways are related to the selfishness behavior adopted. In details, if the selfish behavior is maximization of the macro-agent's energy by an evolution way and each macro-agent has a corresponding population, then we can generate new individuals by using a mean operator on the individuals from two populations which two macro-agents corresponds respectively. Accordingly, the two outcome of the cooperation operation will be resulted: One is that still two macro-agents are generated; the other is that the two macro-agents are synthesized into one macro-agent. We expect to obtain a macro-agent containing all sub-functions finally, so we prefer the latter.

ALGORITHM 3. Macro-Agent Evolutionary Model

Suppose $f(\boldsymbol{x})$ is decomposed as $f(\boldsymbol{x}) = \sum_{i=1}^{m} f_i^s(\boldsymbol{x}_i^s)$. The size of initial population is *PopSize* and *PopSize*≥*m*.

Step 1: Initialize the macro-agent population $Pop^0 \leftarrow \mathcal{MA}_i$, i=1, 2, …, *PopSize*, where $\mathcal{MA}_i$ is initialized as: $\mathcal{MA}_i(\boldsymbol{x}^s) \leftarrow \{\boldsymbol{x}_k^s\}$, $\mathcal{MA}_i(f^s(\boldsymbol{x}^s)) \leftarrow \{f_k^s(\boldsymbol{x}_k^s)\}$, $\mathbb{E}(\mathcal{MA}_i) \leftarrow -f_k^s(\boldsymbol{x}_k^s)$. Herein, if $i \le m$, then $k=i$; otherwise k is a random integer from 1 to m; $t \leftarrow 0$;

Step 2: Optimize each macro-agent in Pop^0 with the approach adopted in the selfishness behavior, $t \leftarrow t+1$;

Step 3: If termination criteria are reached, output the result and stop;

Step 4: Select randomly a macro-agent $\mathcal{MA}_1$ from Pop^t, and determine randomly its behavior between the selfishness and the cooperation; If the selfishness, then optimize its sub-function set by the approach adopted, and delete $\mathcal{MA}_1$ from Pop^t and add it to Pop^{t+1}, go to Step 6; otherwise go to Step 5;

Step 5: Select a macro-agent $\mathcal{MA}_2$ randomly, if $\mathcal{MA}_1$ and $\mathcal{MA}_2$ are homogeneous macro-agents, then perform the competition behavior, and delete $\mathcal{MA}_1$ are $\mathcal{MA}_2$ from Pop^t and add the wined macro-agent to Pop^{t+1}; If $\mathcal{MA}_1$ and $\mathcal{MA}_2$ are incomplete heterogeneous macro-agents, the perform the cooperation behavior, and delete $\mathcal{MA}_1$ are $\mathcal{MA}_2$ from Pop^t and add the new generated macro-agent to Pop^{t+1};

Step 6: If there exist other macro-agents in Pop^t, then go to Step 4; otherwise let $t \leftarrow t+1$, go to Step 3;

In the MacroAEM model, each sub-function at first forms a macro-agent in Step 1, while the rest macro-agents in the initial population randomly select a sub-function. Thus we guarantee that the copy number of each macro-agent is not less than 1. Due to the energies of macro-agents in initialization are low, Step 2 can ensure that each macro-agent has been optimized sufficiently. From the viewpoint of the global optimization, one of termination criteria can be that any two macro-agents are complete homogeneous macro-agents, that is, the global optima can be obtained by assembling the optima of each macro-agent. Steps 4 - 6 can be considered as a coevolutionary process that macro-agents with low energy are eliminated, and macro-agents are also cooperated to generate more potential macro-agents. In addition, sub-functions which macro-agents correspond are optimized sufficiently.

To compare with other algorithms, a function evaluation is considered as calculating all sub-functions of the original functions once. Thus, the terminal condition of the MacroAEM model can be a limited computational cost or a predefined precision, which are the common criteria in the evolutionary computation field.

Actually, the MacroAEM is a generic model, so the ways of function decompositions, optimization method used etc. depend on the practical problems. In next section, based on the MacroAEM, we propose a specific algorithm, called Hierarchical Multi-Agent Genetic Algorithm (HMAGA).

4 Hierarchy Multi-Agent Genetic Algorithm

Dividing the complex system into several layers is an effective measure of human dealing with the complex system. There exist a number of complex systems with hierarchy in business, engineering and science. For instance, university consists of academies, while academy consists of departments. Generally speaking, the intensity of interaction inside subsystem is greater than that of interaction among subsystems. For the complex function optimization problems with high dimension, we can separate them into different layers for reducing the complexity. Accordingly, we can divide the complex function into a number of small-scale sub-functions, then divide the sub-functions into smaller-scale sub-functions in turn. Each sub-function forms a macro-agent in despite of scale of sub-functions. Finally, we get a number of macro-agents with a hierarchical (tree) structure. In what follows, we use Rosenbrock function as an example to show the hierarchical decomposition.

$$\begin{aligned} &f(\boldsymbol{x}) = \textstyle\sum_{i=1}^{n-1} f_i(x_i,\ x_{i+1}),\ \boldsymbol{x} \in \mathcal{R}^n \\ &f_i(x_i,\ x_{i+1}) = 100(x_{i+1} - x_i^2)^2 + (x_i - 1)^2, \\ &\quad -5.12 \le x_i \le 5.12 \end{aligned} \tag{34}$$

where Rosenbrock function is a well-known benchmark function for testing the algorithms. The optimum is located at $\boldsymbol{x}^*$=[1, 1, ..., 1] with $f(\boldsymbol{x}^*)$=0.

Equation (3) gives the hierarchy decomposition of Rosenbrock function with n=9, shown in Fig.3. From (34) and (35), and Fig.3, it can be seen that the f_{00} (0th layer) contains 8 sub-functions (f_1~f_8), f_{11} and f_{12} (first layer) contain 4 sub-functions respectively, and so forth. In addition, the numbers of variables contained by the sub-functions in the first, second and third layers are 5, 3 and 2, respectively. Compare to the number of variables (n=9) of the original function, these sub-functions in first, second and third layers have a low dimension, thus the search space of the algorithm is reduced.

$$\begin{cases} f_{00} = f_1 + \cdots + f_8 \end{cases} \quad \begin{cases} f_{11} = f_1 + \cdots + f_4 \\ f_{12} = f_5 + \cdots + f_8 \end{cases} \quad \begin{cases} f_{21} = f_1 + f_2 \\ f_{22} = f_3 + f_4 \\ f_{23} = f_5 + f_6 \\ f_{24} = f_7 + f_8 \end{cases} \quad \begin{cases} f_{31} = f_1 \\ f_{32} = f_2 \\ f_{33} = f_3 \\ f_{34} = f_4 \\ f_{35} = f_5 \\ f_{36} = f_6 \\ f_{37} = f_7 \\ f_{38} = f_8 \end{cases} \tag{35}$$

Since each macro-agent represents a sub-function, macro-agents have the same hierarchical structure. From the viewpoint of the hierarchical structure, it is advantaged to optimize the macro-agents from bottom up. This is because that when the optimization process reaches the highest layer, namely 0th layer, the sub-function becomes the original function. Herein, we use MAGA as an optimization approach adopted by the macro-agent's selfishness behavior. Therefore, we at first optimize the macro-agents in the lowest layer (3rd layer) through MAGA, then synthesize these macro-agents into the macro-agents in higher layer by using of cooperation behavior. We iterate the process above until we finish optimizing the macro-agents in the 0th layer. This method is called Hierarchical Multi-Agent Genetic Algorithm (HMAGA). In HMAGA, the size of initial population equals to the number of the macro-agents in the lowest layer. For convenience, each macro-agent sets the same values of parameters of MAGA. The cooperation behavior used in HMAGA synthesizes two macro-agents in the ith layer into a new macro-agent in the (i-1)th layer, which is given below.

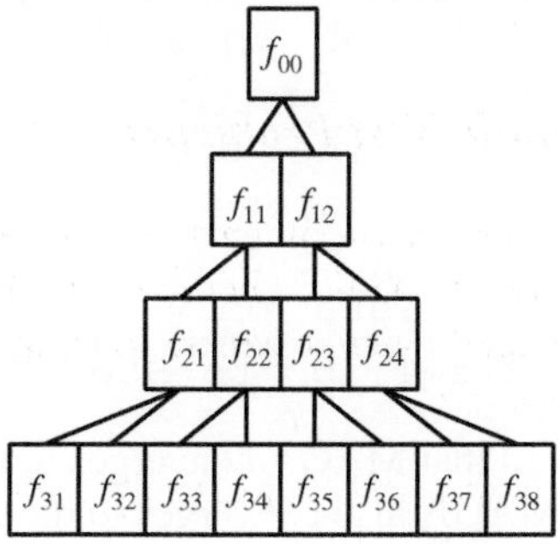

Fig. 3 The hierarchical decomposition of Resenbrock function

Suppose $\mathcal{MA}_1$ and $\mathcal{MA}_2$ are synthesized into $\mathcal{MA}$, then we have $\mathcal{MA}(\boldsymbol{x}^s)\leftarrow\mathcal{MA}_1(\boldsymbol{x}^s)\bigcup\mathcal{MA}_2(\boldsymbol{x}^s)$, $\mathcal{MA}\left(f^s(\boldsymbol{x}^s)\right)\leftarrow\mathcal{MA}_1\left(f^s(\boldsymbol{x}^s)\right)\bigcup\mathcal{MA}_2\left(f^s(\boldsymbol{x}^s)\right)$. Let L^1, L^2 and L are the corresponding agent lattices of $\mathcal{MA}_1$, $\mathcal{MA}_2$ and $\mathcal{MA}$, respectively, the sizes of L^1, L^2 and L are all $L_{size}\times L_{size}$. Let the agents on i^{th} row and j^{th} column of L^1, L^2 and L are $L_{ij}^1=(l_{ij,1}^1,l_{ij,2}^1,\ldots,l_{ij,n}^1)$, $L_{ij}^2=(l_{ij,1}^2,l_{ij,2}^2,\ldots,l_{ij,n}^2)$ and $L_{ij}=(l_{ij,1}, l_{ij,2},\ldots,l_{ij,n})$, respectively, then L_{ij}, $i, j=1,\ldots, L_{size}$ is generated by (36):

$$\begin{cases} l_{ij,k}=l_{ij,k}^1 & \left(l_{ij,k}\in\mathcal{MA}_1(\boldsymbol{x}^s)\right) \text{ and} \\ & \left(l_{ij,k}\notin\left(\mathcal{MA}_1(\boldsymbol{x}^s)\bigcap\mathcal{MA}_2(\boldsymbol{x}^s)\right)\right) \\ l_{ij,k}=l_{ij,k}^2 & \left(l_{ij,k}\in\mathcal{MA}_2(\boldsymbol{x}^s)\right) \text{ and} \\ & \left(l_{ij,k}\notin\left(\mathcal{MA}_1(\boldsymbol{x}^s)\bigcap\mathcal{MA}_2(\boldsymbol{x}^s)\right)\right) \\ l_{ij,k}=\alpha l_{ij,k}^1+(1-\alpha)l_{ij,k}^2 & l_{ij,k}\in\left(\mathcal{MA}_1(\boldsymbol{x}^s)\bigcap\mathcal{MA}_2(\boldsymbol{x}^s)\right) \end{cases} \tag{36}$$

Where k=1, 2, …, n and α is a random number from 0 to 1.

Algorithm 4 describes the details of HMAGA.

ALGORITHM 4. Hierarchical Multi-agent Genetic Algorithm

Suppose the hierarchical decomposition of the original function has been obtained.

Step 1: Construct the macro-agents population Pop^0 that each sub-function in lowest layer forms a macro-agent, and initialize randomly the agent lattices which each macro-agent corresponds in Pop^0, $t\leftarrow 0$;

Step 2: If termination criteria are reached, output the result and stop;

Step 3: Optimize each macro-agent by using MAGA (described in Algorithm 1) in Pop^t;

Step 4: If there exist two incomplete heterogenous macro-agents $\mathcal{MA}_1$ and $\mathcal{MA}_2$ in Pop^t, then perform cooperation behavior on $\mathcal{MA}_1$ and $\mathcal{MA}_2$, and generate a new macro-agent $\mathcal{MA}$. Determine the agent lattice of $\mathcal{MA}$ according to (36), and delete $\mathcal{MA}_1$ and $\mathcal{MA}_2$ from Pop^t and add $\mathcal{MA}$ to Pop^{t+1}, go to Step 4;

Step 5: If there exists only one macro-agent, then move it to Pop^{t+1}, $t\leftarrow t+1$, go to Step 2;

In the above, we have proved that MAGA converges to the global optima theoretically. Theorem 4 shows HMAGA also converges to the global optima.

Theorem 4: Hierarchical multi-agent genetic algorithm converges to the global optimum.

Proof: Suppose $f(\boldsymbol{x})$ can be decomposed as $f(\boldsymbol{x})=\sum_{i=1}^{m} f_i^s(\boldsymbol{x}_i^s)$. According to Algorithm 2 and Step 1 of Algorithm 3, each $f_i^s(\boldsymbol{x}_i^s),\ i=1,2,\ldots,m$ forms a macro-agent, labeled as $\mathcal{MA}_i^I$.

Divide $\boldsymbol{x}_i^s,\ i=1,2,\ldots,m$ into m' groups, labeled as $\boldsymbol{X}_i^g,\ i=1,2,\ldots,m'$, which are shown in (37)~(40):

$$1 \le m' \le m \tag{37}$$

$$\boldsymbol{X}_i^g=\left\{\boldsymbol{x}_{i_j}^s \,\middle|\, j=1,\ 2,\ \ldots,\ |\boldsymbol{X}_i^g|\right\},\quad i=1,\ 2,\ \ldots,\ m',\quad i_j\in\{1,\ 2,\ \ldots,\ m\} \tag{38}$$

$$\bigcup_{i=1}^{m'}\boldsymbol{X}_i^g=\left\{\boldsymbol{x}_k^s \,\middle|\, k=1,\ 2,\ \ldots,\ m\right\} \tag{39}$$

$$\begin{gathered}\forall \boldsymbol{X}_i^g,\ \boldsymbol{X}_k^g,\ i,k\in\{1,2,\ldots,m'\},\ i\ne k,\\ \left(\boldsymbol{X}_i^g\cap\boldsymbol{X}_k^g=\varnothing\right)\ \text{ and }\ \left(\bigcup_{j=1}^{|\boldsymbol{X}_i^g|}\boldsymbol{x}_{i_j}^s\right)\cap\left(\bigcup_{l=1}^{|\boldsymbol{X}_k^g|}\boldsymbol{x}_{k_l}^s\right)=\varnothing\end{gathered} \tag{40}$$

namely, $\boldsymbol{X}_i^g,\ i=1,\ 2,\ \ldots,\ m'$ form m' macro-agents, and among them, any two are complete heterogeneous macro-agents, labeled as $\mathcal{MA}_i^U,\ i=1,\ 2,\ \ldots,\ m'$, where $\mathcal{MA}_i\left(f^s(\boldsymbol{x}^s)\right)=\left\{f_{i_j}^s(\boldsymbol{x}_{i_j}^s)\,\middle|\, j=1,\ 2,\ldots,\ |\boldsymbol{X}_i^g|\right\}$, $\mathcal{MA}_i(\boldsymbol{x}^s)=\left\{\boldsymbol{x}_{i_j}^s \,\middle|\, j=1,\ 2,\ \ldots,\ |\boldsymbol{X}_i^g|\right\},\ i=1,\ 2,\ \ldots,\ m'$.

According to (40), for any $\mathcal{MA}_i^I$ and $\mathcal{MA}_j^I$, if $\mathcal{MA}_i^I(\boldsymbol{x}^s)\subseteq\boldsymbol{X}_{i'}^g$, $\mathcal{MA}_j^I(\boldsymbol{x}^s)\subseteq\boldsymbol{X}_{j'}^g$ and $i'\ne j'$, then $\mathcal{MA}_i^I$ and $\mathcal{MA}_j^I$ are two complete heterogeneous macro-agents. From the Step 4 of Algorithm 3, we can see that the cooperation behavior is only performed on two incomplete heterogenous macro-agents. Therefore, for any two incomplete heterogeneous macro-agents, $\mathcal{MA}_1$ and $\mathcal{MA}_2$, after perform the cooperation behavior, we obtain a new macro-agents $\mathcal{MA}$ and have

$$\mathcal{MA}_1(\boldsymbol{x}^s)\subseteq\boldsymbol{X}_i^g,\ \mathcal{MA}_2(\boldsymbol{x}^s)\subseteq\boldsymbol{X}_i^g,\ \mathcal{MA}(\boldsymbol{x}^s)\subseteq\boldsymbol{X}_i^g,\quad i=1,\ 2,\ \ldots,\ m' \tag{41}$$

namely, candidate vector sets of $\mathcal{MA}_1$, $\mathcal{MA}_2$ and $\mathcal{MA}$ belong to the same group.

Therefore, according to Steps 2 - 5 of Algorithm 3, we can see that $\mathcal{MA}_i^I,\ i=1,2,\ldots,m$ will finally evolve into $\mathcal{MA}_j^U,\ j=1,2,\ldots,m'$. In this case, Algorithm 3 will not perform Step 4, but optimizes the each macro-agent $\mathcal{MA}_j^U$ by using MAGA. Because there are no conflict variables among candidate vector sets of macro-agents $\mathcal{MA}_j^U,\ j=1,2,\ldots,m'$ and we have proved that MAGA converges to the global optima, HMAGA converges the global optima.

Theorem 5: The maximal time complexity of HMAGA is $O(n^a)$ $(a>1)$.

Proof: We have shown that the time complexity of MAGA is $O(n^a)$. Herein we only consider the case of a>1, because when a<1, the original function can be easily solved by MAGA. Without loss of generality, suppose that the number of the macro-agents in the lowest layer is $m = 2^l$ (l+1 is the number of hierarchical decomposition of the original function) and each macro-agent has the same complexity. Suppose $\mathcal{T}(n)$ and $\mathcal{T}_h(n)$ are running time that MAGA and HMAGA need respectively. Due to the time complexity of MAGA is $O(n^a)$, we have $\mathcal{T}(n) = bn^a$. We calculate $\mathcal{T}_h(n)$ from following two types of the decomposable functions,

(i) For separable function,

$$\mathcal{T}_h(n) = 2^l \mathcal{T}(n/2^l) = \frac{1}{(2^l)^{a-1}} \mathcal{T}(n) \tag{42}$$

(ii) For nonseparable function,

As we know, there are (l+1) layers from the lowest layer to the highest layer since m=2^l. The time that HMAGA needs is the sum of time that HMAGA consumes on each layer, thus we have

$$\begin{aligned}\mathcal{T}_h(n) &= \mathcal{T}(n) + 2\mathcal{T}(n/2) + \cdots + 2^l \mathcal{T}(n/2^l) \\ &= bn^a(1 + \tfrac{2}{2^a} + \cdots + (\tfrac{2}{2^a})^l) \\ &= \frac{1-(\frac{2}{2^a})^l}{1-\frac{2}{2^a}} \mathcal{T}(n) < \frac{2^a}{2^a-2} \mathcal{T}(n)\end{aligned} \tag{43}$$

Cases (i) and (ii) show that the maximal time complexity of HMAGA is $O(n^a)$. □

Experimental Studies

We use Rosenbrock function given in (34) to test the performance of HMAGA. Even for n=2, the fitness surface of Rosenbrock function exhibits a narrow ridge, in addition, all variables need to be changed at the same time in a certain manner for obtaining a better solution, which make the optimization methods difficult to approach the optima [15, 21]. We decompose Rosenbrock as the following form:

$$f(\boldsymbol{x}) = \sum_{i=1}^{\lceil (n-1)/\kappa \rceil} f_i$$

$$f_i = \begin{cases} \sum_{j=\kappa(i-1)+1}^{\kappa i} 100(x_{j+1} - x_j^2)^2 + (x_j - 1)^2 & i < \lceil (n-1)/\kappa \rceil \\ \sum_{j=\kappa(i-1)+1}^{n-1} 100(x_{j+1} - x_j^2)^2 + (x_j - 1)^2 & i = \lceil (n-1)/\kappa \rceil \end{cases} \tag{44}$$

where κ is a integer called the decomposition granularity of the function.

A. Comparison between HMAGA and MAGA on Rosenbrock Function with 10~1,000 Dimensions

We firstly investigate the performance of HMAGA along with the dimension of Rosenbrock function, and compare the results with those of MAGA. The range of n is in [10, 1,000]. We get 28 sample points from the range of [10, 200] and [200, 1,000] by a step of 10 and 100 respectively. The decomposition granularity κ is 10. The following termination criterion is defined: if $f_{min}\neq 0$, $|f_{best}-f_{min}|<\varepsilon\cdot|f_{min}|$ or if $f_{min}=0$, $|f_{best}|<\varepsilon$, where f_{best} represents the best solutions found from the beginning to current generation and f_{min} represents the global optima. To be consistent, $\varepsilon=10^{-4}$ is used for both two algorithms. We perform 10 independent runs for each algorithm on each dimension sample point and record the mean number of function evaluations. Fig.4 gives the mean number of function evaluations of both HMAGA and MAGA. From Table 5 and Fig.4, it can be seen that the number of function evaluations of HMAGA is much smaller than that of MAGA, especially when $n>200$.

B. Performance of HMAGA on Rosenbrock Function with 1,000~50,000 Dimensions

In order to study the scalability of HMAGA along the problem dimension, we apply HMAGA on Rosenbrock function with 1,000~50,000 dimensions in this experiment. We first get 17 sample points from the range of [10^3, 10^4] and [10^4, 10^5] in step of 1,000 and 5,000 respectively. Plus 28 points sampled from the range of [10, 1,000], we have obtained 45 sample points totally. The initial decomposition granularity κ is 10. We perform 10 independent runs on each sample points. Table 6 and Fig.5(a) gives the number of function evaluations of HMAGA on each sample points. From Table 6 and Fig.5(a), HMAGA needs a smaller number of evaluations. Even when the number of dimensions is as high as 50,000, HMAGA only need 10^6 evaluations. With the dimension increasing, the number of evaluations increases slowly. According to a statistical analysis, within 200 dimensions, the time complexity of HMAGA is $O(n^{0.6333})$, while $n>200$, the time complexity of HMAGA is $O(n^{0.1729})$. Thus it can be seen that HMAGA has a small time complexity. The number of evaluations increases slower on high dimension than on low dimensions the reason is that with the const κ, higher the dimension is, the lager the number of layers is, but, good candidate sub-solutions has been obtained in the low layers, thus HMAGA only needs a small computation in the high layers. It demonstrates that the computation cost of HMAGA mainly consumes on the low layers. At the same time, it also demonstrates that the hierarchy decomposition is effective. For comparison, Fig.5(b) gives the time complexity of MAGA, $O(n^{1.4378})$.

Table 5 The numbers of function evaluations of HMAGA and MAGA on Rosenbrock function with 10~1,000 dimensions

n	HMAGA	MAGA	n	HMAGA	MAGA
10	39 768	49 279	150	208 097	3 365 243
20	55 721	112 143	160	218 664	2 885 332
30	70 713	201 839	170	226 209	4 511 409
40	82 230	308 824	180	234 931	6 421 595
50	98 142	396 678	190	232 816	5 999 797
60	113 969	544 860	200	227 620	6 744 891
70	129 545	809 087	300	268 380	8 551 755
80	135 563	965 923	400	375 147	12 716 615
90	153 244	1 111 073	500	339 310	14 809 126
100	156 930	1 469 140	600	340 251	16 113 665
110	169 072	1 594 160	700	401 143	15 910 634
120	190 292	1 958 490	800	367 292	21 252 325
130	203 041	2 319 289	900	380 625	21 568 594
140	185 169	3 062 236	1,000	398 601	28 482 792

C. Performance Analysis of Parameter κ

We analyze the effect of parameter κ on the performance of HMAGA. Let n=201, Table 7 and Fig.6 shows the mean number of evaluations and the mean time of HMAGA on 10 independent runs along with κ= 4, 7, 13, 25 and 50. From Table 7 and Fig.6(a), it can be seen that with κ decreasing, the number of evaluations of HMAGA decreases, because sub-functions in the low layers are easier to optimize for smaller κ. In addition, we know that the computation cost mainly consumes on the low layers, therefore small κ will lead to a small computation. On the other

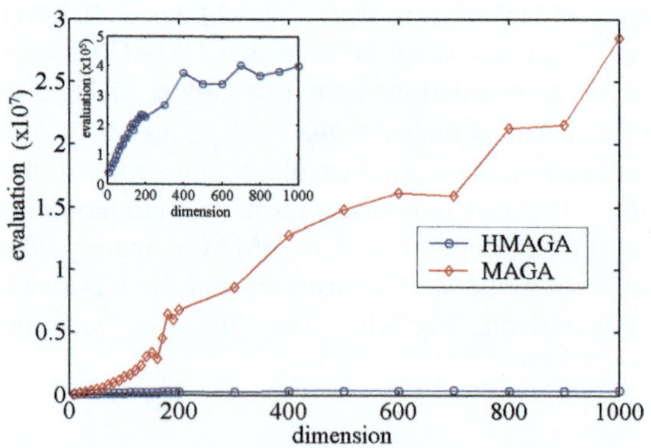

Fig. 4 Comparison between HMAGA and MAGA on Rosenbrock function with 10~1000 dimensions

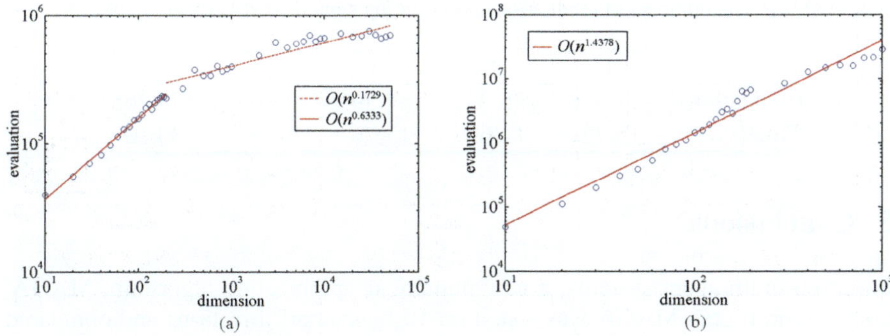

Fig. 5 The time complexities of (a) HMAGA and (b) MAGA

Table 6 The numbers of function evaluations on Rosenbrock function with 2,000~50,000 dimensions

n	2,000	3,000	4,000	5,000	6,000	7,000
HMAGA	487 971	614 902	560 843	601 076	629 034	702 076
n	8,000	9,000	10,000	15,000	20,000	25,000
HMAGA	627 274	657 284	667 731	722 445	683 866	692 096
n	30,000	35,000	40,000	45,000	50,000	
HMAGA	756 596	703 233	659 463	680 174	697 401	

hand, from Table 7 and Fig.6(b), we can see that the minimal consuming time is not obtained on the smallest κ=4, but on κ=13. The reason is that in hierarchy decomposition, smaller κ is, the larger the number of layers is, hence the time consumed on the synthesis of populations from low layer to high layer will be larger. From this viewpoint, κ can not be too small.

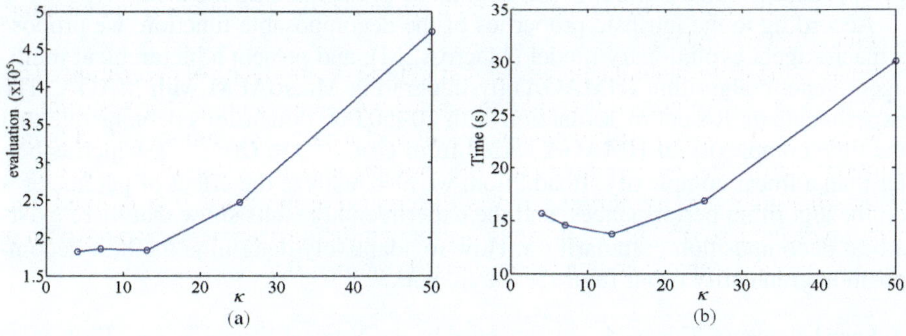

Fig. 6 The mean number of (a) evaluations and (b) consuming time of HMAGA on different κ

Table 7 The numbers of evaluations and running time of HMAGA along with κ

κ	4	7	13	25	50
Evaluations	181 976	185 831	184 395	247 637	473 265
Time(s)	15.72	14.62	13.76	16.93	30.20

5 Conclusions

Based on multi-agent systems, a new numerical optimization algorithm, MAGA, has been proposed. MAGA was tested on 10 benchmark functions and compared with four famous algorithms, FEP [19], OGA/Q [14], BGA [20] and AEA [13]. The experiments on functions with 30 dimensions and 20~1000 dimensions indicated that MAGA outperforms the four algorithms. In order to study the scalability of MAGA along the problem dimension, MAGA was used to optimize the 10 functions with 1000~10,000 dimensions. The results indicated that MAGA can obtain high quality solutions at a low computation cost even for the functions with 10,000 dimensions. For example, MAGA only used thousands of evaluations to optimize f_3, f_7 and f_9. For the 10 functions with 20~10,000 dimensions, the complexity of MAGA varied from $O(n^{0.02})$ to $O(n^{0.80})$, which is better than a linear complexity. Especially for the unimodal functions, the complexity of MAGA is only about $O(n^{0.1})$.

MAGA obtains a good performance for both function optimizations and the approximation of linear systems. This benefits mainly from the model of the agent lattice and the behaviors of agents. In nature, the real natural selection only occurs in a local environment, and each individual can only interact with those around it. In the agent lattice, each agent can only sense its local environment, and its behaviors of competition and cooperation can only take place between the agent and its neighbors. Thus, an agent transfers its information to their neighbors, and the information can be shared by all agents after a process of diffusion. As can be seen, this process is more close to the real evolutionary process in nature than that of traditional GAs. In MAGA, no global control is needed at all, and each agent is independent to some degree, which is in favor of maintaining the diversity.

According to the intrinsic properties of the decomposable function, we propose a macro-agent evolutionary model (MacroAEM), and present a hierarchical multi-agent genetic algorithm (HMAGA) by integrating MacroAEM with MAGA. The experiments on Rosenbrock function with 10~50,000 dimensions demonstrate that the time complexity of HMAGA varied from $O(n^{0.1729})$ to $O(n^{0.6333})$, which is better than a linear complexity. In addition, we also analyze the effect of parameter κ on the algorithm performances, and the experimental results show that there exists a best decomposition granularity κ. How to adaptively determine the best decomposition granularity is our future research work.

Acknowledgments. This work was supported by the National Natural Science Foundation of China under Grant 60872135, and the Program for New Century Excellent Talents in University of China under Grant NCET-06-0857.

References

[1] Russell, S., Norvig, P.: Artificial intelligence: a modern approach. Prentice- Hall, New York (1995)

[2] Jennings, N.R., Sycara, K., Wooldridge, M.: A roadmap of agent research and development. In: Jennings, N.R., Sycara, K., Georgeff, M. (eds.) Autonomous Agents and Multi-Agent Systems Journal, vol. 1(1), pp. 7–38. Kluwer Academic Publishers, Boston (1998)

[3] Ferber, J.: Multi-Agent Systems: An Introduction to Distributed Artificial Intelligence. Addison-Wesley, New York (1999)

[4] Liu, J.: Autonomous Agents and Multi-Agent Systems: Explorations in Learning, Self-Organization, and Adaptive Computation. World Scientific, Singapore (2001)

[5] Liu, J., Tang, Y.Y., Cao, Y.C.: An evolutionary autonomous agents approach to image feature extraction. IEEE Trans. Evol. Comput. 1(2), 141–158 (1997)

[6] Liu, J., Jing, H., Tang, Y.Y.: Multi-agent oriented constraint satisfaction. Artificial Intelligence 136(1), 101–144 (2002)

[7] Whitley, D.: Cellular genetic algorithms. In: Belew, R.K., Booker, L.B. (eds.) Proc. Fifth Int. Conference on Genetic Algorithms, p. 658. Morgan Kaufmann, San Mateo (1993)

[8] Nakashima, T., Ariyama, T., Yoshida, T., Ishibuchi, H.: Performance evaluation of combined cellular genetic algorithms for function optimization problems. In: Proc. 2003 IEEE Int. Symposium on Computational Intelligence in Robotics and Automation, Kobe, Japan, vol. 1, pp. 295–299 (2003)

[9] Folino, G., Pizzuti, C., Spezzano, G.: Parallel hybrid method for SAT that couples genetic algorithms and local search. IEEE Trans. Evol. Comput. 5, 323–334 (2001)

[10] Holland, J.H.: Adaptation in nature and artificial system. MIT Press, Cambridge (1992)

[11] Goldberg, E.: Genetic Algorithms in Search, Optimization & Machine Learning. Addison-Wesley, Reading (1989)

[12] Mitchell, M.: An Introduction to Genetic Algorithms. MIT Press, Cambridge (1996)

[13] Pan, Z.J., Kang, L.S.: An adaptive evolutionary algorithms for numerical optimization. In: Yao, X., Kim, J.H., Furuhashi, T. (eds.) SEAL 1996. LNCS, vol. 1285, pp. 27–34. Springer, Heidelberg (1997)

[14] Leung, Y.W., Wang, Y.: An orthogonal genetic algorithm with quantization for global numerical optimization. IEEE Trans. Evol. Comput. 5(1), 41–53 (2001)

[15] Kazarlis, S.A., Papadakis, S.E., Theocharis, J.B., Petridis, V.: Microgenetic algorithms as generalized hill-climbing operators for GA optimization. IEEE Trans. Evol. Comput. 5, 204–217 (2001)

[16] Rudolph, G.: Convergence analysis of canonical genetic algorithms. IEEE Trans. Neural Networks, Special Issue on Evolutional Computing 5(1), 96–101 (1994)

[17] Dinabandhu, B., Murthy, C.A., Sankar, K.P.: Genetic algorithm with elitist model and its convergence. International Journal of Pattern Recognition and Artificial Intelligence 10(6), 731–747 (1996)

[18] Iosifescu, M.: Finite Markov Processes and Their Applications. Wiley, Chichester (1980)

[19] Yao, X., Liu, Y., Lin, G.: Evolutionary programming made faster. IEEE Trans. Evol. Comput. 3, 82–102 (1999)
[20] Mühlenbein, H., Schlierkamp-Vose, D.: Predictive models for the breeder genetic algorithm. Evolutionary Computation 1(1), 25–49 (1993)
[21] Liang, Y., Zhou, C., Li, S.: Optimization of Rosenbrock's function based on genetic algorithms. Journal of Software 8(9), 701–708 (1997)

An Agent Based Evolutionary Approach for Nonlinear Optimization with Equality Constraints

Abu S.S.M. Barkat Ullah*, Ruhul Sarker, and Chris Lokan

Abstract. To represent practical problems appropriately, many mathematical optimization models require equality constraints in addition to inequality constraints. The existence of equality constraints reduces the size of the feasible space, which makes it difficult to locate feasible and optimal solutions. This paper shows the enhanced performance of an agent-based evolutionary algorithm in solving Constrained Optimization Problems (COPs) with equality constraints. In the early generations of the evolutionary process, the agents use a new learning process that is specifically designed for handling equality constraints. In the later generations, the agents improve their performance through other learning processes by exploiting their own potential. The performance of the proposed algorithm is tested on a set of well-known benchmark problems including two new problems. The experimental results confirm the improved performance of the proposed algorithm.

1 Introduction

Over the last few decades, Evolutionary Algorithms (EAs) have been widely adopted to solve optimization problems due to their flexibility and adaptability to the task at hand, in combination with their robust performance and global search

Abu S.S.M. Barkat Ullah
School of Engineering and Information Technology (SEIT), UNSW@ADFA,
Canberra ACT 2600, Australia
Tel.: +61-2-62688168; Fax: +61-2-62688681
e-mail: barkat@adfa.edu.au

Ruhul Sarker
School of Engineering and Information Technology (SEIT), UNSW@ADFA,
Canberra ACT 2600, Australia
e-mail: r.sarker@adfa.edu.au

Chris Lokan
School of Engineering and Information Technology (SEIT), UNSW@ADFA,
Canberra ACT 2600, Australia

* Corresponding author.

R.A. Sarker and T. Ray (Eds.): Agent-Based Evolutionary Search, ALO 5, pp. 49–76.
springerlink.com

capability [1, 2]. EAs perform very well for unconstrained problems compared to their constrained counterpart. However, most real-world decision processes require solving optimization problems which may involve different types of constraints. The traditional search operators of EAs (such as crossover and mutation) are blind to constraints. As a consequence, the candidate solutions generated by these operators may violate certain constraints [3]. Hence constraint handling is one of the major concerns when applying EAs to solve constrained optimization problems. The main challenge in constrained optimization is optimizing the objective function while still ensuring non-violation of the constraints. Over the past decade, various constraint-handling techniques have been proposed in evolutionary optimization [3-7]. These techniques can be grouped [7] as: penalty functions, special representations and operators, repair algorithms, separation of objectives and constraints, and hybrid methods.

However, even with the existence of many constraint-handling techniques, EAs still have inherent difficulty in dealing with equality constraints when solving the constrained problems. The feasible space of the problems with equality constraints becomes tiny in comparison to the whole search space. It becomes very hard for traditional EAs to find feasible and optimal solutions in such problems [8, 9]. Furthermore, traditional EAs suffer from slow convergence to locate a precise enough solution because of their failure to exploit local information, and face difficulties in solving multi-modal problems that have many local solutions within the feasible space. To improve the performance of EAs, different hybridization of algorithms has been introduced in recent times.

In solving constrained optimization, an improved performance can be achieved by hybridizing the evolutionary algorithms with the Local Search (LS) techniques; this is known as Memetic Algorithms (MAs). MAs have been successfully applied across a wide range of problem domains such as combinatorial optimization [10-15], optimization of non-stationary functions [16], and multi-objective optimization [17-20]. They converge to high quality solutions as well as search more efficiently than their conventional counterparts [21]. MAs are inspired by Dawkins' notion of a meme [22], defined as a unit of information that reproduces itself as people exchange ideas. In this paper, the concept of MAs follows the model of adaptation in natural systems, where an individual of a population may be improved thorough self learning along with the evolutionary adaptation of the population [23, 24]. Some theoretical and empirical investigations on MAs can be found in [21, 23, 25-30]. One of the critical issues regarding the performance of MAs is the selection of appropriate local search methods while hybridizing local search with GAs. If the selection of LS is not appropriate for a particular problem then MAs may not perform well; the performance may even be worse than GAs alone [25, 27, 31]. Many types of local searches are available in the literature but it is very difficult to know which type is appropriate for a particular problem.

Agent-based computation introduces a new paradigm for conceptualizing, designing and implementing intelligent systems, and has been widely used in many branches of computer science [32]. The agents are discrete individuals, situated in an environment, having a set of characteristics and rules to govern their behavior and interactions. They sense the environment and act on it in pursuit of a set of goals or tasks for which they are designed [33]. Agent properties, such as autonomy, communication, proactiveness, learning and reactivity, can potentially be engaged to enhance the performance of EAs. Reciprocal influences and cross-fertilizations of Multi-agent Systems (MAS), as population of agents and evolutionary models, is a promising approach to develop flexible computing techniques based on a population that can proactively evolve and emerge [34].

To mitigate the shortcoming of MAs mentioned above, and to design an efficient technique for handling equality constraints, in this paper we propose an Agent-based Memetic Algorithm (AMA). In AMA, each agent represents a solution vector. Now we define the population of AMA as the population of agents, where each agent contains a solution vector. The agents carry out cooperative and competitive behaviors, and select the appropriate local search adaptively to find optimal solutions for the problem in hand. The agents cooperate and compete through the well known simulated binary crossover (SBX) proposed by Deb and Agrawal [35], and our designed Life Span Learning Process (LSLP). The life span learning processes are basically local search techniques. As we generally see in GAs, an individual in a generation produces offspring and the offspring may be mutated to change the genetic materials. In reality, beside reproduction, an individual learns and gains experiences in different ways during its life time. This process is represented by the proposed LSLPs in this paper.

In this paper, a new LSLP is designed specially to handle the equality constraints in solving nonlinear optimization problems. In any optimization problems with equality constraints, to satisfy the condition of feasibility and optimality, the solution points must lie on each and every equality constraint. That means it might be possible to find an optimal solution by simply exploring an equality constraint function where the constraint function contains all the variables. This common knowledge encourages us to design a new LSLP for dealing with equality constraints. The basic idea is to reach a point on the equality constraint from the current position of an individual solution and then explore on the constraint landscape. That means an individual would explore only a portion of the entire search space. In practical problems, an equality constraint may not contain all the variables which require exploring the landscapes of other equality constraints. To the best of our knowledge, this approach for handling equality constraints does not appear in the literature.

This new LSLP, along with four other LSLPs, are applied with the evolutionary process. The other LSLPs are based on several local and directed search procedures. An agent chooses a LSLP as a local search operator self-adaptively. The new LSLP is particularly worth using early, to speed the search towards the feasible space. For that reason it is used exclusively for the first N generations (N chosen arbitrarily to be 100), and thereafter it is not used at all and the other LSLP's then refine the solution.

The preliminary framework of AMA was reported in [9, 36, 37]. This paper presents an improved version of AMA, where the agents are capable of dealing with the equality constraints more efficiently. To test the performance of the algorithm, we have selected a set of well-known test problems from [4, 38] including two new problems [39] with equality constraints, and compared the results with different algorithms [40, 41]. The comparisons show that the results are of improved quality with low computational time. This paper also analyzes the effect of new learning process for handling the equality constraints.

The rest of this paper is organized as follows. Section 2 describes the Agent-based Evolutionary algorithms. Section 3 discusses the new LSLP for handling the equality constraints. Section 4 presents the proposed Agent-based memetic algorithm. Section 5 provides computational experience, results of the proposed approach on benchmark problems, and compares with other algorithms. Finally, Section 6 concludes the paper and provides future research directions.

2 Agent-Based Evolutionary Algorithms

The agent-based evolutionary algorithms can be of two different approaches. In the first, the multi-agent system incorporates an evolutionary algorithm as a search technique to improve certain functionality of some selected agents or the entire system [42]. For example, a MAS may contain different type of agents and one or more types of these agents deal with some optimization tasks. The second type is population based evolutionary algorithms (EAs), for solving complex decision problems, where the individuals of the population are defined as agents [37, 43, 44]. There is a long debate about the second type whether we should call it an '*agent based system*' or what does it add to EAs by naming it an '*agent based EA*'?

In this section, we will discuss how an agent-based EA can be different from an independent EA.

Here we define the agents a bit differently. The individuals in the population of EAs are not agents rather, based on the individual's belief and learning experiences, each agent stands (or supports) for one of the individuals of the population. A population of agents is endowed with a set of individualistic and social behaviors, in order to explore a local environment within the solution space. The combination of agents' local view exploration with EAs global search ability would establish a superior balance between exploitation and exploration when solving complex optimization problems. In fact, when we define the individual as an agent we can bring anything (like rules for communication, cooperation and competition, intelligence, memory, and learning) onboard which the traditional EAs do not deal with. Davidsson *et al.* [45] indicated that we must capitalize the strength of two approaches in a new hybrid method as they complement each other. We now discuss the agent activities which we can consider with EAs in the context of optimization problem solving.

2.1 Environment of Agents

The environment includes the agent's social network structure, and the size of the neighborhood for interactions. The network topology usually includes ring, two-dimensional lattice, random small-world, and star type. However, we may consider any special structure such as self-organizing network. In optimization problem solving, the two-dimensional lattice-like structure is widely used. There are some similarities of this structure with cellular genetic algorithms [46]. The neighborhood size controls the amount of interaction and diversity in the entire population [37].

2.2 Behavior of Agents

The agents can be cooperative or competitive. Cooperative agents share information with their neighboring agents, whereas competitive agents compete with the neighboring agents. The quality of an agent is represented by fitness value or energy [44]. An agent with a higher fitness value has a better chance of survival in its neighborhood. De Jong [47] stated that the agent behavior is basically a combination of "nature and nurture", those are both inherited and learned components.

2.3 Learning of Agents

De Jong [47] indicated that evolution operates at the population level while "lifetime learning" occurs at the individual (agent) level. The agents learn throughout their life span which improves their quality (fitness value) [37]. This learning process can be chosen by the individual agents independently. For example, in optimization problem solving, the local search techniques could be learning processes for an agent. Vasile and Locatelli [48] indicated that each agent performs a sequence of actions at every generation according to their defined behavior such as inertia, follow-the-tail, random-step, linear blending and quadratic blending. These basically represent local search techniques, which we label as learning processes in this paper.

2.4 Reasoning Capability of Agents

It is well known that to move in the direction of producing the fastest rate of improvement in the fitness value is not always best. So an agent must find reasoning for its next move. An agent may apply either quantitative or qualitative judgment (or both) based on its own belief, social interaction, knowledge and intelligence. In optimization problem solving, the choice of self-learning process may be based on either simple rules or chosen adaptively (like adaptation to environment changes) [37]. The adaptation process also requires some rules or procedures to follow. To make a rational choice of self-learning process, the knowledge must be retrieved, reused, revised and retained. That means it requires systematic archiving of intelligent information. Bajo and Corchado [49] defined the knowledge revising process as the reasoning cycle.

The reasoning capability of agents will make a clear difference between the agent-based EAs and EAs alone. It is known that the incorporation of MAS with EAs would increase the computational time per generation of the new algorithm. However it is expected that a carefully designed hybrid algorithm (MAS plus EAs) would not only improve the quality of solution but also reduce the overall computation time, as is the case for memetic algorithms [50].

Recently, a number of agent-based hybrid evolutionary algorithms have appeared in the literature for solving different problems. For example, Dobrowolski *et al.* [51]

used an evolutionary multi-agent system for solving unconstrained multi-objective problems. Siwik *et al.* [43] developed a semi-elitist evolutionary multi-agent system and they solved so-called *MaxEx* multi-objective problem [43]. Zhong *et al.* [44] used a multi-agent genetic algorithm (MAGA) for solving unconstrained global numerical optimization problems. Barkat Ullah *et al.* [36, 37] used agent-based memetic algorithms for solving constrained optimization problems.

3 Agent-Based Memetic Algorithms

In AMA, we have incorporated the agent concept with memetic algorithms, where an agent stands for a candidate solution in the population. The goal of each agent is to improve its fitness while satisfying constraints. Following the natural adaptation process, in the proposed AMA the agents improve their fitness by selecting a self learning technique, together with the evolutionary adaptation of the population.

The agents are arranged in a lattice-like environment E of size $M \times M$ (where M is an integer). The agents communicate with their surrounding neighbor agents, and interchange information with them through comparison and the crossover operator. The LSLPs are applied to improve the quality of individual agents. As shown in Figure 1, if the goal is achieved the process stops, otherwise the modified agents go through the same process.

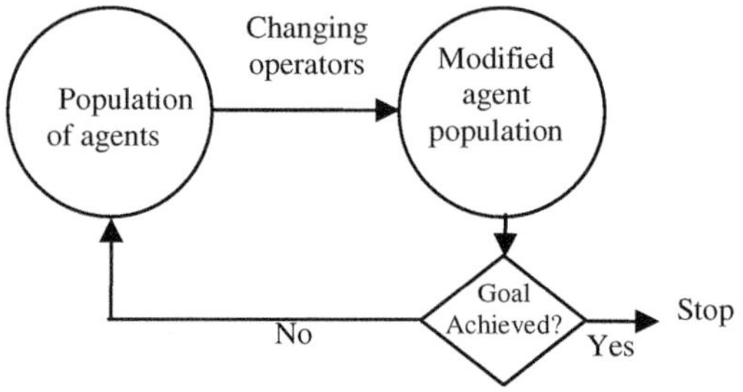

Fig. 1 Agent-based Memetic Algorithm

The overlapped small neighborhoods of the lattice-like environment help in exploring the search space because the induced slow diffusion of solutions through the population provides a kind of exploration (diversification), while exploitation (intensification) takes place inside each neighborhood by crossover operations [52]. We consider that neighborhoods can have four or eight neighbors. For four neighbors we consider the surrounding four agents to the Left, Right, Above and Below in the lattice. For eight neighbors, we consider these same four neighbors, as well as Above-Left, Above-Right, Below-Left and Below-Right. In this paper we use a "*Combined neighborhood Approach*" [37], where the agents consider either four neighbors and eight neighbors interchangeably (i.e. each approach is used for a certain number of generations e.g. 25% of the maximum number of generations alternately) for better performance.

3.1 Crossover

In the proposed AMA, we have used Simulated Binary Crossover (SBX) [35]. SBX operator performs well in solving problems having multiple optimal solutions with a narrow global basin, and has been used in different applications successfully [36, 53, 54].

When this crossover operator is applied on the solution of an agent $A_{i,j}$ (located in location (i,j) in the lattice), the agent searches for its best neighbor agent to mate. The better offspring from these two, denoted as $N_{i,j}$, is stored in a pool. After completing the crossover in a given generation, the fitness of each $N_{i,j}$ $(1 \leq i,j \leq M)$ is compared with its parent $A_{i,j}$. If the new agent's solution's fitness is better than its parent then it takes the place of $A_{i,j}$ and $A_{i,j}$ dies. Otherwise $A_{i,j}$ would survive.

3.2 Life Span Learning Process

Throughout the evolutionary process, after performing crossover a certain percentage of the agents are selected for a LSLP. The idea is that their local learning, together with evolution in the population as a whole, will produce better solutions.

In this paper we introduce a new LSLP for the equality constraints (we call that LSLP type 1). Through this LSLP an agent moves from its current position towards the curvature of an equality constraint, or if it is already there it explores on an equality constraint, by changing one or two variables. We discuss the detail of this LSLP below.

This new LSLP is used in the early stage of the evolution process. Four other LSLPs [37] are then used at the later stage of the evolution process for fine tuning. Among these four different LSLPs, the first one is totally random in nature, the second is restricted random, the third is gradient-based, and the last is directed search. The random LSLPs ensure diversity, and the directed searches try to move towards a better solution which is not necessarily in the individual's locality. The pseudo codes are given later in this section.

3.2.1 New Learning Process for Handling Equality Constraints

The hurdles in solving constrained optimization problems (COPs) arise from the challenge of searching a huge variable space in order to locate good quality feasible points [55]. The size of the feasible space within the search space may depend on the type of the constraints involved in the problem. For example, the existence of equality constraints significantly reduces the size of the feasible space which makes it difficult for the algorithms to find the feasible points [8, 9]. Many traditional EAs [6, 56] convert the equality constraints $h_j(\boldsymbol{X}) = 0$ into inequality constraints $-\delta \leq h_j(\boldsymbol{X}) \leq \delta$ (where δ is a small tolerance value) to increase the feasible space temporarily. Still they may fail to achieve either feasible or good quality solutions.

Here, we introduce a new search process for handling equality constraints. To explain the technique, let us consider an optimization model with all or a subset of constraints of equality type. We now present three propositions related to equality constraints.

Proposition 1: For a given constrained optimization model, where all or a subset of the constraints are of equality type, a feasible solution cannot be found without satisfying any of the equality constraints. We assume that the solution point under consideration satisfies all inequality constraints (if any).

Proof: By definition, a feasible solution point must satisfy all the constraints. To satisfy an equality constraint, the point must be on that constraint.

Proposition 2: A feasible and optimal solution point must lie on each and every equality constraint.

Proof: To satisfy all the equality constraints, a feasible point must lie on all the equality constraints. By definition, the best feasible point is the optimal solution.

Proposition 3: It is possible to find a feasible and optimal solution point by simply searching on an equality constraint function landscape when the function contains all the variables and is continuous.

Proof: As the feasible and optimal point must lie on all equality constraints, by simply moving on a constraint (i.e. points which satisfy the constraint), which involves all the variables, one may be able to reach the optimal solution.

As all variables may not exist in a certain equality constraint and there is no guarantee of having a continuous function, finding a better solution by simply walking on the equality constraint may not work. We use the above arguments to design a new LSLP operator for handling equality constraints in solving optimization problems, as follows:

Choose an equality constraint randomly, and a randomly selected individual.

1. If the equality constraint is not satisfied in this individual, change only one variable so that the constraint is satisfied.
2. If the individual satisfies the constraint, choose two variables and modify them in a way that the resulting point still satisfies the constraint.

The first move would help to reduce overall constraint violation, and the second move would help to increase diversity of individuals in the population.

It is not always possible to satisfy the constraints by changing a single variable. The variable may violate its boundary limit. To ensure the simplicity of the algorithms in that case usually we restrict the variable to the boundary. Sometimes the required value of the variable might be an imaginary number, in that case we assign a random number between its boundary ranges, as shown in the pseudo code.

It is also very complex to move on an equality constraint function landscape. To explore on the constraint landscape we may need to change several variables involved in that constraint. We may need to increase the value of some variables while decreasing the value of some other variables in order to remain on the same equality constraint but at a different point. This is very simple when the constraint involves one or two variables, but becomes complex with more variables. To reduce this complexity in that case, we select randomly only two variables that are involved in the selected equality constraint and then make a small random change on the first variable. The second variable is modified in such a way that the solution remains on the constraint line.

Figure 2 shows an example of how the new LSLP works. Consider a nonlinear optimization problem, consisting of two equality constraints, which can be defined as follows:

Minimize $f(\boldsymbol{X})$, where $\boldsymbol{X} = [x_1, x_2]$, $\boldsymbol{X} \in R^n$ is a set of 2 variables of the solution.
Subject to,

$$h_1(\boldsymbol{X}) = 0,$$
$$h_2(\boldsymbol{X}) = 0,$$
$$\boldsymbol{X} \geq 0.$$

For an example of the first type of move, assume the graphical representation of the problem is like Fig 2. The two equality constraints $h_1(X) = 0$ and $h_2(X) = 0$ intersect at two points. Since only the two intersection points of the constraints satisfy both constraints, we have only two feasible solution points for this problem. As it is a minimization problem consider the objective function value $f(X)$ is optimal on the lower intersection. The new designed LSLP randomly selects an equality constraint $h_1(X) = 0$ for an individual solution. If the solution does not satisfy the constraint (i.e. it does not lie on the line satisfying the equality constraint), the LSLP selects a random variable involved in that constraint, e.g. x_2. Then x_2 is changed so that $h_1(X) = 0$, as shown in Fig 2.

Figure 3 shows the second type of move. If the constraint is satisfied the LSLP will choose two variables x_1 and x_2 and modify them in a way that the resulting point is still on the constraint. This helps to search for the optimum.

Pseudo code: LSLP for Handling Equality Constraints (LSLP Type 1)

Let $\boldsymbol{X}=[x_1,x_2,\ldots,x_n]$ be the current solution vector, CV_j = constraint violation for $h_j(\boldsymbol{X})=0$; $j=1,2,\ldots,q$ with solution vector $\boldsymbol{X}$ and $\boldsymbol{X}_j$ presents a vector that contains the variables involved in constraint $h_j(\boldsymbol{X})=0$. $rnd(.,.)$ is a uniform random number generator.
Select any of the equality constraints $h_j(\boldsymbol{X})=0;\ j=1,2,\ldots,q$ randomly.

If $(CV_j \neq 0)$

Select a random variable $\{ x_a | x_a \in \boldsymbol{X}_j \}$.

Calculate x'_a so that $CV_j=0$;

If x'_a is an imaginary number, set $x'_b = rnd(\underline{x_a}, \overline{x_a})$

If ($x'_a > \overline{x_a}$) set $x'_a = \overline{x_a}$

if $x'_a < \underline{x_a}$; set $x'_a = \underline{x_a}$

Else

Select a random variable $\{ x_a | x_a \in \boldsymbol{X}_j \}$.

Set $x'_a = x_a + rnd(-0.1,0.1) \times x_a$

Select a random variable $\{ x_b | x_b \in \boldsymbol{X}_j, a \neq b \}$.Calculate x'_b such that $C_1 x'_a \oplus C_2 x'_b = C_1 x_a \oplus C_2 x_b$, where $\oplus$ represents any mathematical operators (e.g. $+, -, \times, /$), C_1, C_2 are constants.

End.

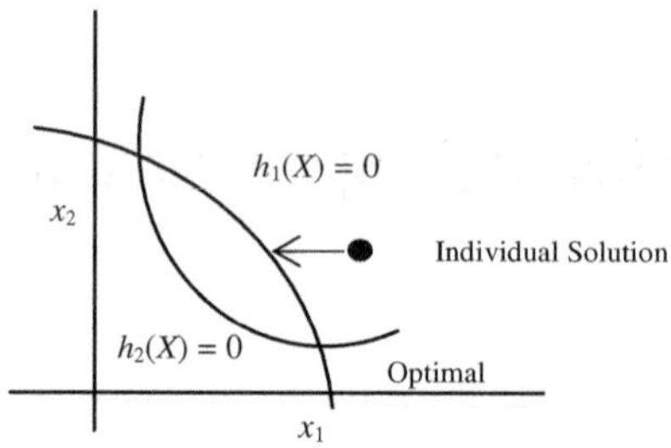

Fig. 2 New LSLP (when the solution does not satisfy the equality constraint)

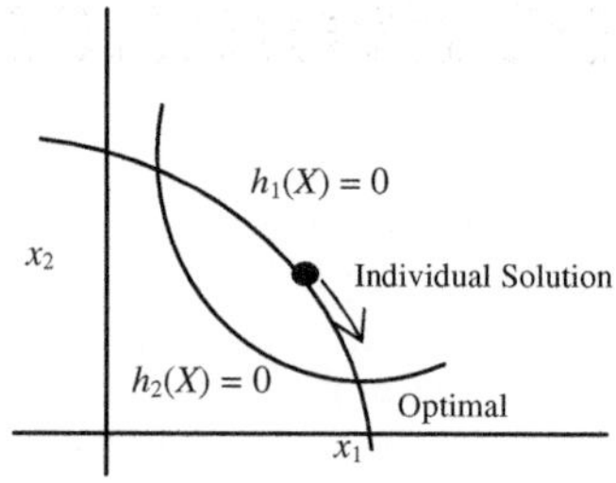

Fig. 3 New LSLP (when the solution satisfies the equality constraint)

3.2.2 Pseudo Code of the Other LSLPs

Let an agent $A_{i,j}$, residing at cell location (i,j) with solution vector of n variables $[a_1,a_{2,\ldots,}a_n]$, be selected for LSLP. Let m be the maximum number of learning steps, and Δ is a positive small value for perturbation.

LSLP Type 2

Step 1. Choose a variable r randomly from n variables

Step 2. Calculate Δ, add / subtract Δ with the variable value a_r and evaluate fitness, and detect in which direction the fitness of the solution vector improves.

Step 3. For t= 1 to m do

Step 3.1. Change the variable value a_r with Δ according to the direction found in step 2.

Step 3.2. If the fitness deteriorates go to step 1 else go to step 3.

[End of Step 3 Loop]

Step 4. Continue the previous steps (1 to 3) until all n variables are modified and then go to step 5.

Step 5. Stop.

LSLP Type 3

Step 1. For all variables r =1 to n do

Step 2. CalculateΔ, add / subtract Δ with the variable a_r and evaluate fitness, and detect in which direction the fitness of the solution vector improves.

Step 3. For $t = 1$ to m do

Step 3.1. Change the variable a_r with Δ according to the direction found in step 2.

Step 3.2. If the fitness deteriorates, go to step 1 else go to step 3.

[End of Step 3 Loop]

[End of Step 1 Loop]

Step 4. Stop.

LSLP Type 4

Step 1. For all variables $r = 1$ to n do

Step 2. Calculate Δ, add / subtract Δ with the variable a_r and evaluate fitness, and detect in which direction the fitness of the solution vector improves.

Step 3. Change the variable a_r according to the direction found in step 2. Find the improvement of the fitness for this change.

[End of Step 1 Loop]

Step 4. Rank the variables based on their effect on the fitness improvement.

Step 5. For all n variables starting from highest rank do

Step 5.1. For $t = 1$ to m do

Step 5.2. Change a_r as $a_r = a_r \pm \Delta$; based on the direction found in step 2.

Step 5.3. If the fitness deteriorates go to step 5, otherwise go to step 5.1

[End of Step 5.1 Loop]

[End of Step 5 Loop]

Step 6. Stop

LSLP Type 5

Step 1. Find the agent with best fitness in the current generation with solution vector $[b_1,b_2,\ldots,b_n]$.

Step 2. For all variables $r = 1$ to n do

Step 2.1. For $t= 1$ to m do

Step 2.2. Calculate Δ.

Step 2.3. If $(a_r > b_r)$ then $a_r = a_r - \Delta$;

Step 2.4. If $(a_r < b_r)$ then $a_r = a_r + \Delta$;

[End of Step 2.1 Loop]

[End of Step 2 Loop]

Step 3. Stop.

The type 2 and 3 LSLPs of the agent try to exploit the existing solution vector by attempting to change the variables. The type 4 attempts to move the solution vector in the direction of gradient and the last type of LSLP leads the agents towards the current best solution. The last two LSLPs try to make the agents converge; the first two maintain diversity. Details of these four LSLPs can be found in [37].

To select an appropriate LSLP among these LSLPs (type 2-5), the agents check a number of simple rules. Initially all the agents are assigned different types of LSLPs randomly with Improvement Index (*II*) zero. Here, *II* indicates the rate of fitness improvement made by a particular LSLP type assigned to an agent. A positive value of

II indicates that fitness is improved by the LSLP, while a negative value indicates deterioration of fitness. When selecting a LSLP, an agent checks the parents' type of LSLPs and *II* values. The parents may have different type of LSLPs associated with *II* values. The offspring will choose the LSLP which has the higher *II* value. Since an agent selects a LSLP based on the knowledge experienced by the parents the adaptation level of the algorithms is Local-level adaptation.

3.3 The Algorithm

The main steps of the proposed algorithm are follows:

Step 1. Create a random population, which consists of $M \times M$ agents.
Step 2. Arrange the agents in a lattice-like environment.
Step 3. Evaluate the agents individually. If the stopping criterion has been met, go to step 7, otherwise go to step 4.
Step 4. For each agent examine its neighborhood. Select an agent from its neighborhood and perform crossover.
Step 5. Select a certain percentage of agents; During the initial generations apply only the LSLP for equality constraint (type 1), then switch to other LSLPs.
Step 6. Go to step 3.
Step 7. Stop.

AMA involves Meta-Lamarckian learning, rather than Lamarckian or Baldwinian learning. In AMA the improved agent (after applying LSLP) is sent back into the population, which follows Lamarckian learning. Since we are using multiple LSLPs during a MA search in the spirit of Lamarckian learning we can say AMA is following Meta-Lamarckian learning [27].

3.4 Constraint Handling

The constraints are handled indirectly while comparing the individuals in ranking. In the pair-wise comparison the best infeasible individual is assigned worse fitness than the worst feasible individual. As such, when comparing two individuals an infeasible individual is penalized and a feasible individual is rewarded, so the constraints are handled indirectly.

4 Experimental Studies

4.1 Initial Design Experience

During the initial design of the LSLP for equality constraints we experienced some interesting situations. We would like to share some of them with the readers.

According to proposition 2, a feasible and optimal solution point must lie on each and every equality constraint. So if we explore any of the equality constraints it is possible to reach the optimal. In our initial design of the new LSLP for equality constraints, we examined only a single equality constraint, by ignoring all other functions, for finding a good quality solution. However, this approach was not consistently providing benefit, as a single equality constraint may not contain all the variables involved in the model.

In the second attempt, we selected one of the equality constraints randomly and then changed one variable by assigning it a value to reach the constraint. For example let us consider problem g13 constraint $h_2(x) = x_2x_3 - 5x_4x_5 = 0$.

According to the first move of the new LSLP we need to select a random variable (e.g. x_2). The required value of x_2 to reach to the equality constraint is $x_2 = \frac{5x_4x_5}{x_3}$. If the target constraint cannot be reached, due to the variable bounds, we assign the variable its upper/lower bound value (which is close to the calculated value). This works fine for most of the problems. However in some test problems it raises a different issue e.g. in problem B01 for any $x_i = 0$ (lower bound) then objective function value $f(x) = \sum_{i=1}^{10} x_i \left(c_i + \ln \frac{x_i}{\sum_{j=1}^{10} x_j} \right)$ will be infinity. To avoid this, in this case we assign a very low to x_i (e.g. x_i=1E-20).

Sometimes the required value of the variable might be an imaginary number, e.g. in problem g11 where $h(x) = x_2 - x_1^2 = 0$, for any negative value of x_2, the required value of $x_1 = \sqrt{-x_2}$ must be an imaginary number. In that case we assign a random number between its boundary ranges.

Furthermore when we try to explore on an equality constraint we change the values of only two variables, to keep it computationally simple. For example in $h_2(x) = x_2x_3 - 5x_4x_5 = 0$, suppose we select randomly the variables x_2 and x_5. Then we change the value of x_2 as $x_2' = x_2 + rnd(-0.1, 0.1) \times x_2$ and we calculate the new value of x_5' as $x_2x_3 - 5x_4x_5 = x_2'x_3 - 5x_4x_5'$. However this process was not simple after all, as the variable may contain any type of nonlinearity. In particular, a variable with a power of 2 could have negative or positive value and have the same impact on satisfying the constraint. For example in problem g03 for any value x_i (positive or negative value) the result is same as $h_1(x) = \sum_{i=1}^{n} x_i^2 - 1 = 0$. If we always assigned a positive value, or always a negative value, we could find it impossible to find the global optimum. We changed our initial LSLP so that in cases of variables with even powers, we assigned randomly either a positive or negative value. This process consistently provided better solutions.

In our third attempt, we used the new LSLP for only the first 100 generations (arbitrarily chosen) and the other four LSLPs afterward. The reason for this choice was due to the fact that the new LSLP increases the diversity of the population. We observed that this process provided even better results.

4.2 Experimental Results and Discussion

We evaluated the performance of our algorithm using those problems containing equality constraints (g03, g05, g11, and g13) from 13 well known benchmark problems (*g*-series) [38]. These problems were initially studied by Michalewicz and Schoenauer [4], Koziel and Michalewicz [57], and further studied by Runarsson and Yao [38] and others. Mezura-Montes and Coello [8] in their research categorized the test problems by difficulty, and reported g05 and g13 as "Very difficult" and g03 as "Difficult". Two new problems (indicated here as B01-B02) considered in this paper are taken from [39].

Table 1 Characteristics of the Test Problems

Prob	*n*	Obj. Func.	LI	LE	NE	AC	Optimal
g03	10	Polynomial	0	0	1	1	-1.000
g05	4	Cubic	2	0	3	3	5126.498
g11	2	Quadratic	0	0	1	1	0.750
g13	5	Nonlinear	0	0	3	3	0.053950
B01	10	Nonlinear	0	3	0	3	-47.764
B02	3	Quadratic	0	1	1	2	961.715

n= number of variables, LI = Linear Inequalities, NI = Nonlinear Inequalities, LE = Linear Equalities, NE = Nonlinear Equalities, AC = Active Constraints.

These problems include different forms of objective function (quadratic, cubic, polynomial, and nonlinear) and different numbers of variables (n). Problem g03 is a maximization problem and the others are minimization problems. The maximization problem is transformed into an equivalent minimization problem. The main characteristics of the benchmark problems are presented in Table 1 and the detailed mathematical representations are provided in the Appendix.

First we investigated the performance of three algorithms:

- SGA: a simple genetic algorithm with tournament selection, SBX crossover and parameter based mutation operator [6].
- AMA-I: AMA as described in [37].
- AMA-II: AMA with the new LSLP for equality constraints.

The initial solution vectors for the agents were randomly generated within the boundary of each decision variable. For AMA-I and AMA-II the agents are arranged in a lattice-like environment of size $M \times M$, so the number of the agents (i.e. the population size) must be a square number. For each algorithm we used the same population size (100), probability of crossover (P_C = 0.90), and probability of learning/LSLP (P_L= 0.2). For SGA the probability of mutation used was P_M=0.2. The maximum number of generations considered was 3500. For each problem Table 2 shows the results from 30 independent runs for each algorithm.

From Table 2, we can see that SGA could not find any feasible solution for problems g03, g05, g11, and B02. Although SGA is able to find feasible solutions for the other three problems, the best solution obtained is far from the known best. Both AMA-I and AMA-II have solved those problems successfully. AMA-I achieves the optimum in three problems (g03, g11, and B02). AMA-II achieves the optimum in four problems (g03, g11, g13, and B02). If we consider the mean results, AMA-II achieves better mean results than AMA-I in g05, g13, B01, and B02. Though the mean results of AMA-I and AMA-II are same for g03 and g11, AMA-II achieves better standard deviation for those problems.

The results achieved by AMA-II show that the inclusion of the new LSLP has enhanced the ability of AMA in solving constrained optimization problems with equality constraints.

4.3 Comparison with Other Algorithms

In this section, we compare the performance of algorithm AMA-II with others in terms of solution quality and computation time. The algorithms compared are Runarsson and Yao's Evolutionary Strategy (ES) based algorithm [38], multi-objective genetic algorithm NSGA-II by Deb [40], and Multi-objective Constrained Algorithm (MCA) proposed by Ray and Sarker [58]. The available best, median, mean, standard deviation (st.dev.), and worst results, as well as execution time, are given in Table 2. (An 'x' in Table 2 indicates that the algorithm did not find any feasible solution.)

Runarsson and Yao (abbreviated as RY) [38] used an interesting ranking procedure, known as stochastic ranking, in their ES-based algorithm. They solved all the *g*-series problems. As the ES-based algorithm [38] produced the best results known so far for all the test problems (g01-g13) we include its result in Table 2 for comparison. Deb *et al* [40] have proposed a computationally fast and elitist multi-objective evolutionary algorithm, based on a non-dominated sorting approach, called NSGA-II. The difference between the conventional single objective GA and NSGA-II lies with the assignment of fitness of an individual. The fitness of an individual in NSGA-II is based on the non-domination level of an individual. NSGA-II is a well-known and well-accepted algorithm for solving both single- and multi-objective constrained optimization problems.

Multi-objective Constrained Algorithm (MCA) [41, 58], proposed by Ray and Sarker, is a close variant of NSGA-II. It has two major differences, in the selection of parents and the process of population reduction. This algorithm can also be used for solving both single- and multi-objective constrained optimization problems.

Due to the availability of code, we were able to use NSGA-II (from http://www.iitk.ac.in/kangal/codes.shtml) and MCA (by personal correspondence) to solve the problems. We ran both NSGA-II and MCA on all test problems, using the same parameters as for SGA. The performance of each algorithm is shown in Table 2.

From Table 2 it is very clear that the performance of AMA-II is better than NSGA-II for all test problems. MCA achieves better results than AMA-II in only one problem (B01); for the other five problems AMA-II has achieved better results than MCA (though for g03 both AMA-II and MCA achieve same best, mean, median and worst results, AMA-II shows better performance by achieving better standard deviation). RY successfully solved the first four problems, and are well known for their performance in solving the *g* series problems (the results of B01, B02 and computational time for all problems of RY are not available in the literature).

Table 2 Statistics for 30 Independent Runs of Different Algorithms

Prob	Optimal	App.	Best	Median	Mean	St.Dev	Worst	Time(s)
G03	-1.000	SGA	×	×	×	×	×	4.14
		RY	-1.000	-1.000	-1.000	1.90E-04	-1.000*	–
		NSGA-II	-0.616	-0.106	-0.171	1.66E-01	-0.003	5.89
		MCA	-1.000	-1.000	-1.000	5.69E-02	-1.000	284.77
		AMAI	-1.000	-1.000	-1.000	2.76E-06	-1.000	80.86
		AMAII	-1.000	-1.000	-1.000	**1.70E-06**	-1.000	62.87
G05	5126.498	SGA	×	×	×	×	×	3.52
		RY	5126.497	5127.372	5128.881	3.50E+00	5142.472	–
		NSGA-II	×	×	×	×	×	6.18
		MCA	5131.295	5364.948	5450.763	3.27E+02	6111.453	482.95
		AMAI	5127.388	5186.761	5208.999	8.20E+01	5463.927	27.62
		AMAII	5126.499	**5126.985**	5129.054	3.69E+00	**5136.921**	23.31
G11	0.750	SGA	×	×	×	×	×	2.55
		RY	0.750	0.750	0.750	8.00E-05	0.750*	–
		NSGA-II	0.754	0.867	0.878	9.05E-02	0.999	7.82
		MCA	0.760	0.760	0.760	1.43E-04	0.761	195.04
		AMAI	0.750	0.750	0.750	3.35E-03	0.750	14.92
		AMAII	0.750	0.750	0.750	**4.00E-07**	0.750	11.11
G13	0.053950	SGA	0.067448	0.465841	0.501076	2.54E-01	1.000018	4.79
		RY	0.053957	0.057006	0.067543	3.10E-02	0.216915	–
		NSGA-II	0.974036	0.994066	0.991300	8.59E-03	0.999997	6.44
		MCA	0.137313	0.416643	0.455166	1.69E-01	0.999374	224.36
		AMAI	0.056754	0.069702	0.242848	2.15E-01	0.820884	40.72
		AMAII	**0.053950**	**0.055993**	**0.055616**	**8.63E-04**	**0.056540**	35.80
B01	-47.764	SGA	-47.330	-41.528	-41.723	1.94E+00	-38.607	3.47
		NSGA-II	-46.874	-41.384	-41.831	1.64E+00	-39.527	5.53
		MCA	**-47.764**	**-47.764**	**-47.764**	**0.00E+00**	**-47.764**	217.15
		AMAI	-46.542	-46.542	-40.094	1.36E+01	-39.808	57.03
		AMAII	-47.688	-47.688	-47.155	3.09E-01	-46.696	77.93
B02	961.715	SGA	×	×	×	×	×	3.29
		NSGA-II	963.764	973.842	973.671	4.16E+00	978.789	7.30
		MCA	961.870	964.849	965.128	2.35E+00	968.546	245.26
		AMAI	961.715	961.716	965.089	2.43E+00	970.210	13.90
		AMAII	961.715	961.716	**961.716**	**1.21E-03**	**961.720**	15.73

Opt= Optimal, '×'= feasible solution were not found, * indicates though the best, worst, median, and mean results are the same, standard deviation is positive due to rounding error. '–'= No result found in the literature. **Bold** fonts indicate the best result achieved.

For g05, no algorithm achieved the optimum (due to the use of relaxation in equality constraint, RY's best result is less than the optimal). Though the mean result for RY is better than AMA-II, there is no statistically significant difference between the two algorithms' achievement for this problem (for the statistical test we used students t-test and degrees of freedom is considered here 60 and t-tabulated value is 2). The median and worst results for AMA-II are also better than for RY. SGA and NSGA-II haven't found any feasible solutions for g05.

For problem g11 and g03, both RY and AMA-II achieved the optimal result and the same mean and median results. However AMA-II achieved better standard deviation.

For g13, only AMA-II has achieved the optimal result. The achievement of AMA-II is not only statistically significant better than RY but also better than all other algorithms.

We could not compare the execution time against RY, because their code was not available to us; for all the others we can make a fair comparison of execution time because they are all being used in the same environment.

SGA is very simple and quick in solving the problems (on average it needs 3.52 seconds to solve each problem). However SGA provided poor results for two test problems (g13 and B01) and could not solve other four problems.

NSGA-II took on average 6.53 seconds to solve each problem, but it could not find the optimal solution for any problem, or any feasible solution at all for problem g05.

MCA achieved the optimal solution for two problems but required huge computational time, averaging 274.92 sec across all of the problems (high computational time is a drawback of this algorithm, as mentioned by the authors [58]).

In AMA-II, during the initial generations it only uses the new LSLP process and so the *I.I.* calculation and other four LSLPs (involving iterative process) are not used in that period. As a result the new LSLP reduces the execution time of AMA-II by 3.66% compared to AMA-I. The new AMA-II took on average 37.79 sec to solve each problem.

Although the execution time of AMA-II is higher than NSGA-II, with the cost of this additional time AMA-II achieves much better quality solutions than NSGA-II. Considering mean performance, AMA-II solves all problems and (in relative terms) achieves an answer that averages less than one percent above the optimum (5 of 6 are less than 1% from the optimum, one is 3% from the optimum). NSGA-II can't solve one problem, is far from the optimum on two, is 12 or 17% above the optimum on two, and only gets to within 1% of optimum on one problem. The time penalty for AMA-II to achieve these much better results is not great (30 seconds difference on average).

Compared to SGA and NSGA-II, AMA-II has a trade-off of speed against solution quality: AMA-II is slower but gives better results. However AMA-II is better than MCA and AMA-I in both speed and quality of solution.

In summary, these results show that AMA-II not only achieves better quality solutions than the other algorithms, it is also time efficient in solving COPs with equality constraints.

4.4 Effect of the New LSLP

As the results shows, with the inclusion of the new LSLP, AMA-II always performs better than AMA-I. To visualize the internal effect of the new LSLP we present two convergence curves from the results with AMA-I and AMA-II, on problem g13. The objective function of this problem is nonlinear and involves equality constraints. As this problem is very difficult to solve (according to Mezura-Montes and Coello [8]), we choose this problem to show the effect of the new LSLP. Figure 4 shows the best objective function value achieved by the two algorithms at every generation for 1000 generations. To see the effect on the whole population, Figure 5 shows the average objective function of the whole population.

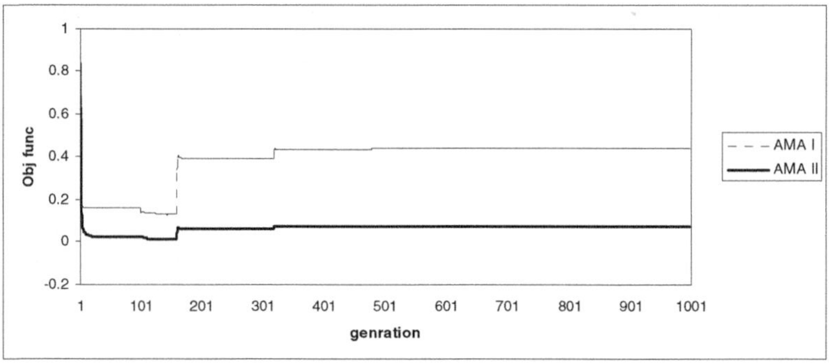

Fig. 4 Convergence curve from the best objective function of AMA-I and AMA-II for problem g13

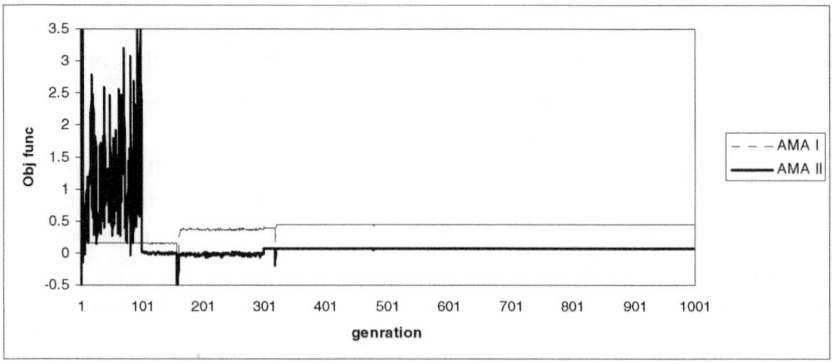

Fig. 5 Convergence curve from the average objective function of AMA-I and AMA-II for problem g13

As the test problems include equality constraints, the feasible space is very tiny. During the initial generation, several individuals in the agent population may have better objective function values but with huge constraint violations. This is why the convergence curves drop from initial high values: although the initial objective function values may be good, the constraints are not satisfied. With time the constraint violation is reduced and objective function values move back towards optimal values, now with low constraint violation.

In most problems, AMA-II converges faster towards the best objective function value. Usually when the new LSLP is applied, it tries to move the agent's existing solution to towards the randomly selected equality constraints. As a result the agents may reach (or nearly reach) different points of different equality constraints. Hence it ensures enough diversity in the population with good quality solutions. That is why AMA-II converges faster than AMA-I in Fig 4. However in some problems, like g03 and g11, both versions of AMA show similar behavior, since both versions achieve the optimal solution in their early generations and solved the problems easily. In Fig 5, for AMA-II, during the first 100 generations (while the new LSLP is working) the average of the objective function values of the population is very high with a high rate of change. This is due to the effect of the increased diversity by the new LSLP. When the new LSLP stops (after 100 generations) the average objective function improves slowly with the other four LSLPs. On the other hand AMA-I has lower diversity and converges slowly. The same behavior is seen for most other problems. We can say the new LSLP helps AMA-II to converge faster, with better quality solutions, than AMA-I.

4.5 Effect of Probability of Using LSLP

The probability of LSLP (P_L) is another important parameter of the algorithm. For low value of P_L a small number of agents are selected for LSLP, which allows lower diversity. On the other hand higher value of P_L allows more agents to use LSLP, which provides more diversity. Figure 6 shows the diversity of the population, for the first 500 generations for problem g13, with different probability of LSLP (we have only shown the diversity using P_L values of 0.05, 0.1, 0.15 to see the effects clearly). We use the average Euclidian distance among the all solution vectors in the population to measure the diversity. In the first 100 generations the agents use LSLP type 1 (designed for handling equality constraints), which ensures enough diversity during that period. Afterwards we can see that higher P_L allows higher diversity in the population. The behavior is same for the other test problems.

Diversity is an important issue for the performance of any population-based search algorithm. Lack of diversity means the algorithm may not perform well; on the other hand, over-diversification causes slower convergence. To see the effect of P_L on the performance of AMA-II, we have run it 30 times on each problem with different values of P_L (0.05, 0.1, 0.15, 0.20, 0.25, and 0.3). For some problems, the algorithm may achieve the optimal solution as an outlier with any of these values of P_L. However if we consider the mean and standard deviation of the results we can see the effect of P_L. From our experiment we realize that for most problems the mean and standard deviations improve with an increase of P_L up to a point, then the quality degrades due to the over diversification. For g05, g13, and B01, AMA-II achieves the best mean results with $P_L = 0.20$. With $P_L = 0.25$ it has achieved best mean for B02. For some easy problems like g03 and g11, the algorithm achieves almost the same results with any P_L. Sometimes a high value of P_L may over diversify the population, shown by high standard deviations in the results.

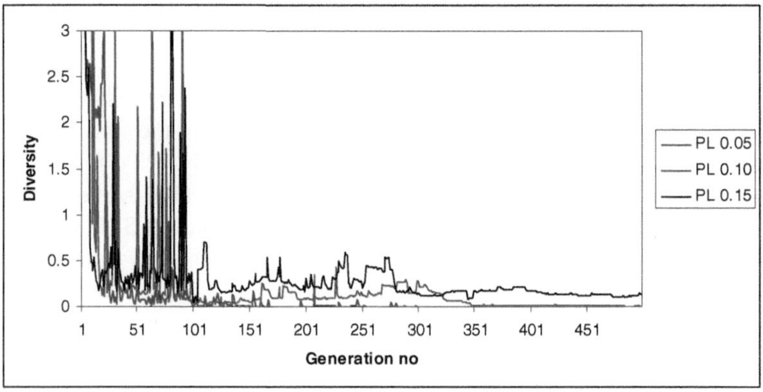

Fig. 6 Effect of Probability of LSLP (P_L) on population diversity for problem g13

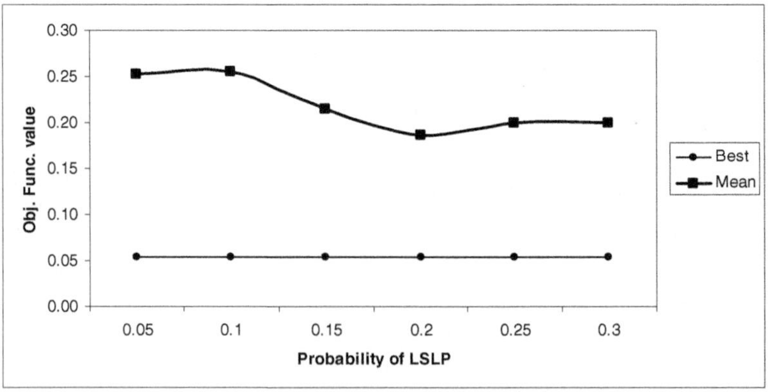

Fig. 7 Effect of Probability of LSLP (P_L) on problem g13. Probability of LSLP vs. achieved best and mean results

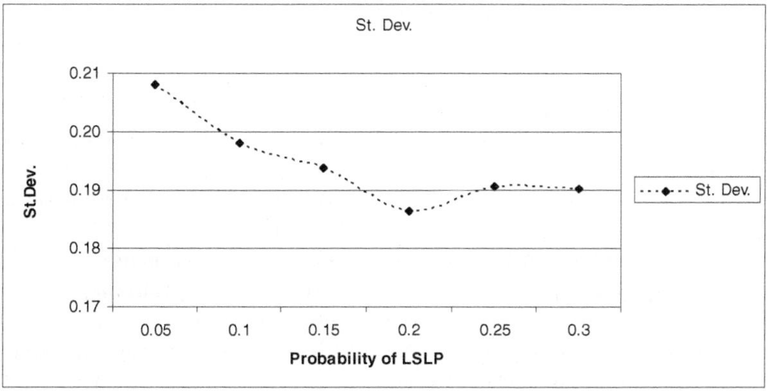

Fig. 8 Effect of Probability of LSLP (P_L) on problem g13. Probability of LSLP vs. St.Dev. of achieved results

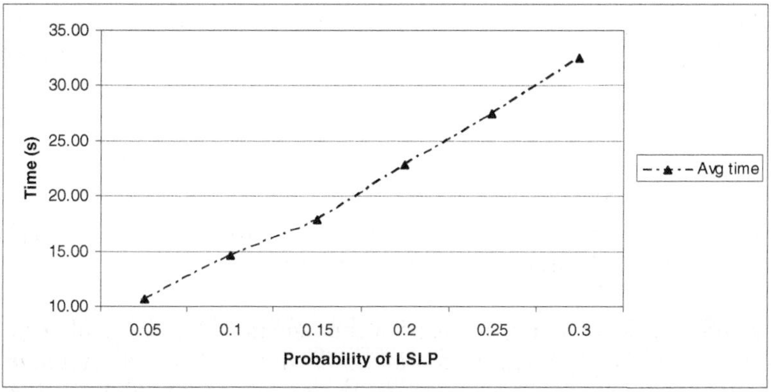

Fig. 9 Effect of Probability of LSLP (P_L) on problem g13. Probability of LSLP vs. Average time required

Figure 7 shows the best and the mean performance of AMA-II on problem g13 as P_L varies; Figure 8 shows standard deviation, and Figure 9 shows average execution time, as P_L varies. It shows by increasing P_L both the mean and standard deviations improve gradually, but once P_L exceeds 0.2 the results are not improving. This is because higher P_L causes slow convergence in the population and so the performance degrades at the same level of generations. That indicates that increasing P_L can improve the performance of the algorithm, however, after a certain point there will be no significant improvement.

We allow more agents to apply LSLP by increasing the probability of P_L. This causes extra time to perform the additional task. The required time is increasing in proportion to the increase of P_L as shown in Fig 9.

So far we have executed the all algorithms up to 350,000 fitness evaluations. One question may arise is whether the other methods could find as good solutions as AMA-II if they were allowed to run for enough generations (i.e. allowed for more fitness evaluations). To investigate this we executed the other algorithms (SGA, NSGA-II, MCA and AMA-I) for 500, 000 fitness evaluations. We found that SGA improved its performance, by achieving feasible solutions in problem g3, g11 and B02. The new mean result for B02 is 0.55% from the optimum. However the mean results of g03 and g11 are still 98.26% and 15.46% away from the optimum. With the additional fitness evaluations, AMA-I also improved a bit in g13. However the best result achieved by AMA-I is still 0.396% and the mean result is 311.8% away from the optimum result. MCA has achieved 0.14% better mean result in B02 than before, which is still 0.22% away from the optimum. It also improved its mean result for problem g13 by 2.72%, however the mean result is 720.74% away from the optimum. NSGA-II could not improve its performance even with the additional fitness evaluations. This shows that the other methods are still so unlikely to find an optimal solution (even with the additional fitness evaluations) because the feasible space of these problems is so small, that an idea like this new LSLP is essential in order to have a realistic chance of finding optimal solutions in practice.

5 Conclusions

This paper has introduced a new learning process to handle equality constraints. We have extended AMA, which uses memetic algorithms and agent-based concepts, and is now capable to solve nonlinear optimization problems with equality constraints more efficiently.

The results show the enhanced performance achieved by the new AMA in terms of solution quality and computational time. The constraint handling techniques used here do not need any penalty functions or additional parameters.

The performance of the proposed algorithms is compared with other algorithms using benchmark problems and the comparison shows promising performance.

The effect of the new learning process and probability of using the learning process is also analyzed. The experimental results show that faster convergence can be achieved with the new learning process. The probability of learning (P_L) is also an important parameter. The performance of the algorithm improves with the increase of P_L up to certain point; beyond that point, high values of P_L may over diversify the population and slow down the convergence.

In future work, we shall investigate the performance of our algorithm in solving more complex problems and will analyze the dynamics of different operators and parameters in solving those problems.

6 Appendix

6.1 g03

Minimize [59]:

$$f(x) = \left(\sqrt{n}\right)^n \prod_{i=1}^{n} x_i$$

subject to:

$$h_1(x) = \sum_{i=1}^{n} x_i^2 - 1 = 0$$

where $n = 10$ and $0 \le x_i \le 1\ (i = 1,\ldots,n)$. The global maximum is at $x^* = 1/\sqrt{n}\ (i = 1,\ldots,n)$ where $f(x^*) = -1.000$.

6.2 g05

Minimize [60]:

$$f(x) = 3x_1 + 0.000001x_1^3 + 2x_2 + 0.000002/3x_2^3$$

subject to:

$$g_1(x) = -x_4 + x_3 - 0.55 \le 0,$$
$$g_2(x) = -x_3 + x_4 - 0.55 \le 0,$$
$$h_3(x) = 1000\sin(-x_3 - 0.25) + 1000\sin(-x_4 - 0.25) + 894.8 - x_1 = 0,$$
$$h_4(x) = 1000\sin(x_3 - 0.25) + 1000\sin(x_3 - x_4 - 0.25) + 894.8 - x_2 = 0,$$
$$h_4(x) = 1000\sin(x_4 - 0.25) + 1000\sin(x_4 - x_3 - 0.25) + 1294.8 = 0.$$

where $0 \le x_1 \le 1200, 0 \le x_2 \le 1200, -0.55 \le x_3 \le 0.55$, and $-0.55 \le x_4 \le 0.55$. The best known solution is x^* = (679.9453, 1026.067, 0.1188764, -0.3962336) where $f(x^*)$=5126.4981.

6.3 g11

Minimize [61]

$$f(x) = x_1^2 + (x_2 - 1)^2$$

subject to

$$h(x) = x_2 - x_1^2 = 0,$$

where $-1 \le x_1 \le 1$ and $-1 \le x_2 \le 1$. The optimum solution is $x^* = \left(\pm 1/\sqrt{2}, 1/2\right)$ where $f(x^*) = 0.750$.

6.4 g13

Minimize [60]

$$f(x) = e^{x_1 x_2 x_3 x_4 x_5}$$

Subject to:

$$h_1(x) = x_1^2 + x_2^2 + x_3^2 + x_4^2 + x_5^2 - 10 = 0$$

$$h_2(x) = x_2 x_3 - 5 x_4 x_5 = 0$$

$$h_3(x) = x_1^3 + x_2^3 + 1 = 0$$

Where $-2.3 \le x_i \le 2.3$ (i = 1, 2) and $-3.2 \le x_i \le 3.2$ (i = 3, 4, 5). The optimum solution is $x^* = (-1.7171422400\ 3, 1.5957212404\ , 1.8272502406\ , -0.76365988191286, -0.76365986736498)$ where $f(x^*) = 0.05394151404$.

6.5 B01

Minimize [39]:

$$f(x) = \sum_{i=1}^{10} x_i \left(c_i + \ln \frac{x_i}{\sum_{j=1}^{10} x_j} \right)$$

Subject to:

$$h_1(x) = x_1 + 2x_2 + 2x_3 + x_6 + x_{10} - 2 = 0$$

$$h_2(x) = x_4 + 2x_5 + x_6 + x_7 - 1 = 0$$

$$h_3(x) = x_3 + x_7 + x_8 + 2x_9 + x_{10} - 1 = 0$$

Where $0 \le x_i \le 10$ (i = 1, …, 10) and $c_1 = -6.089$, $c_2 = -17.164$, $c_3 = -34.054$, $c_4 = -5.914$, $c_5 = -24.721$, $c_6 = -14.986$, $c_7 = -24.1$, $c_8 = -10.708$, $c_9 = -26.662$, $c_{10} = -22.179$. The best known solution is at x^* = (0.0406684113216282, 0.147721240492452, 0.783205732104114, 0.00141433931889084, 0.485293636780388, 0.000693183051556082, 0.0274052040687766, 0.0179509660214818, 0.0373268186859717, 0.0968844604336845) where $f(x^*) = -47.7648884594915$.

6.6 B02

Minimize [39]:

$$f(x) = 1000 - x_1^2 - 2x_2^2 - x_3^2 - x_1 x_2 - x_1 x_3$$

subject to:

$$h_1(x) = x_1^2 + x_2^2 + x_3^2 - 25 = 0$$

$$h_2(x) = 8x_1 + 14x_2 + 7x_3 - 56 = 0$$

where the bounds are $0 \le x_i \le 10$ (i = 1, 2, 3). The best known solution is at x^* = $(3.5121281261, 0.2169875104, 3.5521785492)$ where $f(x^*) = 961.715022289961$.

References

[1] Ho, P.Y., Shimizu, K.: Evolutionary constrained optimization using an addition of ranking method and a percentage-based tolerance value adjustment scheme. Information Sciences 177, 2985–3004 (2007)
[2] Back, T., Hammel, U., Schwefel, H.P.: Evolutionary computation: comments on the history and current state. IEEE Transactions on Evolutionary Computation 1, 3 (1997)
[3] Chootinan, P., Chen, A.: Constraint handling in genetic algorithms using a gradient-based repair method. Computers & Operations Research 33, 2263–2281 (2006)
[4] Michalewicz, Z., Schoenauer, M.: Evolutionary algorithms for constrained parameter optimization problems. Evolutionary Computation 4, 1–32 (1996)
[5] Venkatraman, S., Yen, G.G.: A Generic Framework for Constrained Optimization Using Genetic Algorithms. IEEE Transactions on Evolutionary Computation 9, 424–435 (2005)
[6] Deb, K.: An efficient constraint handling method for genetic algorithms. Computer Methods in Applied Mechanics and Engineering 186, 311 (2000)
[7] Coello Coello, C.A.: Theoretical and numerical constraint-handling techniques used with evolutionary algorithms: a survey of the state of the art. Computer Methods in Applied Mechanics and Engineering 191, 1245–1287 (2002)
[8] Mezura-Montes, E., Coello, C.A.C.: A Numerical Comparison of Some Multiobjective-Based Techniques to Handle Constraints in Genetic Algorithms, Dept. Ing. Eléct., México (2002)
[9] Barkat Ullah, A.S.S.M., Sarker, R., Lokan, C.: An Agent-based Memetic Algorithm (AMA) for Nonlinear Optimization with Equality Constraints. In: The 2009 IEEE Congress on Evolutionary Computation (CEC 2009), Norway (2009)
[10] Merz, P., Freisleben, B.: Genetic local search for the TSP: new results. In: IEEE International Conference on Evolutionary Computation, pp. 159–164 (1997)
[11] Cheng, R., Gen, M.: Parallel machine scheduling problems using memetic algorithms. In: IEEE International Conference on Systems, Man, and Cybernetics, pp. 2665–2670 (1996)
[12] Burke, E.K., Smith, A.J.: A memetic algorithm to schedule planned maintenance for the national grid. Journal of Experimental Algorithmics 4 (1999)
[13] Tang, J., Lim, M.H., Ong, Y.S., Er, M.J.: Solving large scale combinatorial optimization using PMA-SLS. In: Proceedings of the 2005 conference on Genetic and evolutionary computation. ACM Press, Washington (2005)
[14] Merz, P., Freisleben, B.: Fitness landscape analysis and memetic algorithms for the quadratic assignment problem. IEEE Transactions on Evolutionary Computation 4, 337–352 (2000)
[15] Alkan, A., Ozcan, E.: Memetic algorithms for timetabling. In: The 2003 Congress on Evolutionary Computation, pp. 1796–1802 (2003)
[16] Vavak, F., Fogarty, T., Jukes, K.: A genetic algorithm with variable range of local search for tracking changing environments. In: Ebeling, W., Rechenberg, I., Voigt, H.-M., Schwefel, H.-P. (eds.) PPSN 1996. LNCS, vol. 1141, pp. 376–385. Springer, Heidelberg (1996)
[17] Knowles, J., Corne, D.: A comparative assessment of memetic, evolutionary and constructive algorithms for the multi-objective d-msat problem. In: GECCO 2001 Workshop Program, pp. 162–167 (2001)

[18] Knowles, J.D., Corne, D.W.: M-PAES: a memetic algorithm for multiobjective optimization. In: Proceedings of the 2000 Congress on Evolutionary Computation, pp. 325–332 (2000)
[19] Hu, X., Huang, Z., Wang, Z.: Hybridization of the multi-objective evolutionary algorithms and the gradient-based algorithms. In: The 2003 Congress on Evolutionary Computation, pp. 870–877 (2003)
[20] Knowles, J., Corne, D.: Memetic Algorithms for Multiobjective Optimization: Issues, Methods and Prospects. In: Recent Advances in Memetic Algorithms, pp. 313–352 (2005)
[21] Tang, J., Lim, M., Ong, Y.: Diversity-adaptive parallel memetic algorithm for solving large scale combinatorial optimization problems. Soft Computing - A Fusion of Foundations, Methodologies and Applications 11, 873–888 (2007)
[22] Dawkins, R.: The selfish gene. Oxford University Press, New York (1976)
[23] Krasnogor, N., Smith, J.: A tutorial for competent memetic algorithms: model, taxonomy, and design issues. IEEE Transactions on Evolutionary Computation 9, 474–488 (2005)
[24] Moscato, P.: On Evolution, Search, Optimization, Genetic Algorithms and Martial Arts Towards Memetic Algorithms. Caltech Concurrent Computation Program Report 826, California Institute of Technology, Pasadena, CA, U.S.A (1989)
[25] Hart, W.E.: Adaptive Global Optimization With Local Search. vol. PhD thesis: Univ. California, San Diego, CA (1994)
[26] Krasnogor, N.: Studies on the Theory and Design Space of Memetic Algorithms. vol. Ph.D. Thesis, University of the West of England (2002)
[27] Ong, Y.S., Keane, A.J.: Meta-Lamarckian learning in memetic algorithms. IEEE Transactions on Evolutionary Computation 8, 99–110 (2004)
[28] Merz, P., Freisleben, B.: A comparison of memetic algorithms, tabu search, and ant colonies for the quadratic assignment problem. In: Proceedings of the Congress on Evolutionary Computation, p. 2070 (1999)
[29] Goldberg, D.E., Voessner, S.: Optimizing Global-Local Search Hybrids. In: Proceedings of the Genetic and Evolutionary Computation Conference, pp. 220–228 (1999)
[30] Tang, J., Lim, M.H., Ong, Y.S.: Adaptation for parallel memetic algorithm based on population entropy. In: Proceedings of the 8th annual conference on Genetic and evolutionary computation. ACM Press, Seattle (2006)
[31] Davis, L.: Handbook of Genetic Algorithms. Van Nostrand Reinhold, New York (1991)
[32] Ferber, J.: Multiagent systems as introduction to distributed artificial intelligence. Addison-Wesley, Reading (1999)
[33] Stan, F., Art, G.: Is It an agent, or just a program?: A taxonomy for autonomous agents. In: Jennings, N.R., Wooldridge, M.J., Müller, J.P. (eds.) ECAI-WS 1996 and ATAL 1996. LNCS, vol. 1193, pp. 21–35. Springer, Heidelberg (1997)
[34] Chira, C., Gog, A., Dumitrescu, D.: Exploring population geometry and multi-agent systems: a new approach to developing evolutionary techniques. In: Proceedings of the 2008 GECCO conference companion on Genetic and evolutionary computation, Atlanta, GA, USA, pp. 1953–1960 (2008)
[35] Deb, K., Agrawal, R.B.: Simulated Binary Crossover for Continuous Search Space. Complex Systems 9, 115–148 (1995)
[36] Barkat Ullah, A.S.S.M., Sarker, R., Cornforth, D., Lokan, C.: An Agent-based Memetic Algorithm (AMA) for Solving Constrained Optimization Problems. In: IEEE Congress on Evolutionary Computation, CEC 2007, pp. 999–1006 (2007)

[37] Barkat Ullah, A.S.S.M., Sarker, R., Cornforth, D., Lokan, C.: AMA: a new approach for solving constrained real-valued optimization problems. In: Soft Computing - A Fusion of Foundations, Methodologies and Applications, August 11 (2008)

[38] Runarsson, T.P., Yao, X.: Stochastic ranking for constrained evolutionary optimization. IEEE Transactions on Evolutionary Computation 4, 284 (2000)

[39] Himmelblau, D.M.: Applied Nonlinear Programming. McGraw-Hill, USA (1972)

[40] Deb, K., Pratap, A., Agarwal, S., Meyarivan, T.: A fast and elitist multiobjective genetic algorithm: NSGA-II. IEEE Transactions on Evolutionary Computation 6, 182 (2002)

[41] Ray, T., Sarker, R.: Genetic algorithm for solving a gas lift optimization problem. Journal of Petroleum Science and Engineering 59, 84–96 (2007)

[42] Liu, H., Frazer, J.H.: Supporting evolution in a multi-agent cooperative design environment. Advances in Engineering Software 33, 319–328 (2002)

[43] Siwik, L., Kisiel-Dorohinicki, M.: Semi-elitist Evolutionary Multi-agent System for Multiobjective Optimization. In: Alexandrov, V.N., van Albada, G.D., Sloot, P.M.A., Dongarra, J. (eds.) ICCS 2006. LNCS, vol. 3993, pp. 831–838. Springer, Heidelberg (2006)

[44] Zhong, W., Liu, J., Xue, M., Jiao, L.: A multiagent genetic algorithm for global numerical optimization. IEEE Transactions on Systems, Man and Cybernetics, Part B 34, 1128–1141 (2004)

[45] Davidsson, P., Persson, J., Holmgren, J.: On the Integration of Agent-Based and Mathematical Optimization Techniques. In: Agent and Multiagent Systems: Technologies and Applications, pp. 1–10 (2007)

[46] Nakashima, T., Ariyama, T., Yoshida, T., Ishibuchi, H.: Performance evaluation of combined cellular genetic algorithms for function optimization problems, pp. 295–299 (2003)

[47] De Jong, K.A.: Evolving intelligent agents: A 50 year quest. IEEE Computational Intelligence Magazine 3, 12–17 (2008)

[48] Vasile, M., Locatelli, M.: A hybrid multiagent approach for global trajectory optimization. Journal of Global Optimization (2008)

[49] Bajo, J., Corchado, J.: Multiagent Architecture for Monitoring the North-Atlantic Carbon Dioxide Exchange Rate. In: Marín, R., Onaindía, E., Bugarín, A., Santos, J. (eds.) CAEPIA 2005. LNCS (LNAI), vol. 4177, pp. 321–330. Springer, Heidelberg (2006)

[50] Hasan, S.M.K., Sarker, R., Essam, D., Cornforth, D.: Memetic Algorithms for Solving Job-Shop Scheduling Problems. In: Memetic Computing. Springer, Heidelberg (2008) (in press)

[51] Dobrowolski, G., Kisiel-Dorohinicki, M., Nawarecki, E.: Evolutionary multiagent system in multiobjective optimisation. In: Proc. of the IASTED Int. Symp. on Applied Informatics (2001)

[52] Alba, E., Dorronsoro, B.: The exploration/exploitation tradeoff in dynamic cellular genetic algorithms. IEEE Transactions on Evolutionary Computation 9, 126–142 (2005)

[53] Deb, K.: Multi-Objective Optimization Using Evolutionary Algorithms. John Wiley & Sons, Inc., Chichester (2001)

[54] Barkat Ullah, A.S.S.M., Sarker, R., Cornforth, D.: An Evolutionary Agent System for Mathematical Programming. In: Kang, L., Liu, Y., Zeng, S. (eds.) ISICA 2007. LNCS, vol. 4683, pp. 187–196. Springer, Heidelberg (2007)

[55] Barkat Ullah, A.S.S.M., Sarker, R., Cornforth, D.: Search space reduction technique for constrained optimization with tiny feasible space. In: Proceedings of the 10th annual conference on Genetic and evolutionary computation, Atlanta, GA, USA, pp. 881–888 (2008)
[56] Elfeky, E.Z., Sarker, R.A., Essam, D.L.: A Simple Ranking and Selection for Constrained Evolutionary Optimization. In: Wang, T.-D., Li, X., Chen, S.-H., Wang, X., Abbass, H.A., Iba, H., Chen, G.-L., Yao, X. (eds.) SEAL 2006. LNCS, vol. 4247, pp. 537–544. Springer, Heidelberg (2006)
[57] Koziel, S., Michalewicz, Z.: Evolutionary algorithms, homomorphous mappings, and constrained parameter optimization. Evolutionary Computation 7 (1999)
[58] Sarker, R., Ray, T.: Multiobjective Evolutionary Algorithms for solving Constrained Optimization Problems. In: International Conference on Computational Intelligence for Modelling, Control and Automation (CIMCA 2005), Vienna, Austria. IEEE Press, USA (2005)
[59] Michalewicz, Z., Nazhiyath, G., Michalewicz, M.: A Note on Usefulness of Geometrical Crossover for Numerical Optimization Problems. In: 5th Annual Conference on Evolutionary Programming, Cambridge, MA, San Diego, CA, pp. 305–312 (1996)
[60] Hock, W., Schittkowski, K.: Test Examples for Nonlinear Programming Codes. LNEMS. Springer, Heidelberg (1981)
[61] Koziel, S., Michalewicz, Z.: Evolutionary Algorithms, Homomorphous Mappings, and Constrained Parameter Optimization. Evolutionary Computation 7, 19–44 (1999)

Multiagent-Based Approach for Risk Analysis in Mission Capability Planning

Lam T. Bui*, Axel Bender, Michael Barlow, and Hussein A. Abbass

Abstract. In this chapter, we propose a multiagent-based approach for risk analysis in military capability planning. A hierarchical system is introduced that has two layers: an Option Production Layer (OPL) to find all possible options for the given planning problem, and a Risk Tolerance Layer (RTL) in which DMs' acceptance of risk is evolved. The OPL uses metaheuristic techniques such as evolutionary algorithms to deal with multi-objectivity of a class of NP-hard resource investment problems, called the Mission Capability Planning Problem (MCPP), under the presence of risk factors. This problem has at least two inherent conflicting objectives: minimizing the cost of investment in resources as well as optimizing the makespan of plans. The framework allows for the addition of a risk-based objective to the problem in order to support risk assessment during the planning process. The RTL is run by a multi-agent system which simulates the risk attitudes of DM. The system determines different types of attitudes towards risk with each type applying to a sub-set of MCPP solutions. The goal of each agent is to maximize its risk tolerance levels with respect to a given subset of solutions determined in the OPL. Risk tolerance levels are used as surrogates for risk attitudes. The hierarchical system is flexible in terms of using a feedback mechanism when necessary. The RTL uses information from the OPL and can itself return some hyper-information to guide the OPL further. In a case study, we use a mission planning scenario to validate our proposal. The results from this study demonstrate the advantage of our proposed system. A diverse set of agents was found; hence different types of options can be grouped and offered to the decision-makers.

Lam T. Bui · Michael Barlow · Hussein A. Abbass
School of SEIT, University College, University of New South Wales, Canberra, Australia
e-mail: lam.bui07@gmail.com

Axel Bender
Defence Science and Technology Organisation (DSTO), Edinburgh,
South Australia, Australia

* Corresponding author.

R.A. Sarker and T. Ray (Eds.): Agent-Based Evolutionary Search, ALO 5, pp. 77–96.
springerlink.com

1 Introduction

Under risk consideration, planning systems in general offer many different option sets. Each set might convey an attitude of a decision-maker (DM) towards risk. Clearly, it is worthwhile to identify all possible combinations of objectives and risk attitudes in order to provide DM with flexibility when evaluating these options. This is the main focus of the work reported in this chapter.

It is important to note that risk is not a new concept. Vast literature exists on risk assessment and management in different domains such as financial markets [40], engineering [51, 27], defence [7] and security [49]. However, in the literature there is no unifying definition of the term 'risk' (see, for example, the differences in [22, 39]). Risk assessment identifies points of vulnerability, the likelihood of a potential problem, the possible frequency of risk, and the extent of potential damage. Various approaches have been devised such as probabilistic risk assessment in engineering [9], value-at-risk in financial market [40], and multi-agent based systems for defense and security applications [45, 7].

In this chapter, we are interested in the area of military planning for capabilities of future missions. Risk might arise from various sources of hazards and uncertainty. Indeed, for mission capability planning, uncertainty is unavoidable; for example, the appearance of risk factors such as enemy actions, bad weather, and potentially hostile civilians are difficult to determine, and so are the consequences of a delay in executing a task (that might cause the collapse of the whole plan). We expect our work to provide a computational methodology supporting risk analysis in the domain of defense risk management [44, 23, 17, 47]. Based on our previous work investigating a special class of planning problem called Mission Capability Planning Problem (MCPP) [14], we propose to build a two-level hierarchical system for risk analysis. At the first level, a planning engine is used to produce options for a particular planning problem. The second layer is in charge of managing DM's attitudes towards risk.

For the first layer, the problem is initially analyzed in the context of a military capability planning process and then mathematically formulated as a multi-objective resource investment problem (a special class of the *project scheduling problem*). Three objectives are proposed comprising cost of capabilities, execution time of the plan, and level of associated risk. A general risk-based framework is introduced in order to deal with this problem. In this way, the framework allows for the progressive integration of risk assessments throughout the planning process rather than as a post-analytical step. At each time step, the first layer will provide a set of non-dominated solutions (or options).

For the second layer, we propose to design a multi-agent system to manage the risk attitudes. Here risk attitude is approximated by the surrogate of risk tolerance. It describes the range of risk values that are acceptable to a DM. Each agent represents a risk attitude. Agents can interact with each other cooperatively or competitively. However, they all have the same goal, namely to maximize their risk tolerance level while trying to minimize the total cost of the subset of options that they appreciate the most amongst all possible solutions. Furthermore, the system offers a possible method for sending feedback from the agent layer to the first layer for further guidance.

The remainder of the chapter is organized as follows: an overview of project scheduling is presented in Section 2. The problem context and formulation are given in Section 3. The proposed framework is introduced in Section 4. Experimental studies are presented in Section 5. The last section is devoted to the conclusion of the work and lessons learnt.

2 Background

2.1 Project Scheduling Problems

2.1.1 Overview

Scheduling and resource allocation (SRA) is a decision-making support process involving assignment and allocation of limited resources to tasks (sometimes referred as jobs, operations or activities) over time under certain constraints. It also needs to deal with defining which tasks are to be executed at a specific time [37]. Metaheuristic techniques providing approximate solutions are often used for SRA because many scheduling problems are NP-hard [31]. In these situations the majority of exact methods fails to obtain the optimal solutions [4].

Here we summarize the main development of research on project scheduling in general. There are usually more than one type of resources (machine, people, money, etc). Traditionally, this problem is considered in a situation when there are scarce resources for scheduling, and is usually referred to as *resource constraint project scheduling problem* – RCPS. In RCPS, tasks are characterized by several aspects such as task duration and resources required. Tasks can be involved in several execution modes and are constrained by a precedence graph. Only one mode is performed at a time. Algorithms for RCPS are expected to find an optimal schedule that minimizes the processing time (called the makespan). In general, this is an NP-hard problem [10]. There are a number of practical approaches that can be transformed to a RCPS, including production sequencing, timetabling, and flight scheduling.

In [33], the author observed that problems related to RCPS were usually addressing two matters: resource and time. Therefore, he proposed the following categorization:

- **Problem of scare resources:** This is the conventional RCPS. Given a limitation K on resources, the problem is to find the time slots to process tasks. The objective is to minimize the makespan. This is sometimes referred to as *budget problem* [12]
- **Problem of scare time:** Given limitation T on processing time, the problem is to find the time slots to process tasks that minimize the resource cost. This

is refereed to as *resource investment problem*. In some other work [12], it is considered part of a special class called *resource leveling* problems.

The problem of RCPS is even more interesting when the above two problems are considered together and results in a problem of time-cost trade-off where the objective is to find a *vector of durations* for tasks that either minimize time or cost. Different values of time factors will give different optimal values for cost, and vice versa. A detailed RCPS review with more than 200 relevant research papers can be found in [12].

In general, RCPS is a mixed integer programming problem. Therefore, conventional linear programming methods, such as the simplex method, are not suitable. Researchers have actively developed a number of approaches. Broadly speaking, these approaches can be divided into two categories: exact and approximate methods. In exact methods, the final solution will be the optimal solution. Some typical approaches are branch and bound [20], Lagrangian relaxation [21], and dynamic programming [16]. However, exact methods usually face an issue of scalability and are limited to problems with task number ≤ 60 [4]. Therefore, approximation methods are preferred instead, including heuristic/ meta-heuristic techniques such as priority-based, truncated branch and bound, sampling techniques, local search techniques, tabu search, simulated annealing, scatter search, and evolutionary algorithms [30, 26, 32]. For stochastic project scheduling problems, several approaches have been proposed such as dynamic PERT networks [5], or MDP-based Q-learning [2].

It is surprising that although RCPS is inherently multi-objective, the literature of RCPS is dominated by work that considers only single objectivity [43, 48]. There exists a number of possible objectives for RCPS such as time, cost, resource balancing, robustness, etc. These objectives might conflict with each other to varying degree. It is clear that for multi-objective optimization, finding good (optimal) solutions is not the only consideration, but also addressing the trade-off between objectives among obtained solutions.

Perhaps, the most common approach for dealing with multi-objective RCPS (and also for machine scheduling [34]) is the weighted-sum method [1]. For this, objectives are summed according to a predefined vector of weights. However, this approach faces several systematic issues, such as objective scaling and convexity of the Pareto optimal front. Moreover, it is not easy to perform a trade-off analysis. Alternatively, Pareto-based approaches are used to obtain a set of trade-off solutions in a single run [48, 8]. Therefore, these approaches seem to be more suitable for solving multi-objective RCPS.

2.1.2 Resource Investment Problems

In this section, we extend our brief review to include the resource investment problem (RIP), which is closely related to our proposed problem. It is a special class of RCPS where there is no limitation on the resource capabilities. Instead, there is a limitation on the processing time. The primary objective is to minimize the cost of available resources. It still requires for scheduling tasks to satisfy a time constraint; but with an objective of optimizing the resource vector. Note that this is also an NP-hard problem [33]. Unlike RCPS, which is extensively covered in the research literature, there doesn't seem to be too much published work on RIP.

Mohring [33] pioneered the field with an exact approach based on graph algorithms. It is similar to a two-stage method where for each level of resource capabilities, the problem is transformed to an RCPS. Likewise, Demeulemeester [19] also proposed a two-stage algorithm using a method of cutting plane and a version of branch and bound algorithm. Several heuristic approaches have been introduced, such as scatter search [52], genetic algorithms [42], the polynomial priority-rule based method [36], and the multi-mode priority-rule based approach [28]. Note that when referring to RIP, we mean the static RIP where tasks are associated with a predefined set of required resources and a duration. In fact, it is worthwhile to note that the time-cost trade-off problem, as mentioned above, can be considered a stochastic version of RIP. For this problem, there are usually no deterministic requirements of resources for tasks, except boundaries for each task; and further, the task's duration is a probabilistic function of allocated resources. The question is what are task durations that satisfy the time-cost trade-off?

As with RCPS, RIP is inherently associate with many objectives. However, we are not aware of any work on RIP dealing with multi-objectivity and trade-off analysis using meta-heuristic techniques.

3 Mission Capability Planning

3.1 Overview of Capability Planning Process

Capability development (or *capability planning*) is an abstract concept that is performing at the visionary level and is about making future plans with hypothesized scenarios, but not making future decisions [38]. Capability planning is considered a part of *defense planning*. In the military, *mission planning* is also a popular term. Although capability planning needs to be considered under certain future missions, the term mission planning is usually used for *operations planning* and at the implementation level; it is about generating *courses of actions* and therefore supports making decisions at the present time.

Similar to operations planning [41, 29, 11, 8], capability planning can be applied to every level of missions: strategic, operational, or tactical. Note that this hierarchy is also applied to the business world where an organization or company needs to establish visionary plans for all levels of its business. However, for the business world, operational and tactical levels are swapped in the hierarchy.

The initial point in capability planning is the quantification of capabilities. In a general sense, a capability is viewed as a capacity to be, to process, to do, or to affect something [38, 3]. In defense, capability is understood as the power to achieve a particular operational effect. The operational effects might be described in terms of the nature of the effect given an environment, time, and duration. In capability planning, all these factors are synthesized for the future, not the present. The question is what factors constitute a capability of a system, organization, or a military? In the military, they can be understood as organization structures, facilities, personnel, and other resources, and all of these are referred to as Fundamental Inputs to Capability (FIC). Note that capability is partitionable. Each capability can be broken into several sub-capabilities [46].

Capability planning, in its simplest form, is the determination of FIC requirements. However, it is worthwhile to note that in a broad context a comprehensive capability planning process will generally be partitioned into different phases. Australian Defence Force's (ADF's) documents determine that each system capability has a life cycle of five phases [3]: *Needs* (identifying the capability gaps), *Requirements* (identifying and developing affordable options that satisfy the needs — this also requires risk assessment for each option and might involve simulation and evaluation for each risk factor, estimating cost and schedule for plans, preparing management and acquisition plans, and obtaining approvals), *Acquisition* (capability procurement and transition into service), *In Service* (capability operation, its support and possible modifications), and *Disposal* (withdrawal from service, disposal or redeployment of capability, depending on the nature of the individual capability input).

It is quite clear from the process that risk assessment is essential in order to make decisions on selecting capability options. There are many risk factors that will have impacts on future military missions such as enemies, terrains, weather, hostile civilians, etc. To support risk analysis, it is important to have a taxonomy defining all possible risks, such as [24].

Further, capability is clearly an abstract concept. It is determined via consideration of what needs to be done in order to achieve the objectives of the mission. In this context capability planning is about determining requirements in terms of cost, force, function, and performance to guarantee the capability, and we situate our research on capability planning for future missions at the *operational level*. At this level, campaigns and major operations are planned, conducted, and sustained to accomplish strategic objectives within theaters or areas of operations. Activities link tactics and strategy by establishing operational objectives needed to accomplish the strategic objectives, sequencing events to achieve the operational objectives, and initiating actions and applying resources to bring about and sustain those events [11].

For planning at the operational level (either operations planning or capability planning), we need to deal with a sequence of tactical events (usually referred to as tasks) associated with a *precedence relation* and solve the requirements for resources (capabilities) from tasks. The difference between capability and operations planning is that for operations planning, the available resources are limited and the problem is how to assign the resources (carried by tactical units) to the tasks to get the maximal benefit of assignment and the minimal assignment-cost (mainly the time of execution).

To date, there are many conceptual approaches to capability planning. These approaches differ in the way the problem of planning is viewed [35] and include: threat-based, scenario-based, capability-based, budget-based planning. They are not stand-alone but are usually integrated. Currently, TTCP [http://www.dtic.mil/ttcp/], a joint program among AU, UK, US, CA, and NZ, has proposed a methodology centered around a capability-based approach, while taking into account all other approaches. It is very clear that, regardless of the selection of approaches, scenarios are essential in capability planning. This is also our approach in tackling the capability planning problem.

3.2 Mission Capability Planning Problem

As mentioned above, one of the potential problems in the capability planning process is the determination of FIC requirements for capabilities in the *Requirements* phase. We thus focus this book chapter on this problem. From the computational point of view, there are not many differences between planning and scheduling [8]. Both are about making a plan for performing tasks. Depending on the problem domain, the planning can be broken down into different categories. In defense, this includes both operations planning that requires scheduling for performing a mission with given capabilities and at a specific time (normally, the current time); and capability planning that is about making a visionary plan for preparing resource capabilities for a future mission.

There are analogies between capability planning problem (CPP) and RIP. CPP is normally a large-scale instance of RIP where there is no limitation on resources and time. CPP not only focuses on order and time to process tasks, but also on the allocation scheme so that the costs of resource investment become minimal and the impact of plans is well-controlled. In the simplest form of CPP, there are at least two obvious objectives: minimizing makespan and minimizing resource cost. They are inherently conflicting; if all required resources are satisfied at the same time, makespan will be minimal, however, the cost will be highest. If risk is used as another objective, it somewhat conflicts with makespan and resource cost. Clearly, the more time allowed or the more resources invested, the less probability of risk occurring.

Here, we propose to investigate an instance of CPP, called the Mission Capability Planning Problem - MCPP, as follows:

- **Mission capability:** Combat engagement in future operational missions
- **Planning level:** Operational level
- **Capability inputs:** Force structure and related resources
- **Criteria:** Time, cost, and associated risk.

To our knowledge, there has not been any work dedicated to this problem, especially no work on using multi-objective metaheuristic techniques.

3.3 Mathematical Formulation of MCPP

The problem formulation is described as follows:

- **Inputs:**
  - A set V of N tasks: $V = V_1, V_2, ..., V_N$, these are non-pre-emptive. Each V_i will have:
    - Task durations d_i
    - A vector r_i of required resources by tasks: $r_i = r_{ij}, j = 1, \ldots, M$ (M is the number of resource types).
  - A network G, called activity-on-node network, of tasks where nodes and arcs represent the tasks and the precedence relations, respectively: $G = (V, E)$. $Pred(j)$ defines a set of direct predecessors, while $Succ(j)$ is the set of direct successors of task j.
  - A vector c of costs $c = c_1, c_2, ..., c_M$, c_k is the cost of a unit of resource k.

- **Parameters:**
  - Indices of the tasks: $I = I_1, I_2, ..., I_N$ (or a schedule ts that is translated from I: $ts = ts_1, ts_2, ..., ts_N$)
  - Vectors of resource capabilities: $R = R_1, R_2, ..., R_M$

- **Constraints:**
  $ts_i + d_i \leq ts_j, \forall j$, and $\forall i \in Prec(j)$
  $\sum_{j=1}^{N} r_{ij}(t) \leq R_i \; \forall i$ and t
- **Objective functions:**
  - Makespan: Minimization of $ts_{\overline{N}} + d_{\overline{N}}$ ($\overline{N}$ is the last task to be scheduled)
  - Resource Investment: Minimization of $\sum_{k=1}^{M} c_k R_k$
  - Risk level: Minimization of probability of mission failure

- **Outputs:**
  Sets of trade-off options (schedules and vectors of resources). Each option has
  - A vector R of M resource capabilities $R = (R_1, R_2, ..., R_M)$.
  - A schedule ts and an assignment scheme of resources to tasks over time.

4 A Multiagent-Based Framework

Multi-agent techniques have been popular computational tools for many problem domains ranging from simulations, controls to optimization [50]. An agent is equipped with a sensor enabling it to perceive the surrounding environment; hence allowing it to interact with others. Multi-agent systems (MAS) are able to distribute computation among member agents and to keep track of information in all local areas occupied by agents. Because of this ability, multi-agent systems are very suitable tools for modeling group behaviors [6, 13, 53] as well as population-based search in the optimization domain [54]. This is why we select MAS for our work.

4.1 General Framework

We propose to design a hierarchical framework with two layers. The motivation of the two-layer system is that our previous method usually offers a wide range of solutions and it is quite difficult to classify them for the decision making process. So it is better to superimpose an analytical component that carries out a meta-analysis. The lower layer is designed to find all possible options for the planning problem. That is why we call it the "*options production layer*" — OPL. This layer is mainly based on our previously proposed design [14]. Since the problem is a multi-objective one, at any time step, it passes on a set of non-dominated solutions (options) to the higher layer for further analysis.

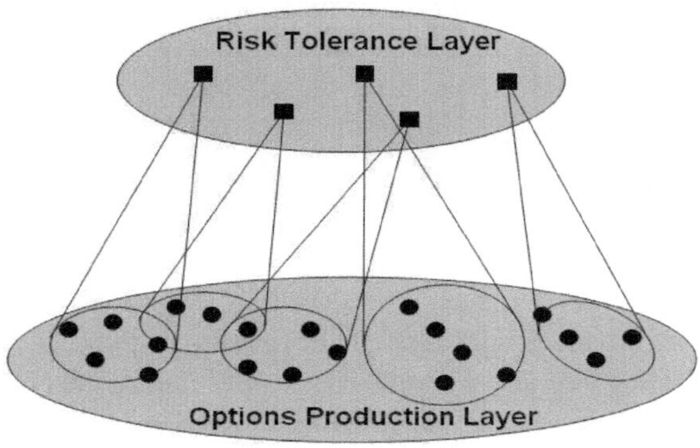

Fig. 1 General framework for our multiagent based framework

Since the options provided by OPL have different risk values, it is difficult for the DM to select the final possible options. This requires them to express their attitudes towards risk that is represented by risk tolerance levels. However, it is a challenging task since the options are trade-offs in which an option might be better in terms of optimizing cost and time, but might pose a high risk of failure. The question then is which options are acceptable to the DM in terms of risk, but with less cost of resources? In general, options might be clustered into many sub-groups under different levels of risk acceptance and resource costs. We come up with an approach of using multi-agent technique for finding these levels of risk acceptance. We name the layer "*risk tolerance layer*" — RTL. The general framework is depicted in Figure 1.

4.2 Options Production Layer — OPL

In this layer, a given future mission will undergo a period of *pre-processing* where the mission is decomposed into a series of tasks depending on the given scenarios. Basically, each task is considered a tactical event (sub-mission). Military staff usually perform the task decomposition and pre-determine the resources required by each task. Also, a precedence relationship between tasks is determined. Further, it is required to identify all possible risk factors associated with the mission; here a predefined risk taxonomy can support the risk identification process. This is a difficult process because it relies heavily on intelligence sources and generally is performed during the set up of scenarios.

After this period, a planning engine is started to search for capability options, such as total resources that lead to an optimal schedule for task execution. This is an iterative process until the obtained plans satisfy mission objectives. Here, risk assessment is performed via estimating a level of risk for each plan and the degree to which this risk is addressed describes another measure of the plan's quality. Therefore, the evaluation of an option (or plan) is based on three aspects (as mathematically specified in Section 3.3):

- *Resource cost* is the cost of all resources that will be used for the mission.
- *Execution time* is the time to execute the mission given the planned resources.
- *Associated risk* is assessed based on the effect of risk factors on the success (failure) of the mission. Details will be given in the section on risk simulation (Section 4.4).

4.3 Risk Tolerance Layer — RTL

Running this layer is a multi-agent system. Each agent is in charge of a risk tolerance level (that is modeled by a lower bound of risk value and an interval). It has the goal of maximizing risk tolerance, while keeping the cost of the subset of options that belong to it at a minimum. An option is considered to belong to an agent if its risk value falls into the interval of the agent's risk tolerance. Note that in practice there might be many factors contributing to the DM's perception of risk. These factors need to be quantified before applying our system. To demonstrate the concept, we use the cost of options as the factor affecting DM.

Because of multi-objectivity, after a period of cooperation and competition, there will be a set of agents that offer different trade-off subsets of options. Here we propose to use an evolutionary multi-agent structure, although any kind of multi-agent system can be valid. The group of agents can be either allowed to evolve after completing the task of the OPL, or run alongside with the population of the options generated in the OPL. In this study we choose the latter approach and we believe that with this approach, the agents can keep track of some good information that might otherwise be lost during the evolution of the OPL. The first approach thus might not be able to recover. In addition, in the selected approach, information that might be relevant for option production can conveniently be fed back from the RTL to the OPL as and if required.

4.4 Risk Simulation

Determining the risk to a mission is a difficult task and riddled with a high degree of uncertainty. In particular, for a combat mission, which by its core nature is a complex system, tasks are not isolated. They are connected via a precedence constraint. A risk event occurring within a single task might trigger catastrophic consequences for others. Here, we propose to use simulation methods to estimate risk. There are several approaches such as multi-agent based simulators to simulate risk factors and measure their effects on mission success. However, in this chapter, we make the case that the risk factor is the variation of the execution time (the duration) of each task. For this, we try to address the question of what is the probability of failure of a plan given that the variability of duration follows a normal distribution. In this case, Monte Carlo simulation is the most popular tool to estimate this probability [25, 18, 15]. We consider the measure of this probability as the risk-based objective in our proposed MCPP.

Although for the sake of simplicity we assume that the distributions of task durations are normal (since it is the most common type in practice and is widely investigated in the literature), it is possible to use any distribution in practice. The failure of a plan is simply considered to be the violation of both precedence and resource constraints. Formally, for a plan p_i, the Monte Carlo simulation is carried out T times of the plan execution with randomly generated task durations. The probability of failure $Prob_i$ is calculated as follows:

$$Prob_i = \frac{\sum_{j=1}^{T} x_j^i}{T} \qquad (1)$$

where x_j^i represents whether the plan violates the constraints or not and is determined as follows:

$$x_j^i = \begin{cases} 1 & \text{if } p_i \text{ violates the constraints} \\ 0 & \text{otherwises} \end{cases} \qquad (2)$$

Note that the starting time for each task is not changed during the simulation.

4.5 *Feedback from Agents to the Solutions*

With our design, it is possible to get feedback from the agent layer to guide option generation in the OPL. If we enable this feedback, we can formulate some additional goals that utilize information of change from both layers. Further, the interaction between the two layers reflects the situation when DM adjust their risk tolerance level based on solution options provided but, at the same time, want to place some preference on the process of finding solutions which is handled by the OPL.

From a design perspective, it is quite straightforward to incorporate the feedback from the agent layer to the solutions. Although it depends on the type of feedback, this feedback is first quantified in terms of numerical values and is then interpreted as either a reward or a penalty. Each solution will be added with this reward/penalty. A particular design will be given in the next section in which a case study is described.

5 Case Study

5.1 Test Scenarios

We adapted our test scenarios from a mission planning problem [14]. Suppose that the future operational mission will be an offensive one with 18 tasks, and that there are eight possible necessary resources with the cost of 23, 4, 14, 1, 7, 18, 9, 10, respectively, per unit . The precedence relation among tasks is depicted in the graph of Fig. 2 in which the mission must be started with tasks 1, 2, 3, 4, 5, 6, and 17. Over time, the execution of tasks must follow the precedence constraint in order to fulfil the mission. Tasks 15, 16, and 18 are the final ones to be executed before ending the mission. Further, each task's parameters (resource requirements and duration) are given in the table on the right of the graph.

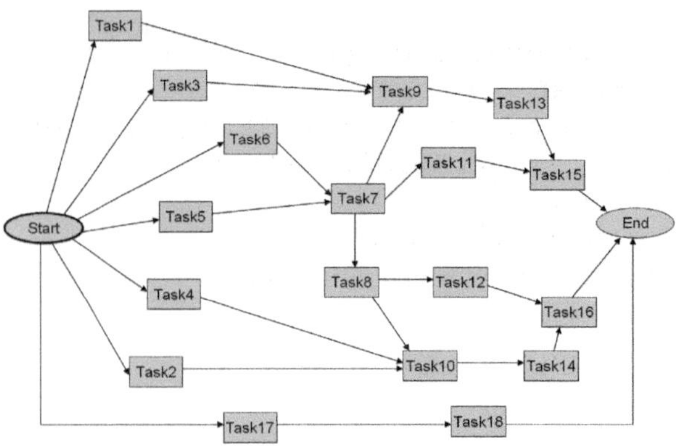

Task No.	Required resources	Duration
1	5 3 10 0 0 8 0 6	30
2	5 3 10 0 0 8 0 6	30
3	0 3 0 0 0 0 0 0	10
4	0 3 0 0 0 0 0 0	10
5	0 3 0 0 0 0 0 0	10
6	0 0 0 10 14 12 0 0	20
7	0 0 0 10 14 12 0 0	15
8	0 0 0 10 14 12 0 0	10
9	5 0 0 0 0 5 0 0	10
10	5 0 0 0 0 5 0 0	10
11	0 0 0 0 0 10 5 0	25
12	0 0 0 0 0 10 5 0	10
13	0 0 0 0 0 8 0 6	20
14	0 0 0 0 0 8 0 6	20
15	0 0 0 20 10 4 0 0	15
16	0 0 0 20 10 4 0 0	15
17	0 0 0 0 0 8 0 4	10
18	0 0 0 8 6 0 4 10	20

Fig. 2 Precedence graph for tasks and parameters

5.2 Parameter Settings

The chromosome size is 26. Initially, we used a population size of 40 and, in a later step, we actually analyzed the effect of population sizing. The crossover and mutation rates were 0.9 and 0.01, respectively. The evolution was terminated after 1000 generations. Further, for the validation process, an elitist GA was used with the same parameters.

The Monte Carlo simulation was allowed to run 1000 times when evaluating a plan. Each task's duration will follows a normal duration with the mean set as the predefined duration used in the first scenario, and the standard deviation set to 1.0. The probability that each task is executed with a time less than or equal the predefined duration is 0.99.

5.3 Results and Discussion

To obtain a better understanding of the proposed system, we ran it for a long period of 1000 loops (generations). We recorded the set of agents that are non-dominated relative to each other regarding two objectives. As analyzed in Section 4, because of multi-objectivity we expect that the system will offer a diverse set of risk tolerance levels. In this case study, we obtained 38 solutions (out of a possible 40). They are visualized in Figure 3.

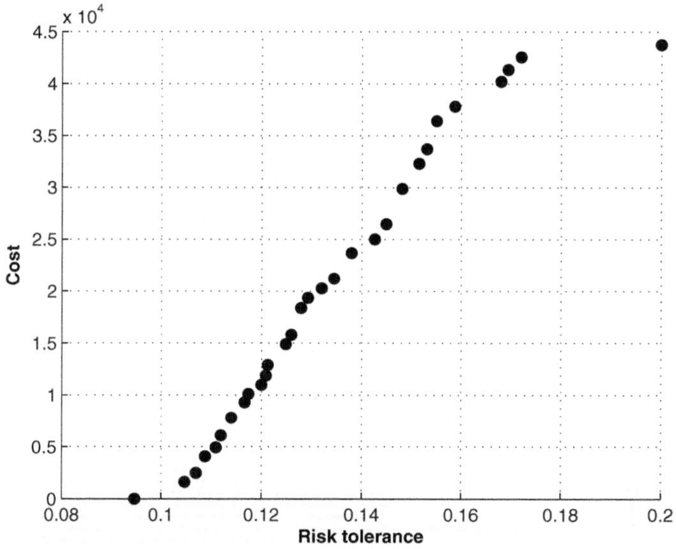

Fig. 3 Set of trade-off agents (visualized in the objective space) after 1000 generations

As can be seen from the figure, there is no single agent that has a zero risk tolerance level. That is because there was no risk-free solution provided from OPL. The non-existence of zero risk tolerance in this case study can be easily validated in real-world applications where it is impossible to find any risk-free plan for a scenario. This is also obvious from Figure 4 where all 38 non-dominated solutions obtained by the OPL have quite small but non-zero values of risk (between 0.02 and 0.1).

Further, the agents indeed found a quite diverse range of risk tolerance. It spreads evenly from the minimal value (around 0.08) to the case-study specific maximum value of 0.2. We note here that it is really difficult to find an agent with maximum risk tolerance level and minimal cost for the options belonging to it. This poses the following questions regarding cost-risk trade-offs: *should we choose a maximal level of risk tolerance*? or; *should we opt for a small interval of risk tolerance but centered around a high value of risk*? (Recall that risk tolerance is represented by

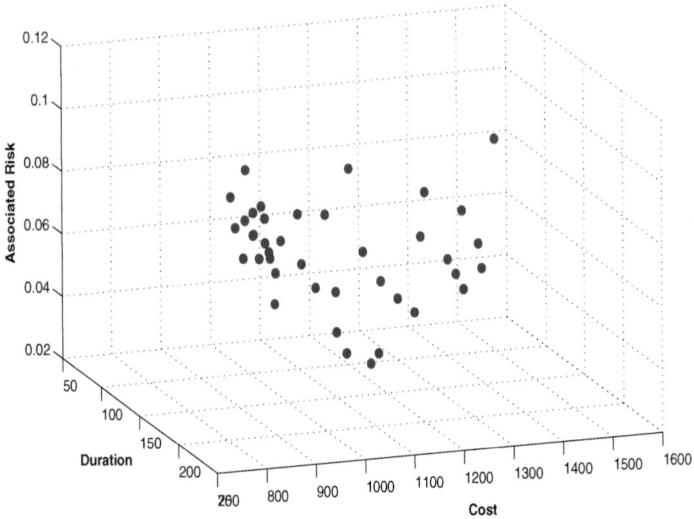

Fig. 4 Non-dominated solutions obtained in the OPL after 1000 generations

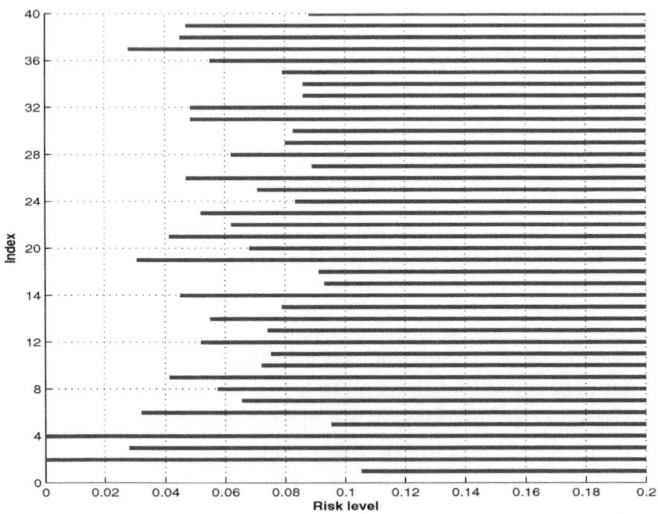

Fig. 5 Risk tolerance levels (a lower bound risk value and an interval) of agents after 1000 generations

two parameters: a lower bound risk value and an interval) The results show that neither of the two strategies is preferred. An agent with maximal level of risk tolerance (0.2) also accepts maximum cost for the solutions belonging to it; i.e. all provided

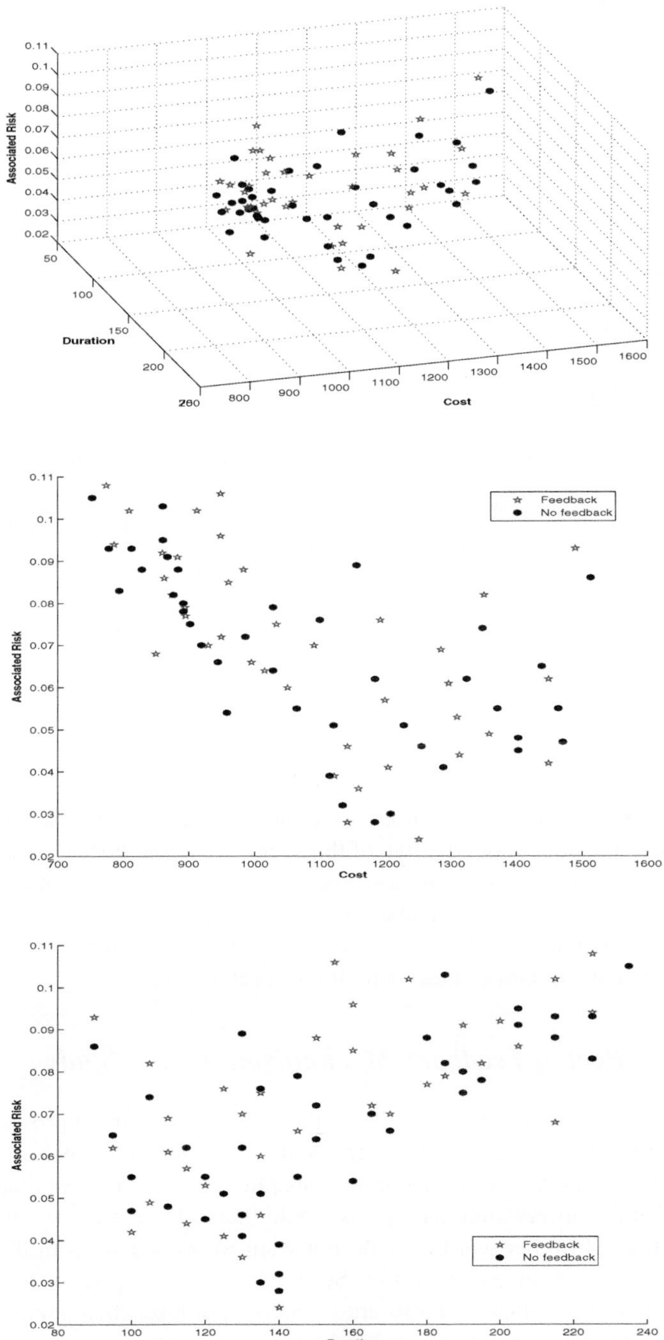

Fig. 6 3D view and 2D projections of non-dominated solutions obtained with and without feedback after 1000 generations

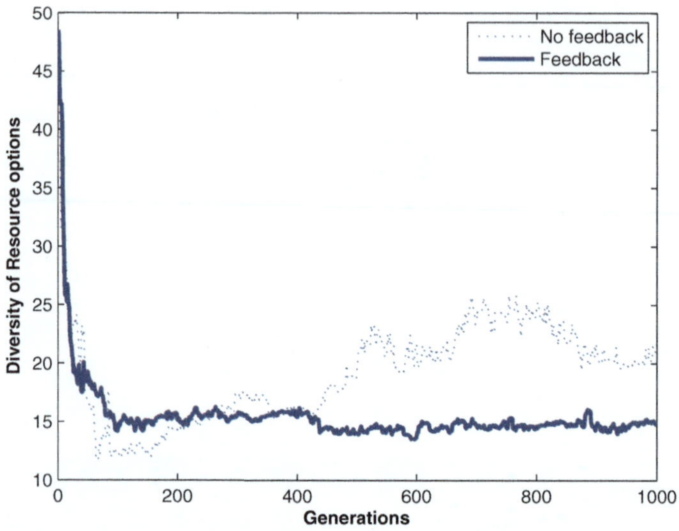

Fig. 7 The diversity of the population of solutions (in terms of resources) over time

OPL solutions are acceptable. On the other hand, an agent with minimal risk tolerance does not accept any of the solutions; the level of tolerance is too low for any solution from the non-dominated set to be acceptable.

Figure 5 shows that the agents tried to extend their risk tolerance as much as possible (towards the upper bound). Two of the agents' risk tolerances spread from 0 to 0.2 resulting in huge cost of resources. Note that the distribution of the lower bound for the agents' risk tolerance is consistent with the spread of agents in the cost-risk trade-off space (Figure 3). With this result, it is straightforward to recommend the risk tolerance level to DMs based on their cost preference.

5.4 The Effect of Feedback Mechanism: A Pilot Study

Feeding back information from the RTL to the OPL is important. It provides a mechanism to apply lessons learnt by the agents and experiences to the process of finding solutions for the planning task. In our demonstration of the basic feedback concept, we start from the observation that agents would like to have their own distinct subset of solutions. In other words, we do not want to see agents with different risk tolerance having the same capabilities. So the feedback we give to the OPL is information about how often solutions are found in option sets belonging to different agents. This information will be used to impose a penalty on solutions that belong with many agents.

It is interesting to see that in objective space, the non-dominated sets obtained with and without feedback are quite similar (Figure 6). Imposing a penalty did not cause the system to worsen. The approach with feedback still obtained solutions that were as good as those found by the non-feedback approach. However, their behavior in decision space is different. To see the difference, we measure the diversity of the population in terms of resources over time. It turns out that the two approaches were quite similar at the early stage of evolution when they both were converging and diversity did reduce rapidly. However, at a later time without feedback, the system had a need to explore around as can be seen from the changes in population diversity (Figure 7). On the other hand, the system under the feedback constraint was smoothly progressing. This satisfies our design purpose that we allow for exploration of good options but have them as distinctly clustered in different groups as possible.

6 Conclusion

In this chapter, we proposed for the first time a quantitative model in support of risk analysis in multi-objective military mission capability planning problems (MCPP). Our model takes into account decision-makers' attitudes to risk. It evolves planning option sets that reflect different cost-capability trade-offs in objective space and, at the same time, maximizes associated risk tolerance levels. Such option sets facilitate cost-capability trade-off negotiations because they have increased likelihood of meeting a variety of decision-makers' tolerances for risk. They are also more likely to be robust because they are less vulnerable to changes which can come about when new decision-makers enter the military capability planning process or when old decision-makers alter their attitudes to risk.

Our quantitative risk analysis model is a hierarchical system that has two layers: an Option Production Layer (OPL) that finds non-dominated options for the given MCPP, and a Risk Tolerance Layer (RTL) in which decision-makers' risk tolerance is evolved. Every MCPP has at least two conflicting objectives: minimization of required investment in resources and minimization of the capability plan's makespan. The OPL uses metaheuristics to deal with these objectives. It also considers a risk-based objective in order to support risk assessment during the planning process.

The RTL is run by a multi-agent system which simulates the risk attitudes of the decision makers. The goal of each agent in the RTL is to maximize its risk tolerance level with respect to a given subset of non-dominated solutions determined in the OPL. These subsets reflect different cost-capability trade-offs. The hierarchical system is flexible because the RTL can feed back information to the OPL in order to guide the option generation process.

We have demonstrated the concept of our risk analysis model in a simple case study. In future work, we expect to use the system in more involved MCPP. We also endeavor to elaborate our modeling of risk attitudes. The current approach of equating risk attitudes to intervals of tolerable risk values is very simplistic and rarely describes situations in the real-world of capability planning.

Acknowledgements

The authors gratefully acknowledge the financial support from the Australian Research Council via the ARC Discovery Grant No. DP0667123.

References

1. Abbasi, B., Shadrokh, S., Arkat, J.: Bi-objective resource-constrained project scheduling with robustness and makespan criteria. Applied Mathematics and Computation 180, 146–152 (2006)
2. Aberdeen, D., Thiébaux, S., Zhang, L.: Decision-theoretic military operations planning. In: Proc. ICAPS, pp. 402–411. AAAI Press, Menlo Park (2004)
3. ADF: Defence Capability Development Manual. Australian Defence Publishing, Canberra, Australia (2006)
4. Alcaraz, J., Maroto, C.: A robust genetic algorithm for resource allocation in project scheduling. Annals of Operations Research 102(4), 83–109 (2001)
5. Azaron, A., Tavakkoli-Moghaddam, R.: Multi-objective time cost trade-off in dynamic PERT networks using an interactive approach. European Journal of Operational Research 180, 1186–1200 (2007)
6. Barlow, M., Easton, A.: Crocadile-an open, extensible agent-based distillation engine. Information & Security 8(1), 17–51 (2002)
7. Barlow, M., Yang, A., Abbass, H.: A Temporal Risk Assessment Framework for Planning A Future Force Structure. In: IEEE Symposium on Computational Intelligence in Security and Defense Applications, CISDA 2007, pp. 100–107 (2007)
8. Belfares, L., Klibi, W., Lo, N., Guitouni, A.: Multi-objective tabu search based algorithm for progressive resource allocation. European Journal of Operational Research 177, 1779–1799 (2007)
9. Bier, V.: Challenges to the Acceptance of Probabilistic Risk Analysis. Risk Analysis 19(4), 703–710 (1999)
10. Blazewicz, J., Lenstra, J., Kan, A.R.: Scheduling subject to resource constraints: Classification and complexity. Discrete Applied Math. 5, 11–24 (1983)
11. Boukhtouta, A., Bedrouni, A., Berger, J., Bouak, F., Guitouni, A.: A survey of military planning systems. In: The 9th ICCRTS Int. Command and Control Research and Technology Symposium, Copenhagen, Denmark (2004)
12. Brucker, P., Drexl, A., Mohring, R., Neumann, K.: Resource constrained project scheduling: Notation, classification, models and methods. European Journal of Operational Research 112, 3–41 (1999)
13. Bui, L.T., Barlow, M.: Performance of path-planning agents in an agent-based distillation. In: Proceedings of SimTecT 2003, SIAA, Adelaide, Australia (2003)
14. Bui, L.T., Barlow, M., Abbass, H.A.: A Multiobjective Risk-based Framework for Mission Capability Planning. New Mathematics and Natural Computation 5(2), 459–485 (2009)
15. Cabeza, A.Z., Ridao, M.A., Ridao, E.F., Camacho, E.F.: Using a risk-based approach to project scheduling: A case illustration from semiconductor manufacturing. European Journal of Operational Research 190(3), 708–723 (2008)
16. Carruthers, J.A., Battersby, A.: Advances in critical path methods. Operations research quarterly 17(4), 359–380 (1966)
17. Center, A.: Risk Management: Multiservice Tactics, Techniques, and Procedures (2001)

18. Cheung, K.M., Ko, A., Dang, V., Heckman, D.: Risk Analysis for Non-Deterministic Mission Planning and Sequencing. In: 2005 IEEE Conference on Aerospace, pp. 1–12 (2005)
19. Demeulemeester, E.: Minimizing resource availability costs in time-limited project networks. Management Science 41(10), 1590–1598 (1995)
20. Demeulemeester, E., Herroelen, W.: A branch-and-bound procedure for the multiple resource-constrained project scheduling problem. Management Science 38(12), 1803–1818 (1992)
21. Drexl, A., Kimms, A.: Optimization guided lower and upper bounds for the resource investment problem. Journal of Operations Reseach Society 52, 340–351 (2001)
22. Fischhoff, B.S., Watson, S.R.S., Hope, C.S.: Defining risk. Policy Sciences 17(2), 123–139 (1984)
23. Gaidow, S., Boey, S.: Australian Defence Risk Management Framework: A Comparative Study. Defense Technical Information Center (2005)
24. Gallagher, B., Pamela, J.: A Taxonomy of Operational Risks. Carnegie Mellon University, Software Engineering Institute (2005)
25. Greenland, S.: Sensitivity Analysis, Monte Carlo Risk Analysis, and Bayesian Uncertainty Assessment. Risk Analysis 21(4), 579–584 (2001)
26. Hindi, K.S., Yang, H., Fleszar, K.: An evolutionary algorithm for resource-constrained project scheduling. IEEE Trans. on Evolutionary Computation 6(5), 512–518 (2002)
27. Holmgren, Å.: Using Graph Models to Analyze the Vulnerability of Electric Power Networks. Risk Analysis 26(4), 955–969 (2006)
28. Hsu, C.C., Kim, D.: A new heuristic for multi-mode resource investment problem. Journal of Operations Reseach Society 56, 406–413 (2005)
29. Kewley, R., Embrechts, M.: Computational military tactical planning system. IEEE Trans. on Systems, Man, and Cybernetics, Part C 32(2), 161–171 (2002)
30. Kolishch, R., Hartmann, S.: Experimental investigation of heuristic for resource constrained project scheduling: An update. European Journal of Operational Research 174, 23–37 (2006)
31. Leung, J.: Handbook of scheduling: algorithms, models, and performance analysis. Chapman & Hall/CRC (2004)
32. Merkle, D., Middendorf, M., Schmeck, H.: Ant colony optimization for resource-constrained project scheduling. IEEE Trans. on Evolutionary Computation 6(4), 333–346 (2002)
33. Mohring, R.: Minimizing costs of resource requirements in project networks subject to a fixed completion. Operations Research 31(1), 89–119 (1984)
34. Nagar, A., Haddock, J., Heragu, S.: Multiple and bi-criteria scheduling: a literature survey. European Journal of Operational Research 81, 88–104 (1995)
35. NATO: Handbook on Long term Defence Planning. St. Joseph Print Group Inc., Canada (2003)
36. Neumann, K., Zimmermann, J.: Resource levelling for projects with schedule-dependent time windows. European Journal of Operational Research 117, 591–605 (1999)
37. Pinedo, M.: Scheduling: Theory, Algorithms, and Systems, 2nd edn. Prentice Hall, NJ (2001)
38. Radford, K.J.: Strategic Planning: an Analytical Approach. Reston Pub. Co., Reston (1980)
39. Renn, O.: Three decades of risk research: accomplishments and new challenges. Journal of Risk Research 1(1), 49–71 (1998)
40. Rockafellar, R., Uryasev, S.: Optimization of conditional value-at-risk. Journal of Risk 2(3), 21–41 (2000)

41. Schlabach, J., Hayes, C., Goldberg, D.: FOX-GA: A genetic algorithm for generating and analysing battlefield courses of action. Evolutionary Computation 7(1), 45–68 (1999)
42. Shadrokh, S., Kianfar, F.: A genetic algorithm for resource investment project scheduling problem, tardiness permitted with penalty. European Journal of Operational Research 181, 86–101 (2007)
43. Slowinski, R.: Multiobjective project scheduling under multiple-category resource constraints. In: Advances in project Scheduling. Elsevier, Amsterdam (1989)
44. Smith, J., Egglestone, G., Farr, P., Moon, T., Saunders, D.: Technical Risk Assessment of Australian Defence Projects. DSTO Information Sciences Laboratory (2004)
45. Torrellas, G.: A Framework for Multi-Agent System Engineering using Ontology Domain Modelling for Security Architecture Risk Assessment in E-Commerce Security Services. In: Proceedings of the Network Computing and Applications, Third IEEE International Symposium on (NCA 2004), vol. 00, pp. 409–412 (2004)
46. TTCP: The guide to capability-based planning. Tech. Rep. TR-JSA-TP3-2-2004, The Technical Cooperation Program (2004)
47. US Department of the Army: FM 5-19: Composite Risk Management (2006)
48. Viana, A., de Sousa, J.P.: Using metaheuristics in multiobjective resource constrained project scheduling. European Journal of Operational Research 120, 359–374 (2000)
49. Willis, H.: Estimating Terrorism Risk. RAND Corporation (2005)
50. Woolridge, M., Wooldridge, M.: Introduction to multiagent systems. John Wiley & Sons, Inc., New York (2001)
51. Yacoub, S., Ammar, H.: A methodology for architecture-level reliability risk analysis. IEEE Transactions on Software Engineering 28(6), 529–547 (2002)
52. Yamashita, D., Armentano, V.A., Laguna, M.: Scatter search for project scheduling with resource availability cost. European Journal of Operational Research 169, 623–637 (2006)
53. Yang, A., Abbass, H.A., Sarker, R.A.: Characterizing warfare in red teaming. IEEE Transactions on Systems, Man, and Cybernetics, Part B 36(2), 268–285 (2006)
54. Zhong, W., Liu, J., Xue, M., Jiao, L.: A multiagent genetic algorithm for global numerical optimization. IEEE Transactions on Systems, Man, and Cybernetics, Part B 34(2), 1128–1141 (2004)

Agent Based Evolutionary Dynamic Optimization

Yang Yan, Shengxiang Yang, Dazhi Wang, and Dingwei Wang

Abstract. Agent-based Evolutionary Search (AES) has attracted a growing interest from the evolutionary computation community in recent years due to its robust ability in solving large scale problems, ranging from online trading, disaster response to financial investment planning. In order to solve these problems, a great variety of intelligent techniques have been developed to improve the framework and efficiency of AES. This chapter investigates an AES algorithm in which the agents are updated and co-evolve to track dynamic optimum by imitating the exhibited feature of living organism. In the proposed algorithm, all agents live in a lattice like environment, where each agent is fixed on a lattice point. In order to increase the predefined energy function, individual agent is designed to compete with its neighbors and also acquire knowledge through cumulative information. For the purpose of maintaining the diversity of the population, random immigrants and adaptive primal dual mapping schemes are incorporated. Simulation experiments on a set of dynamic benchmark problems show the proposed AES algorithm can yield a better performance on dynamic optimization problems (DOPs) in comparison with several peer algorithms.

1 Introduction

Agent based system which is composed of multiple interacting agents has generated lots of excitement in recent years because it has been suggested as a promising

Yang Yan · Dazhi Wang · Dingwei Wang
School of Information Science and Engineering
Northeastern University, Shenyang 110004, China
e-mail: yanyangmail@163.com, wongdz@gmail.com,
dwwang@mail.neu.edu.cn

Shengxiang Yang
Department of Computer Science, University of Leicester
University Road, Leicester LE1 7RH, United Kingdom
e-mail: s.yang@mcs.le.ac.uk

R.A. Sarker and T. Ray (Eds.): Agent-Based Evolutionary Search, ALO 5, pp. 97–116.
springerlink.com

technique for conceptualizing and solving various optimization problems [11, 16, 20, 21, 27]. This promising technique is particularly attractive for solving operational problems in business and logistics environments, such as the production planning problems, transportation and distribution problems, and resource allocation problems [9, 10, 17, 22]. Zhong et al. [39] also integrated agent systems with Genetic Algorithms(GAs) to form a new algorithm for solving the global numerical optimization problem.

In recent years, AES has been applied widely to solve stationary optimization problems where the fitness curve or the objective function maintain unchanged during the searching process of AES. However most real-world optimization problems are dynamic since the coefficients of objective function, environmental parameter and/or constraint conditions maybe change over time. For example, in job scheduling production systems, re-scheduling of the jobs is often required due to change of the specification of jobs or malfunction of the machines. For these DOPs, the goal of an algorithm is no longer to find an optimal solution but to track the moving optima in the search space. In order to enhance the performance of Evolutionary Algorithms (EAs) on DOPs, many researchers have developed a number of approaches, including diversity-increasing methods [8], diversity-keeping methods [2, 14], memory-based approaches [4], and multi-population approaches [3, 5].

In this chapter, a hybrid Agent-based Evolutionary Search algorithm is proposed. In the proposed AES, a population of agents that represent potential solutions are located in a lattice like environment and accomplish their update via competing in a local neighborhood or learning strategy based on the statistical feedback information of the whole system. Similar to other EAs, AES may gradually converge in the search space during the run, especially when the environment has been stationary aldong with time. In order to address such problems, two diversity maintaining, random immigrants and adaptive dual mapping, are introduced into our AES to improve its performance on DOPs.

The rest of this chapter is organized as follows. Section 2 describes the AES algorithm in details. In Section 3, the dynamic testing suite is presented. Section 4 reports the experimental results. Finally, Section 5 concludes this chapter.

2 Proposed Agent Based Evolutionary Search Algorithm

2.1 *The Framework of AES*

Multi-agent system, also called 'self organized system' is a computational system in which multiple interacting intelligent agents work together to solve difficult problems which may be impossible for a single agent. In a nutshell, the agent in a multi-agent system is a physical or virtual entity which has several important characteristics. According to [21], four elements should be defined before an agent-based system is used to solve problem:

- First, it is able to live and act in the environment.
- Second, it is able to sense its local environment.
- Third, it is driven by certain purposes.
- Fourth, it has some reactive behaviors.

As can be seen, the meaning of an agent is quite comprehensive, and what an agent represents is different from each other for specific optimization problems. In this chapter, an AES algorithm is proposed to investigate the dynamic 0-1 optimization problems. Obviously, each agent is a 0-1 array, which represents a candidate solution. The energy value of an agent may be equal to the value of the objective function. More detailed definitions can be described as follows:

Definition 1: An agent L represents a candidate solution to the optimization problem in hand, and can be expressed as a 0-1 array:

$$L = (L^1, L^2, \cdots, L^n),\ L^i = 0\ or\ 1,\ 1 \le i \le n, \tag{1}$$

where n represents the scale of the problem. The value of its energy $E(L)$ is equal to the value of the objective function $f(L)$, that is, $E(L) = f(L)$.

In order to realize the local perceptivity of agents conveniently, the environment is organized as a lattice-like structure. All the agents live in a lattice-like environment, which is called an agent lattice *LL*. The size of environment is $L_{size} \times L_{size}$.

Each agent is fixed on a lattice-point and only interacts with its neighbors. Suppose that the agent located at (i, j) is represented as $L_{i,j}$. And the agents which interact with $L_{i,j}$ could be decided by the apperceive range R_s. Thus, the neighbor agents of $L_{i,j}$ could be calculated as follows:

$$L_{k,l},\ i - R_s \le k \le i + R_s,\ j - R_s \le l \le j + R_s. \tag{2}$$

Due to the ring-shaped lattice structure, if $k < 1$, then $k = k + L_{size}$; if $k > L_{size}$, then $k = k - L_{size}$; if $l < 1$, then $l = l + L_{size}$; if $l > L_{size}$, then $l = l - L_{size}$. The agents that interact with $L_{i,j}$ are called its neighbor $N_{i,j}$.

Therefore, the agent lattice can be represented as the one in Fig. 1. Each cross-lattice represents an agent, the data in a rectangle represents its position in the lattice, and two agents can interact with each other if and only if there is a line connecting them. In traditional genetic algorithms (GAs), those individuals that will generate offspring are usually selected from all individuals according to their fitness. Therefore, the global fitness distribution of a population must be determined. But in nature, a global selection does not exist, and the global fitness distribution cannot be determined either.

In fact, the natural selection in real world only occurs in a local environment, and each individual can only interact with those around it. That is, in some phases, the natural evolution is just a kind of local phenomenon. The information can be shared globally only after a process of diffusion. In the agent lattice, in order to achieve their purposes, agents will compete or cooperate with each other so that they can gain more resources. Since each agent can only sense its local environment, its behaviors of competition and cooperation can only take place among the agent and its

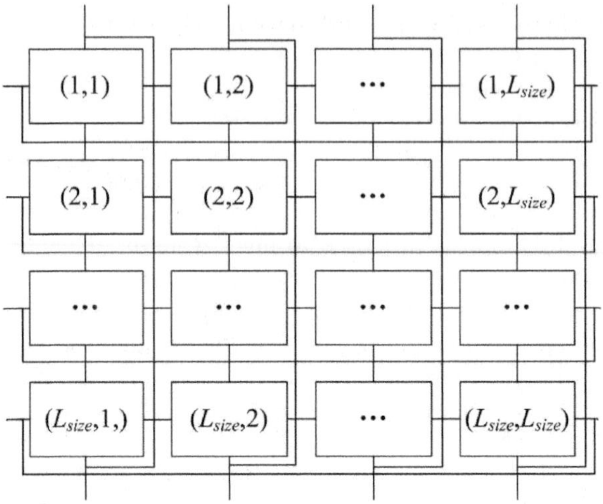

Fig. 1 Model of the agent lattice

neighbors. Simultaneously, the feedback of the best agent can usually influence the behavior of the agents. An agent interacts with its neighbors so that information is transferred to them, and the information feedback from the Lattice can help agents to learn. In such a manner, the information is diffused to the whole agent lattice. As can be seen, the model of the agent lattice is more close to the real evolutionary mechanism in nature than the model of the population in traditional GAs. In summary, the general AES algorithm can be illustrated in Fig. 2.

Procedure general AES algorithm

initialize all the parameters
$t := 0$
initialize a $L_{size} \times L_{size}$ agent lattice $LL(0)$
evaluate every agent in $LL(0)$
repeat
  for each agent L_{ij} in $LL(t)$ **do**
    execute one behavior to L_{ij}
  $t := t + 1$
until a termination condition is met

Fig. 2 Pseudocode for general AES

2.2 Behaviors of Agents

In the proposed AES algorithm, all the agents update and co-evolve by executing a certain behavior. And the ideology of the behaviors is generated by imitating the character of local competitive and learning mechanism of the living organisms.

In nature, several organisms always work together to achieve a common goal. And many algorithm are formulated by simulating these natural phenomena.

In recent years, there has been an increasing concern from the evolution computation community on a class of hybrid EAs, named as memetic algorithms (MAs), which hybridizes local search (LS) methods to refine solution quality with EAs. The term MA is now widely used as a synergy of evolutionary or any population-based approach with separate individual learning or local improvement procedures for problem search. So far, MAs are extensively used for solving many optimization problems, such as scheduling problems [19], [1], travel salesman problems [26], combinatorial optimization problems [12], quadratic assignment problems [23] and other applications [6], [15].

Here, competitive and learning behaviors are defined. For each agent, if its energy is not the best, then it will execute the competitive behavior; if there's no better agent, the statistics-based learning behavior is executed by using the information feedback from the whole multi-agent system. The detailed discussion is given as follows.

2.2.1 Competitive Behavior

The apperceive range of each agent is 1, and the neighbor agents are called competitive neighborhood. Therefore, there are 8 agents in the competitive neighborhood of each agent. For an agent $L_{i,j}$, it will compare its energy with the agents in its neighborhood. If its energy is no less than any agent in its neighborhood, then it can live; else, it will extinct and its position will be replaced by the offspring of the best agent. The details are discussed further as follows.

Assume $L_{i,j} = L^1_{i,j}, L^2_{i,j}, \cdots, L^n_{i,j}$, $L_{max} = L^1_{max}, L^2_{max}, \cdots, L^n_{max}$, and $\forall L \in N_{i,j}, E(L) < E(L_{max})$. If $E(L_{i,j}) < E(L_{max})$, then L_{max} is used to generate an offspring $L'_{i,j} = L'^1_{i,j}, L'^2_{i,j}, \cdots, L'^n_{i,j}$ to replace $L_{i,j}$. There are two kinds of generating methods.

In the first competitive method [39], the information of $L_{i,j}$ and L_{max} is utilized together to generate an offspring $L'_{i,j}$. It randomly chooses some position that $L_{i,j}$ differs from L_{max} to alter the corresponding position in $L_{i,j}$.

$$L'^k_{i,j} = \begin{cases} L^k_{max}, & k \notin D \text{ or } (k \in D \text{ and } random(2) = 0) \\ 1 - L^k_{max}, & \text{otherwise,} \end{cases} \tag{3}$$

where $k = 1, 2, \cdots, n$, set D record the sequence number of the position that $L_{i,j}$ differs from $L_{i,j}$. Namely, $D = \{k | L^k_{i,j} \neq L^k_{max}, k = 1, \cdots, n, \}$. $random(2)$ means to generate 0 or 1 randomly.

The second competitive method is the position based mutation, which is often used in evolutionary computation.

$$L'^{k}_{i,j} = \begin{cases} L^k_{max}, & rand() > \frac{1}{n} \\ 1 - L^k_{max}, & \text{otherwise,} \end{cases} \tag{4}$$

where $k = 1, 2, \cdots, n$, and $rand()$ means to generate a real number between 0.0 and 1.0.

The number of elements in set D actually equals to the Hamming distance between $L_{i,j}$ and L_{max}. When $|D|$ is small, it means the Hamming distance between $L_{i,j}$ and L_{max} is small, and the two agents are similar. Then the probability of using the first competitive pattern to generate better agent is minor, thus we introduce a parameter $D^k \in (0,1)$ to determine which pattern should be used to generate $L'_{i,j}$: If $|D|/n > D^k$, then the first method is adopted; or else the second.

2.2.2 Statistics Based Learning Behavior

The purpose of the learning behavior is to enhance the energy of an agent by statistics based adaptive non-uniform mutation (SANUM) [34] or statistics based adaptive non-uniform crossover (SANUX) [33] with the *elite* (the best agent in the current generation). Because the resource is limited in the environment, only when the energy of an agent is no less than any of the agents in its neighborhood, it can get a chance to learn. The first learning scheme is statistics based mutation and the second is statistics based crossover with the *elite* agent. The schemes are described as follows.

First of all, we decide how to calculate the uniform crossover or mutation probability of a locus by using a statistics-based approach. A global information-based statistics, which has been proved to be effective in enhancing the exploration capability of GAs by Yang [33, 34], is introduced here. Let $p(i)$ denote the crossover or mutation probability of the locus i, f_{ki} denote the frequency of $k's$ in the gene locus i over all the past generations, where $i = 1, 2, \cdots, n$ and k is the allele value for the gene locus. Then, in the binary encoding space, we have $f_{1i} + f_{0i} = 1$, $0 \leq f_{0i}, f_{1i} \leq 1$, and f_{1i} can be regarded as the tendency to '1' for the locus i over all the past populations. SANUM and SANUX make use of this convergence information as feedback to control mutation and crossover by adjusting the probability of each locus. Thus the one-dimension statistic vector $F1 = \{f_{11}, f_{12}, \cdots, f_{1n}\}$ can express the convergence degree of the population from the gene level. Then, $p(i)$ can be calculated from f_{1i} as follows:

$$p(i) = p_{max} - 2 \times |f_{1i} - 0.5| \times (p_{max} - p_{min}), \tag{5}$$

where $|y|$ denotes the absolute value of y, p_{min} and p_{max} denote the minimum and maximum allowable mutation or crossover probability for a locus respectively. For example, for the mutation probability, let $p_{min} = 10^{-4}$ and $p_{max} = \frac{1}{n}$, and for the crossover probability, let $p_{min} = 0.0$ and $p_{max} = 0.5$. Firstly, we calculate the distribution of 1's $f_1(i)$ for each locus i over the lattice, and the crossover and mutation probability $p(i)$ for that gene locus can be obtained. For the global statistics based method, the learning behavior can be adjusted by the convergence degree of the population.

The first learning scheme is the statistics based adaptive non-uniform mutation (SANUM). If L gets a chance to learn, then L is mutated to generate L'. If $E(L') > E(L)$, then agent L is replaced by L'.

Both SANUM and SANUX belong to the class of adaptive mechanisms that occur at the bottom level of mutation and crossover. Traditional bit mutation and crossover keep a constant probability over all the loci. In fact, as the population converges, fewer and fewer offsprings generated by mutating and crossover will survive in the next generation. That is, many operations are wasted on those converged loci. Most of these wasted operations and hence wasted fitness evaluations are saved by SANUM and SANUX through adaptively decreasing $p(i)$ to p_{min} for those converged loci.

We introduce a population index ξ [25] in order to measure its diversity and it can be calculated as follows:

$$\xi = \frac{E_{best} - E_{ave}}{E_{best}}, \qquad (6)$$

where E_{best} and E_{ave} are respectively the best and average energy among the energy values of the agents. Obviously, the index ξ can measure the state of agents via the energy calculation. When ξ decreases from 1 to 0, it means that the agents lose the diversity gradually.

The criterion of selecting a learning scheme is based on the value of ξ as follows.

- If $\xi \leq 0.1$, the first statistics based mutation learning scheme is applied considering the mutation operation can help a converging population to jump from a local optimum.
- If $\xi \geq 0.9$, the second statistics based crossover learning scheme is applied since the crossover operation can accelerate the exploitation to a diversifying population.
- If $0.1 < \xi < 0.9$, one of the two learning schemes is chosen randomly. *Random* means to generate a real number between 0 and 1. If $rand() < 0.5$, the first learning scheme is selected; else, the second one is chosen.

2.3 Two Diversity Maintaining Schemes on AES

In dynamic environments, the fitness landscape could change over time, that is, the current optimum point may become a local optimum and the past local optimum maybe become a new global optimum point. Since a spread-out population can adapt to these changes more easily, two diversity-keeping methods, namely random immigrants (RI) and adaptive dual mapping (ADM), will be introduced into our algorithm framework of AES algorithm to address DOPs.

2.3.1 Random Immigrants Method (RI)

Among the approaches developed for GAs to address DOPs, the random immigrants scheme has been proved to be a beneficial one for many DOPs. The random immigrants scheme aims to maintain the diversity of the population by replacing worst or randomly selected individuals from the population with randomly created individuals [29]. Here, we always choose the worst energy agents to be replaced by randomly generated individuals every generation. In order to avoid that random immigrants disrupt the ongoing search progress too much, especially during the period when the environment does not change, the ratio of the number of random immigrants to the population size is usually set to a small value, e.g., 0.1.

2.3.2 Adaptive Dual Mapping Method(ADM)

Dualism and complementarity are quite common in nature, such as the double-stranded structure in DNA molecules. Inspired by the complementarity mechanism in nature, a primal-dual genetic algorithm has been proposed and applied for DOPs [35]. Elitist is the currently-best solutions in subsequent generation. It preserves currently-best solutions from one generation to the next. The concept of elitism has been introduced into several EAs. Elitism has been considered an effective method for improving the efficiency of an EA[7][18][28]. Effective manipulation of the elite can contribute to better performance of EAs.

In this chapter, we investigate the application of the combination of dualism and elitism into AES. For the convenience of description, we first introduce the definition of dual agent here. Given an agent L, its dual agent is defined as L', where $L'^k = 1 - L^k, k = 1, \cdots, n$. With this definition, an agent *elite* is to evaluate its dual agent ($elite'$) firstly before executing statistics based learning. If its dual agent is evaluated to have more energy, that is, $E(elite') > E(elite)$, then *elite* is replaced by $elite'$.

Given the above discussion, the proposed AES incorporate two diversity maintaining techniques (RI and ADM) is illustrated in Fig. 3.

3 The Dynamic Testing Suite

3.1 Stationary Test Problems

A set of well studied stationary problems, forming a range of difficulty levels for EAs, are selected as the experimental functions to compare the performance of AES with traditional standard GA (SGA), the Primal-Dual GA (PDGA) [35, 30], and the GA with RI (RIGA), where the worst 10% individuals are replaced with random individuals every generation. In this chapter, dynamic test problems will be constructed from these stationary problems by a dynamic problem generator, which is called the XOR generator.

```
Procedure proposed AES algorithm
initialize all the parameters
t := 0
initialize a L_size × L_size agent lattice LL(0)
evaluate every agent in LL(0)
calculate crossover and mapping prob. p(i), i = 1,···,n
repeat
  for each agent L_{i,j} in LL(t) do
    if there exists an agent L_max ∈ N_{i,j} with
        E(L_max) > E(L_{i,j}) then
      calculate |D|/n, where |D| is the Hamming
        distance between L_max and L_{i,j}
      if |D|/n > D^h then
        select the first competitive scheme
      else Select the second competitive scheme
    else
      L'_{i,j} := Dual(L_{i,j})
      if (E(L'_{i,j}) > E(L_{i,j})) then
        L_{i,j} := L'_{i,j}
      if (ξ < 0.1) then
        select the first learning scheme to mutate L_{i,j}
      else if (ξ > 0.9) then
        select the second learning scheme to crossover
          L_{i,j} with elite, which is the best agent so far
      else
        randomly select the first or second learning
          scheme and execute it for L_{i,j}
  calculate ξ
  t := t+1
until a termination condition is met
```

Fig. 3 Pseudocode for the proposed AES algorithm

3.1.1 One-Max Function

This problem simply aims to maximize ones in a binary string. The fitness of a string is the number of ones it contains. A string length of 100 bits will be used for this study. And the unique optimum solution has a fitness of 100.

3.1.2 Royal Road Function

This function was proposed by Michel, Forrest and Holland [24]. It is defined on a 64 bit string consisting of eight contiguous building-blocks (BBs) and each BB has 8 adjacent bits. The fitness of this function is defined as follows:

$$f(x) = \sum_{i=1}^{8} c_i \delta_i(x), \tag{7}$$

where $c_i = 8$ and $\delta_i = \{1, if\, x \in S; 0, otherwise\}$. The optimal for this function is given as $f(111\ldots1) = 64$.

3.1.3 Deceptive Function

Deceptive functions are a family of functions where there exists low-order BBs that do not combine to form the higher-order BBs. In this study, a deceptive function is constructed, which consists of 10 copies of the order-4 fully deceptive function DF2 [32] and has an optimum fitness value of 300. DF2 is defined as follows:

f(0000)=28	f(0001)=26	f(0010)=24	f(0011)=18
f(0100)=22	f(0101)=6	f(0110)=14	f(0111)=0
f(1000)=20	f(1001)=12	f(1010)=10	f(1011)=2
f(1100)=8	f(1101)=4	f(1110)=6	f(1111)=30

3.1.4 Double Deceptive Function

A six rank bipolar Double (Strongly Interrelated) Deceptive function [13] discussed here consists of 10 copies of 6 rank bipolar deceptive function DF1.

$$f(a) = \sum_{i=1}^{n/6} f_{bipolar6}(a_{6i-5}, a_{6i-4}, a_{6i-3}, a_{6i-2}, a_{6i-1}, a_{6i}) \tag{8}$$

DF1 is designed as follows:

$$f_{bipolar6}(a_1, a_2, a_3, a_4, a_5, a_6) = \begin{cases} 0.9, & \text{if } u = 3 \\ 0.8, & \text{if } u = 2 \text{ or } 4 \\ 0, & \text{if } u = 1 \text{ or } 5 \\ 1, & \text{if } u = 0 \text{ or } 6 \end{cases}$$

where u is the number of 1s in the variable. The strongly interrelated deceptive function has an optimum fitness of 10.

3.2 *Generating Dynamic Test Problems*

In this chapter, the dynamic test environments are constructed based on the above stationary functions by using an "XOR" operator [36, 38]. Suppose that the environment is periodically changed every τ generations, the environmental dynamics can be formulated as follows:

$$f(x,t) = f(x \oplus M(k)), \tag{9}$$

where $k = \lceil t/\tau \rceil$ is the period index, t is the generation counter, and $M(k)$ is the XOR mask for period k. Given a value for parameter ρ, $M(k)$ can be incrementally generated as follows:

$$M(k) = M(k-1) \oplus T(k), \tag{10}$$

where $T(k)$ is an intermediate binary template randomly created for period k containing $\rho \times n$ ones. For the period $k = 1$, $M(1)$ is initialized to be a zero vector.

The advantage od this operator is the fitness landscape can still keep the certain properties in the original landscape after a change occurs, e.g., the total number of optima and fitness values of optima though their locations shifted. For example, if we apply a template $T = 11111$ to a 5-bit One-Max function, the original optimal point $x^* = 11111$ becomes the least fit point while the original worst point $x = 00000$ becomes the new optimal point in the changed landscape, but the optimal fitness value (i.e., 5) and the uniqueness of optimum remain invariant.

In this chapter, we have constructed dynamic versions of above stationary problems. In our experiment, the change severity ρ is set to 0.1, 0.5 and 0.9, which is used to examine the algorithms' performance in the changing environments with different severities respectively: from very slight change ($\rho = 0.1$) to moderate variation ($\rho = 0.5$) to intense change ($\rho = 0.9$).

Another dynamics parameter considered is the change frequency, i.e. how often the environment changes. In our experiment the frequency parameter τ is set to 20, 50 and 100 respectively because all algorithms are at quite early searching stage at generation 20, at medium searching stage at generation 50, and at late stage or converged at generation 100. By setting τ to these values, we can inspect the capability of the algorithms adapting to dynamic environment under different level of searching stage.

Thus, we construct a series of 9 different dynamic problems from each stationary test problem. The dynamics parameter settings are summarized in Table 1.

Table 1 The index table for dynamic parameter setting

τ	Environmental Dynamics Index		
20	1	2	3
50	4	5	6
100	7	8	9
$\rho \rightarrow$	0.1	0.5	0.9

4 Experimental Study

4.1 Experimental Setting

For AES algorithm, $L_{size} = 10$. When the mutation probability is computed, p_{min} is set to be 10^{-4}, p_{max} is $1/n$. And $p_{min} = 0.0$ and $p_{max} = 0.5$ are set for crossover probability computation. For SGA, RIGA and PDGA, parameters are set as follows:

the population size *pop_size* is set to 100, one-point crossover with a fixed probability $p_c = 0.6$, bit mutation with the mutation probability $p_m = 0.001$, and fitness proportionate selection with commonly used roulette wheel mechanism. The best N chromosomes among all the parents and children are always transferred to the next generation population. For every generation, the best-of-generation fitness was recorded. And for each run of an algorithm on a dynamic problem, 10 periods of environmental changes are allowed. Each experimental result is averaged over 100 runs with different random seeds.

For dynamic optimization problems, the optimal solution is not unique. Hence, it is not sufficient to compare only optimal solution obtained at last. It is extremely important to compare the performance of algorithms at every generation. The overall offline performance of an algorithm on a DOP is defined as:

$$\overline{F}_{BOG} = \frac{1}{G}\sum_{i=1}^{G}(\frac{1}{100}\sum_{j=1}^{100}\overline{F}_{BOG_{ij}}), \tag{11}$$

where $G = 10\tau$ is the total number of generations for a run and $\overline{F}_{BOG_{ij}}$ is the best-of generation fitness of generation i of run j. The off-line performance $\overline{F}_{BOG}$ is the best-of-generation fitness averaged over 100 runs and then averaged over the data gathering period.

4.2 Experimental Results on DOPs

The experimental results with respect to the mean best-of-generation fitness against evaluations of four EAs on the dynamic test problems are plotted in Fig. 4 to Fig.7 respectively.

The overall offline performance is presented in Fig.8. The corresponding statistical results of comparing algorithms by the one-tailed t-test with 38 degrees of freedom at a 0.05 level of significance are given in Table 2. In Table 2, the t-test result regarding Alg. 1 - Alg. 2 is shown as "s+","s-", "+", "-" or "∼" when Alg. 1 is significantly better than, significantly worse than, insignificantly better than, insignificantly worse than, or statistically equivalent to Alg. 2 respectively. From these tables and figures several results can be observed.

First, when the environment shifted, AES can always outperform other peer EAs on almost all the problems, which has been indicated by the relevant t-test results in Table 2. AES can converge faster, and simultaneously it can trace the optimal solution better in dynamic environments. This result shows the efficiency of our proposed behaviors of the agents in dynamic environments. The competitive and learning behaviors of the agents, and the diversity maintaining schemes help the algorithm adapt well to the most changing environment. When the environment shifts slightly ($\rho = 0.1$), AES significantly perform better over other EAs for the changing periods. Both PDGA and RIGA perform better than SGA, because the primal dual and RI operators can help them adapt to the new environments. But for One Max problem, when the environment changes slightly, the AES can be beaten by

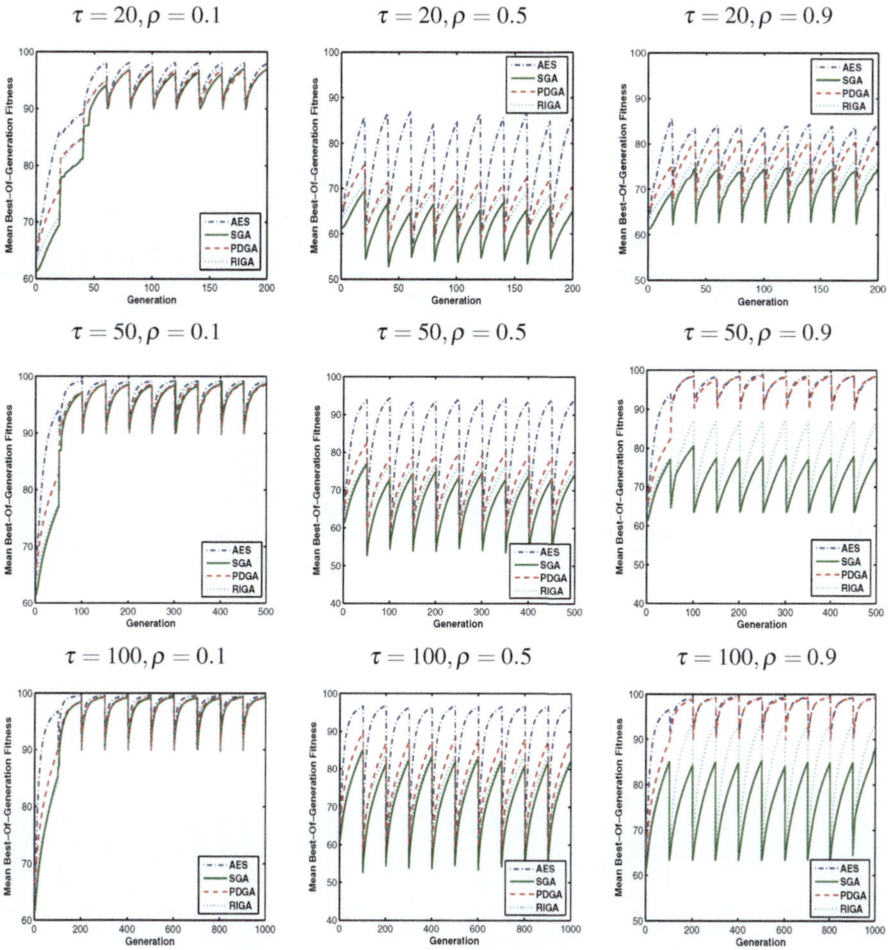

Fig. 4 Dynamic performance of EAs on Dynamic One-Max Problems

SGA and PDGA; When the shift severity is moderate ($\rho = 0.5$), AES keeps performing perfectly well with the best-of-generation fitness staying in the optimum over all dynamic periods. SGA always maintains a lower fitness level than RIGA, PDGA, and AES can do for all environmental periods on all dynamic problems; when $\rho = 0.9$, both AES and PDGA perform much better than SGA and RIGA. It can be shown from the figures that both AES and PDGA can converge fast and make good performance when intense shift happened.

Second, AES algorithm has a more quickly and robust converge capability than SGA, PDGA and RIGA. On the One-Max problem, when the environment shifts slightly, AES performs as well as SGA, PDGA and RIGA; when $\rho = 0.5$, AES keeps performing well with the best-of-generation fitness staying in the optimum over all dynamic periods; when $\rho = 0.9$, both AES and PDGA perform much better

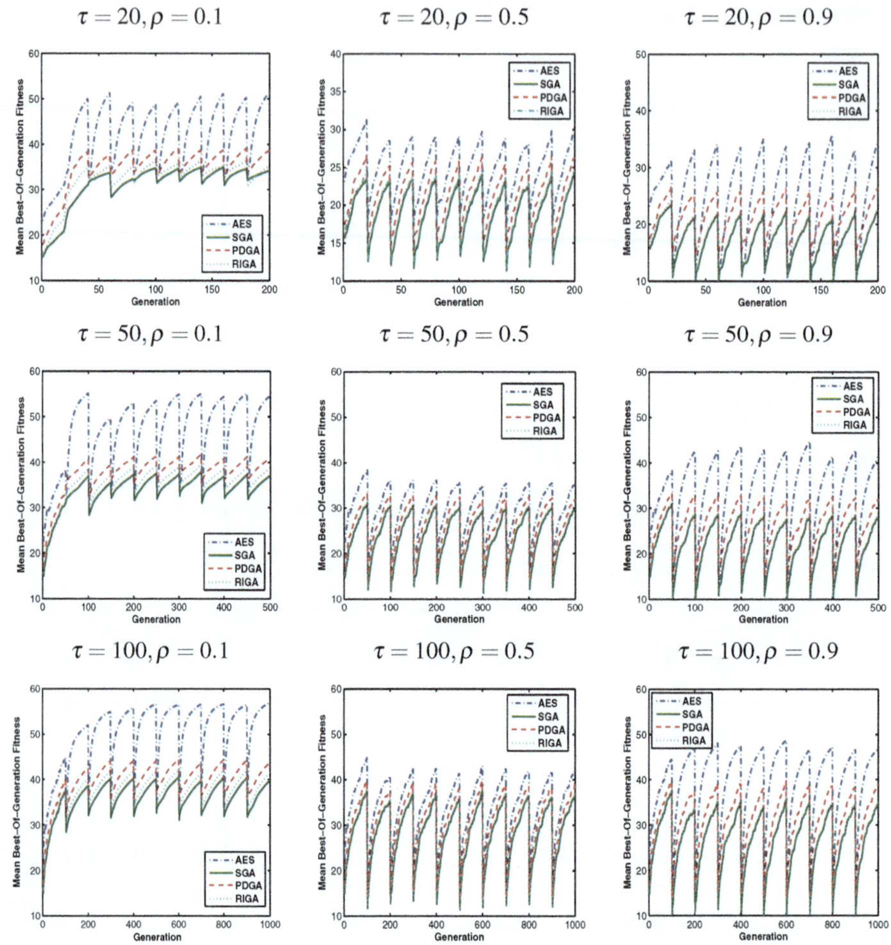

Fig. 5 Dynamic performance of EAs on Dynamic Royal Road Problems

than SGA and RIGA, and also when change took place, it can be seen from the figures that both AESF and PDGA can converge fast and track the optimum when intense shift happens. On the Royal Road Problem and Deceptive Problem, when $\rho = 0.1$ and $\rho = 0.9$, AES outperforms SGA, PDGA and RIGA; and when the environment half shift, both AES and PDGA perform well. The mean best-of-generation fitness first drops a little when changes occur, then rises up quickly to near optimum value. It is because the complementary and dominance mechanisms embedded in AES work perfectly under the condition of extreme environment change. On the the Strongly Interrelated Deceptive problem, AES outperforms SGA, PDGA and RIGA for all periods in all the shifting environments. For AES, whenever the fitness function changes, the statistics based mechanism and diversity maintaining mechanisms in AES rapidly draw back the best individuals in the last generation of previous period. And the lattice construction also helps to suit the new environment. And also,

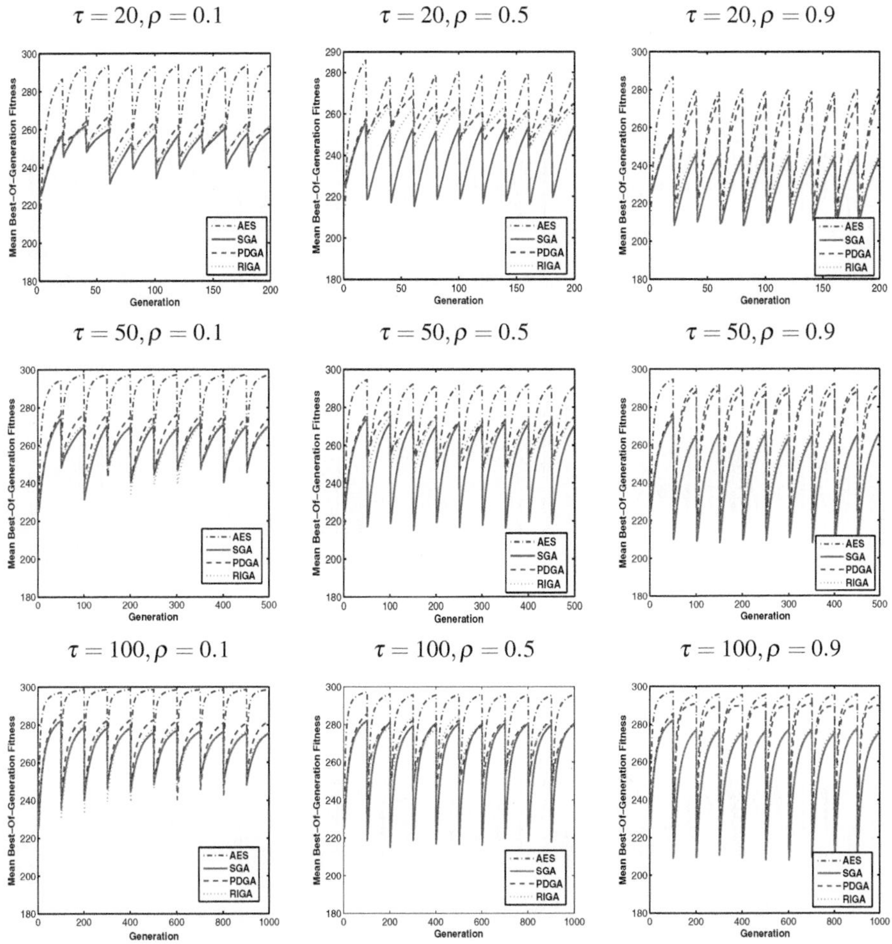

Fig. 6 Dynamic performance of EAs on Dynamic Deceptive Problems

it can be seen from the figures, for all the DOPs, especially when the change is intense, that PDGA has the ability to track the optimal faster than SGA and RIGA. This lies in that dual mapping operator help PDGA adapt to the new environment quickly, maintain enough diversity, and therefore effectively track the new optimum after moderate and intense changes.

Third, the competitive and learning behaviors in AES algorithm make AES have a better exploitation ability than GAs. All the agents accomplish their update via competing in a local neighborhood or learning based on the statistical feedback information of the whole system, so AES algorithm has a rapid convergence capability.

Fourth, the combination of the two diversity maintaining schemes RI and ADM improves the adaptability of AES algorithm in non-stationary environments. The information of the former generation is reserved and the diversity is increased,

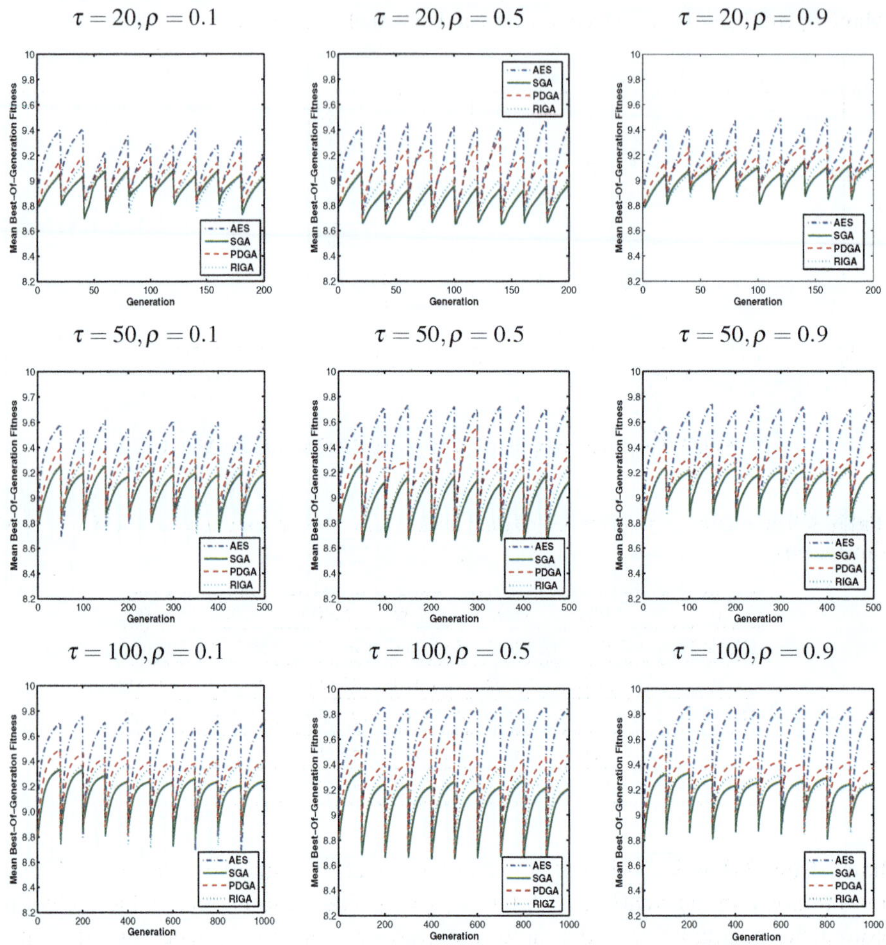

Fig. 7 Dynamic performance of EAs on Dynamic Double Deceptive Problems

therefore, AES algorithm with reserved information and multiplex diversity can adapt to the dynamic environment due to effectiveness of the combined maintaining scheme of RI and ADM.

Finally, the environmental parameters affect the performance of algorithms. The performance of all algorithms can be increased when the value of τ arises from 20 to 50 to 100. It is easy to understand since algorithms have more time to find better solutions before the next change with the increment of the value of τ. The effect of the changing severity parameter ρ is different. For example, if ρ is fixed, the performance curve of AES is always lower when the environment changes with a middle severity degree than that when $\rho = 0.1$ and $\rho = 0.9$.

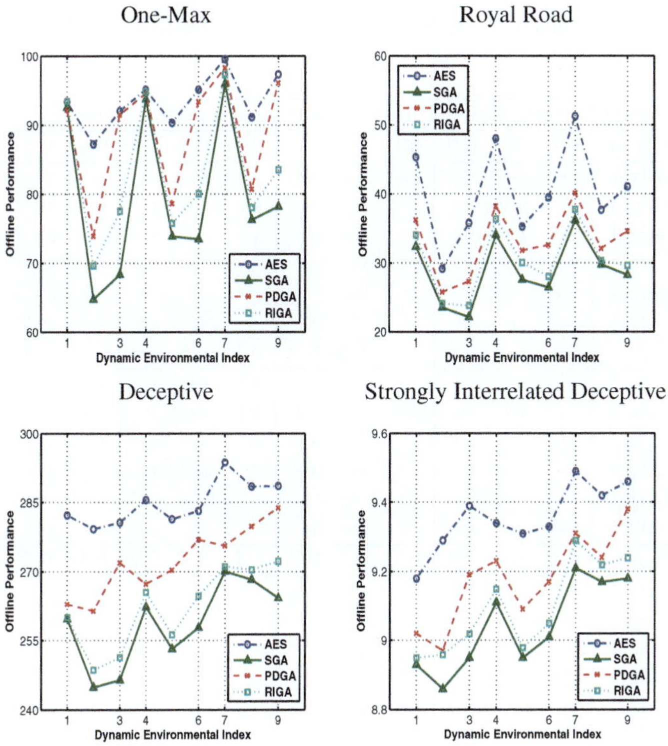

Fig. 8 Overall Offline performance of EAs

5 Conclusions

This chapter investigates the application of an agent-based evolutionary search algorithm (AES) for binary coded problems. Two special behaviors of agents, competing and learning, are proposed. Two diversity schemes, random immigrants and adaptive dual mapping scheme, are integrated into AES in order to improve its performance in dynamic environments. Based on the XOR dynamic problem generator, a series of dynamic test problems were constructed systematically, in the mean time, the proposed AES algorithm is used to solve the dynamic test problems.

From the experimental results, we can draw the following conclusions. First, the competitive and learning behaviors of agents can always help AES algorithm obtain a better performance than the peer GAs can do. Second, both random immigrants scheme and dual mapping method can help increase the population diversity of EAs. Combining them together also improves the performance of AES algorithm in dynamic environments efficiently. Third, some dynamic characteristics of the environments may affect the performance of algorithms.

Table 2 Statistical results of algorithms on dynamic problems (Alg.1−Alg. 4 denotes *AES*, *SGA*, *PDGA* and *RIGA* respectively)

t-test Result	One Max			Royal Road			Deceptive			Double Deceptive		
$\tau = 20, \rho \rightarrow$	0.1	0.5	0.9	0.1	0.5	0.9	0.1	0.5	0.9	0.1	0.5	0.9
Alg.1-Alg.2	+	s+	s+	s+	s+	s+	s+	s+	s+	s+	s+	s+
Alg.1-Alg.3	∼	s+	∼	s+	s+	s+	s+	s+	s+	s+	s+	s+
Alg.1-Alg.4	∼	s+	s+	s+	s+	s+	s+	s+	s+	s+	s+	s+
Alg.2-Alg.3	−	s−	s−	−	s−	s−	−	s−	s−	−	s−	s−
Alg.2-Alg.4	∼	−	s−	−	∼	−	∼	−	−	∼	s−	−
Alg.3-Alg.4	∼	+	s+	+	+	s+	+	s+	s+	+	∼	s+
$\tau = 50, \rho \rightarrow$	0.1	0.5	0.9	0.1	0.5	0.9	0.1	0.5	0.9	0.1	0.5	0.9
Alg.1-Alg.2	+	s+	s+	s+	s+	s+	s+	s+	s+	s+	s+	s+
Alg.1-Alg.3	∼	s+	∼	s+	s+	s+	s+	s+	s+	s+	s+	s+
Alg.1-Alg.4	∼	s+	s+	s+	s+	s+	s+	s+	s+	s+	s+	s+
Alg.2-Alg.3	−	s−	s−	−	s−	s−	−	s−	s−	−	s−	s−
Alg.2-Alg.4	−	−	s−	−	−	−	−	−	s−	−	−	−
Alg.3-Alg.4	∼	+	s+	+	+	s+	+	s+	s+	+	s+	s+
$\tau = 100, \rho \rightarrow$	0.1	0.5	0.9	0.1	0.5	0.9	0.1	0.5	0.9	0.1	0.5	0.9
Alg.1-Alg.2	+	s+	s+	s+	s+	s+	s+	s+	s+	s+	s+	s+
Alg.1-Alg.3	∼	s+	∼	s+	s+	s+	s+	s+	s+	s+	s+	s+
Alg.1-Alg.4	∼	s+	s+	s+	s+	s+	s+	s+	s+	s+	s+	s+
Alg.2-Alg.3	−	s−	s−	−	s−	s−	−	s−	s−	s−	s−	s−
Alg.2-Alg.4	∼	−	s−	−	∼	−	∼	−	s−	s−	−	−
Alg.3-Alg.4	∼	+	s+	+	+	s+	+	s+	s+	∼	∼	s+

Generally speaking, the experimental results indicate that the AES algorithm with the random immigrants and adaptive dual mapping schemes exhibits better performance in dynamic environments. AES algorithm combining some diversity schemes seems to be a good choice to address DOPs.

The work studied in this chapter can be extended in several ways. It is interesting to hybridize AES algorithm with other advanced diversity schemes, e.g., hybrid memory schemes [31, 37]. Another interesting work is to further investigate the idea of AES algorithm to solve other optimization problems with sequential and real encodings in non-stationary environments.

References

1. Annibale, G.D., Leone, R.D., Festa, P., Marchitto, E.: A New Meta-Heuristic for the Bus Driver Scheduling Problem: GRASP Combined with Rollout. In: Proc. of the, IEEE Symposium on Computational Intelligence in Scheduling, pp. 192–197 (2007)
2. Blackwell, T.: Particle swarms and population diversity. Soft Computing 9, 793–802 (2005)
3. Blackwell, T., Branke, J.: Multi swarms, exclusion, and anti-convergence in dynamic environments. IEEE Transactions on Evolutionary Computation 10(4), 459–472 (2006)

4. Branke, J.: Memory enhanced evolutionary algorithms for changing optimization problems. In: Proc. of the 1999 IEEE Congress on Evolutionary Computation, pp. 1875–1882 (1999)
5. Branke, J., Kaußler, T., Schmidth, C., Schmeck, H.: A multi-population approach to dynamic optimization problems. In: Proc. of the 5th International Conference on Adaptive Computing in Design and Manufacturing, pp. 299–308 (2000)
6. Caponio, A., Cascella, G.L., Neri, F., Salvatore, N., Sumner, M.: A Fast Adaptive Memetic Algorithm for Online and Offline Control Design of PMSM Drives. IEEE Transactions on Systems, Man, and Cybernetics-Part B: Cybernetics 37(1), 28–41 (2007)
7. Chang, W.A., Ramakrishna, R.S.: Elitism-Based Compact Genetic Algorithms. IEEE Transactions on Evolutionary Computation 7(4), 367–385 (2003)
8. Cobb, H.G., Grefenstette, J.J.: Genetic algorithms for tracking changing environments. In: Proc. of the 5th Int. Conf. on Genetic Algorithms, pp. 523–530 (1993)
9. Davidsson, P., Wernstedt, F.: A Multi-Agent System Architecture for Coordination of Just-in-time Production and Distribution. In: Proc. of the 17th ACM Symposium on Applied Computing, ACM SAC, UK, pp. 294–300 (2002)
10. Persson, J.A., Davidsson, P.: Integrated optimization and multi-agent technology for combined production and transportation planning. In: Proc. of the 38th Hawaii Int. Conf. on System Sciences, pp. 1–9. IEEE Press, NY (2005)
11. Ferber, J.: Multi-Agent Systems: An Introduction to Distributed Artificial Intelligenc. Addison-Wesley, New York (1999)
12. Gallardo, J.E., Cotta, C., Ferndez, A.J.: On the Hybridization of Memetic Algorithms with Branch-and-Bound Techniques. IEEE Transactions on Systems, Man, and Cybernetics-Part B: Cybernetics 37(1), 77–83 (2007)
13. Goldberg, D.E., Deb, K., Korb, B.: Messey genetic algorithm revisited: studies in mixed size and scale. Complex Systems 4(4), 145–444 (1990)
14. Grefenstette, J.: Genetic algorithms for changing environments. In: Proc. of the 2nd Int. Conf. on Parallel Problem Solving from Nature, pp. 137–144 (1992)
15. Hodgson, R.J.W.: Memetic Algorithms and the Molecular Geometry Optimization Problem. In: Proc. of the, Congress on Evolutionary Computation, pp. 625–632 (2000)
16. Jennings, N.R., Sycara, K., Wooldridge, M.: A roadmap of agent research and development. Autonomous Agents and Multi-Agent Systems Journal 6(4), 317–331 (1998)
17. Karageorgos, A., Mehandjiev, N., Weichhart, G., Hammerle, A.: Agent-based optimization of logistics and production planning. Engineering Application of Artificial Intelligence 16, 335–348 (2003)
18. Kim, J.L., Ellis Jr., R.D.: Permutation-Based Elitist Genetic Algorithm for Optimization of Large-Sized Resource-Constrained Project Scheduling. Journal of Construction Engineering and Management 134(11), 904–913 (2008)
19. Liu, B., Wang, L., Jin, Y.H.: An Effective PSO-Based Memetic Algorithm for Flow Shop Scheduling. IEEE Transactions on Systems, Man, and Cybernetics-Part B: Cybernetics 37(1), 18–27 (2007)
20. Liu, J., Tang, Y.Y., Cao, Y.C.: An evolutionary autonomous agents approach to image feature extraction. IEEE Transactions on Evolutionary Computation 1, 141–158 (1997)
21. Liu, J.: Autonomous Agents and Multi-Agent Systems. In: Explorations in Learning Self-Organization, and Adaptive Computation. World Scientific, Singapore (2001)
22. Manvi, S.S., Birje, M.N.: An agent-based resource allocation model for grid computing. In: Proc. of 2005 IEEE Int. Conf. on Services Computing, pp. 311–314. IEEE Press, NY (2005)
23. Merz, P., Freisleben, B.: A Comparison of Memetic Algorithms, Tabu Search and Ant Colonies for the Quadratic Assignment Problem. In: Proc. of the 1999 Congress on Evolutionary Computation, pp. 2063–2070 (1999)

24. Mitchell, M., Forest, S., Holland, J.H.: The royal road for genetic algorithms: fitness landscape and GA performance. In: Proc. of the 1st Euro. Conf. on Artificial Life, pp. 245–254 (1992)
25. Neri, F., Toivanen, J., Makinen, R.A.E.: An adaptive evolutionary algorithm with intelligent mutation local searchers for designing multidrug therapies for HIV. Applied Intelligence 27(3), 219–235 (2007)
26. Nguyer, H.D., Yoshihara, I., Yamamori, K., Yasunaga, M.: Implementation of an Effective Hybrid GA for Large-Scale Traveling Salesman Problems. IEEE Transactions on Systems, Man, and Cybernetics-Part B: Cybernetics 37(1), 92–99 (2007)
27. Russell, S., Norvig, P.: Artificial Intelligence: A Modern Approach. Prentice-Hall, Englewood Cliffs (1995)
28. Tan, G., Zhou, D., Jiang, B.: MI Dioubate Elitism-based immune genetic algorithm and its application to optimization of complex multi-modal functions. Journal of Central South University of Technology 15(6), 845–852 (2008)
29. Vavak, F., Fogarty, T.C.: A comparative study of steady state and generational genetic algorithms for use in nonstationary environments. In: Fogarty, T.C. (ed.) AISB-WS 1996. LNCS, vol. 1443, pp. 297–304. Springer, Heidelberg (1996)
30. Wang, H., Wang, D.: An improved primal-dual genetic algorithm for optimization in dynamic environments. In: King, I., Wang, J., Chan, L.-W., Wang, D. (eds.) ICONIP 2006. LNCS, vol. 4234, pp. 836–844. Springer, Heidelberg (2006)
31. Wang, H., Wang, D., Yang, S.: Triggered memory-based swarm optimization in dynamic environments. In: Giacobini, M. (ed.) EvoWorkshops 2007. LNCS, vol. 4448, pp. 637–646. Springer, Heidelberg (2007)
32. Whitley, L.D.: Fundamental principles of deception in genetic search. In: Foundations of Genetic Algorithms I, pp. 221–241 (1991)
33. Yang, S.: Adaptive non-uniform crossover based on statistics for genetic algorithms. In: Proc. of the 2002 Genetic and Evolutionary Computation Conference, pp. 650–657 (2002)
34. Yang, S.: Adaptive mutation using statistics mechanism for genetic algorithms. In: Coenen, F., Preece, A., Macintosh, A. (eds.) Research and Development in Intelligent Systems, pp. 19–32. Springer, London (2003)
35. Yang, S.: PDGA: the primal-dual genetic algorithm. In: Abraham, A., Koppen, M., Franke, K. (eds.) Design and Application of Hybrid Intelligent Systems, pp. 214–223. IOS Press, Sydney (2003)
36. Yang, S.: Non-stationary problem optimization using the primal-dual genetic algorithm. In: Proc. of the 2003 Congress on Evolutionary Computation, vol. 3, pp. 2246–2253 (2003)
37. Yang, S.: Memory-based immigrants for genetic algorithms in dynamic environments. In: Proc. of the 2005 Genetic and Evolutionary Computation Conference, vol. 2, pp. 1115–1122 (2005)
38. Yang, S., Yao, X.: Experimental study on population-based incremental learning algorithms for dynamic optimization problems. Soft Computing 9(11), 815–834 (2005)
39. Zhong, W.C., Liu, J., Xue, M.Z., Jiao, L.C.: A multiagent genetic algorithm for global numerical optimization. IEEE Transactions on System, Man, and Cybernetics-Part B: Cybernetics 34(2), 1128–1141 (2004)

Divide and Conquer in Coevolution: A Difficult Balancing Act

Hemant Kumar Singh and Tapabrata Ray

Abstract. In recent years, Cooperative Coevolutionary Evolutionary Algorithms (CCEAs) have been developed as extensions to traditional Evolutionary Algorithms (EAs). CCEAs attempt to solve the optimization problems by decomposing them into subcomponents referred to as *collaborators*. CCEAs have been deemed attractive for certain complex problems (with high number of decision variables), as they can achieve better fitness values than traditional EAs by employing "divide and conquer" strategy. However, their performance can vary from good to bad depending on the choice of collaborators, separability of problem and the underlying recombination scheme. This chapter highlights that a basic CCEA is inadequate to handle a wide variety of problems. Thereafter, a CCEA with adaptive partitioning (CCEA-AVP) has been introduced, which attempts to chose the collaborators adaptively during the search, depending on the relationships between the design variables. Studies have been done on various test functions and the proposed technique has been compared with conventional EA as well as conventional CCEA to highlight the benefits. A number of areas of further research in CCEA are highlighted to fully exploit the benefits of coevolution.

1 Introduction

Evolutionary Algorithms (EA) have demonstrated a great competence in solving various classes of optimization problems in the recent past. They have been quite

Hemant Kumar Singh · Tapabrata Ray
School of Engineering and Information Technology,
University of New South Wales at Australian Defence Force Academy (UNSW@ADFA),
Canberra ACT, Australia
e-mail: h.singh@adfa.edu.au, t.ray@adfa.edu.au

R.A. Sarker and T. Ray (Eds.): Agent-Based Evolutionary Search, ALO 5, pp. 117–138.
springerlink.com

successful as generic optimizers as they are easy to apply, do not require assumptions on the continuity or the slope of the function and are able to deliver a set of non-dominated solutions for multi-objective problems in a single run (as opposed to point-to-point techniques). However, it has also been observed that the performance of EAs tend to degrade for problems with large number of decision variables, which is often referred to as the "curse of dimensionality". Therefore, there has been a growing need for effective mechanisms to deal with such problems of high dimensionality, often referred to as large scale optimization problems.

The notion of collaboration among evolving species was introduced by Potter and De Jong [10], to improve the performance of existing EAs for handling large scale problems. Their approach is referred to as Cooperative Coevolutionary Algorithms (CCEAs). CCEAs try to solve large scale problems by dividing the variable space into multiple populations of smaller dimensions, which are then evolved individually. These individual populations also interact (collaborate) frequently with each other in order to improve global fitness. This way, the modularity of optimization problem can be exploited to generate solutions with higher fitness in relatively less computational expense. The approach is also commonly referred to as *problem decomposition* or *divide and conquer* strategy.

While the concept of evolving subcomponents of a problem independently and co-adaptively sounds natural and attractive, the dynamics of CCEA is far more complex as compared to EA [6]. In a CCEA, in addition to the parameters of EA (viz. population size, number of generations, crossover and mutation rates), decisions regarding variable partitions, number of generations at each subpopulation level and collaboration strategies (single best, random etc) play a significant role in the algorithm's performance. Without a careful choice of these additional parameters, CCEA, in its basic form can result in a large variation in outcome ranging from good to disappointing, as echoed in the studies by Popovici and De Jong [7].

In an attempt to better understand the dynamics of CCEA, empirical studies have focused on various aspects – choice of collaborators [15, 4], interaction frequency [8], sequential and parallel versions of information exchange (referred as update timing) [9], disconnection between the external goal and the internal behavior of CCEA [6], etc. All these studies have highlighted the complexity of the dynamics of CCEA.

On a parallel front, a number of variants of CCEA have beein developed, such as coevolutionary particle swarms [13, 14], coevolutionary differential evolution [3], cooperative coevolutionary models based on self-adaptive neighborhood search differential evolution [16]. The original study [10] and its variants have been applied to a number of single and multi-objective, separable and nonseparable benchmark problems. In a recent work [2], the paradigm of cooperation based optimization has been further extended to Dynamic Multiobjective Optimization problems.

While excellent performance of CCEAs have been reported in many studies, very few highlight the difficulties of using CCEAs for black box functions. Recently, Ray and Yao [11] have proposed a CCEA based on adaptive partitioning and highlighted a number of cases where traditional CCEAs may not be able to deliver competitive results. The CCEA proposed by them tries to identify the collaborators (subcomponents) adaptively during the search and partitions the decision variable space accordingly. In this chapter, the correlation based partitioning proposed by them has been further improved and investigated on a number of problems with large number of variables. Thereafter, comparisons have been made between CCEA with adaptive variable partitioning (CCEA-AVP) and fixed partition CCEAs in order to highlight the improvements in performance.

Rest of the chapter is organized as follows: The background of the basic CCEA is discussed in Section 2. The proposed algorithm (CCEA-AVP) is described in Section 3. Numerical experiments and related discussions are presented in Section 4. The Summary of findings and further scope of studies have been discussed in Section 5.

2 Background

2.1 Basic CCEA

Cooperative Coevolutionary Algorithm (CCEA) was proposed by Potter and De Jong [10][1]. Basic CCEA is outlined in Algorithm 1. It comprises the following 3 major steps:

1. Decompose the complex decision space into multiple subsystems of smaller dimensions. In basic CCEA, the population is partitioned into s (>1) subpopulations, containing equal number of variables. The variables to be grouped into any particular subpopulation are chosen randomly at each CCEA generation.
2. Evolve each subsystem individually for a specified number of *subevolve* generations using a traditional EA. Since the complete set of variables are needed in order to evaluate a global function value, collaborators are needed from other subpopulations in order to evolve any individual subpopulation. Two strategies were suggested in [10]:
   a. Single best collaboration strategy: The subpopulation under consideration is combined with variable values corresponding to the best solution obtained so far. The subpopulations are evolved in a round-robin fashion, and the best function value is continuously updated.
   b. Random/best collaboration strategy: The subpopulation is combined with either the best *or* random solution from remaining subpopulations, whichever has the better global fitness value.
3. After evolving subpopulations for a specified number of generations, the evolved subpopulations are integrated back together, so that the subpopulations can co-adapt and interact in order to obtain high global fitness value.

[1] Potter and De Jong had used the term CCGA for their GA- based algorithm. CCEA has also been referred to as CCA in certain studies [15].

Algorithm 1. Basic CCEA

Require: N_G (Number of CCEA Generations), N_V (Number of variables for the problem), N_P (Population size), *CR* (Crossover Rate), *MR* (Mutation Rate), Distribution index of Crossover and Distribution index of Mutation.

Initialize Population *pop*
Set $Soln_{best}$ as the best individual in the population
for $cgen = 2$ to N_G **do**
  Split pop_{cgen} into s partitions $pop_{1,cgen}, pop_{2,cgen} \cdots pop_{s,cgen}$. {Each partition consisting of randomly assigned N_V/s variables. Let the $\mathbf{V_j}$ denote the indices of variables contained in a partition j}
  for $i = 1$ to s **do**
    Construct $spop_i$ by combining variable values in partition i with best/random variable values from other partitions
    for $subgen = 2$ to N_{SG} **do**
      $childpop_{i,subgen} \leftarrow$ Subevolve $(spop_{i,subgen-1})$
      Evaluate $(childpop_{i,subgen})$
      $CP \leftarrow$ Sort $(childpop_{i,subgen} + spop_{i,subgen-1})$
      $spop_{i,subgen} \leftarrow$ Reduce (CP)
    end for
    Set $pop_{cgen}(\mathbf{V_i}) = spop_i$
    Update $Soln_{best}$ {Update best solution}
  end for
end for

2.2 Why Are CCEAs Attractive?

As suggested by various studies in the past [10, 16], for many large scale problems, CCEAs can provide much better solutions than traditional Evolutionary Algorithms. While dealing with large number of variables, evolving different modules of a problem individually can expedite convergence rate appreciably as compared to traditional EAs, which attempt to evolve all variables simultaneously. A simple illustration to this effect for four popular scalable benchmark problems (viz. Rastrigin, Schwefel, Rosenbrock and Ackely (details given in Table (4))) is shown in Table 1 (For 50 Dimension problem) and Table 2 (for 100 Dimension problems). Both the algorithms were run for 100,000 evaluations. The results shown are average of 15 independent runs. For CCEA, single best collaboration strategy has been used[2]. The number of *subevolve* generations used is 10. Results have been obtained using 5 and 10 equal partitions in CCEA. It can be seen that for both 50D and 100D problems, CCEA is able to achieve results which are at times orders of magnitude better than EA. Figures 1 and 2 show convergence of the function values (averaged over 15 runs), which indicates the faster convergence rate of CCEA over traditional

[2] Although single best collaboration strategy makes the search greedy, we have used it consistently throughout our studies, in order to limit the variation in results due to stochastic selection of collaborators. In our subsequent studies, we wish to explore the random collaboration strategy as well.

EA. For the presented studies, Non-dominated Sorting Algorithm II (NSGA-II) [1] has been used as the representative EA. Same EA has also been used as an underlying mechanism for CCEA, *i.e.* Simulated Binary Crossover and Polynomial mutation are used for evolving solutions during the search. The ranking of solutions is done using non-dominated sorting (which equates to sorting based on function fitness values for single objective functions).

Table 1 Results using CCEA (fixed partitions) and EA for 50D Rastrigin, Schwefel, Rosenbrock and Ackley

	EA		CCEA			
			5 partitions		10 partitions	
	Mean	S.D.	Mean	S.D.	Mean	S.D.
Rastrigin	23.508	3.273	0.0164	0.0055	0.0123	0.0043
Schwefel	1254.723	385.828	150.103	137.738	7.978	30.597
Rosenbrock	85.012	40.344	70.957	33.683	95.233	37.976
Ackley	0.935	0.164	0.00287	0.00389	0.0259	0.00486

Table 2 Results using CCEA (fixed partitions) and EA for 100D Rastrigin, Schwefel, Rosenbrock and Ackley

	EA		CCEA			
			5 partitions		10 partitions	
	Mean	S.D.	Mean	S.D.	Mean	S.D.
Rastrigin	183.366	13.231	6.720	1.764	0.556	0.526
Schwefel	5296.503	433.643	923.844	279.744	309.944	213.665
Rosenbrock	260.094	51.086	149.952	44.674	138.206	58.0476
Ackley	4.185	0.102	0.269	0.0216	0.108	0.0171

Apart from obtaining better function values, CCEA also has an additional advantage of saving on the cost of evolving the solutions. This is because for each new solution generated during the run, conventional EA has to perform recombination and mutation operators on each variable, whereas CCEA has to deal with much less number of variables as only a part of the population is evolved at a given point of time. For problems with large number of variables, this may result in a considerable computational overhead, as mentioned in [10]. However, this effect may not be a major concern in computational complexity of the algorithm if cost of evaluating a solution itself is considerably higher than recombination and mutation operators.

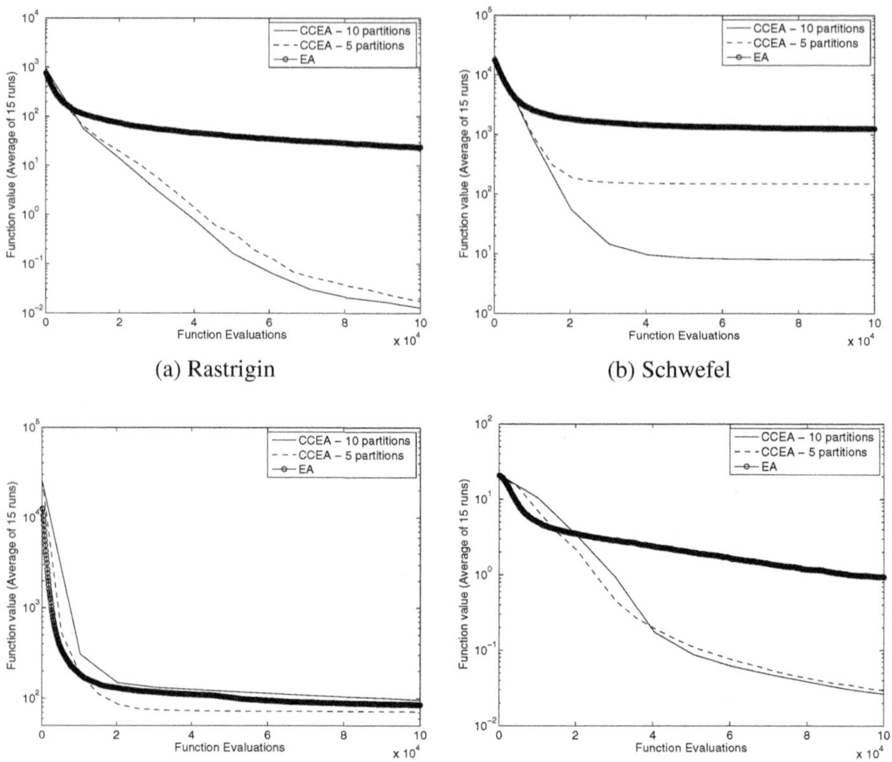

(a) Rastrigin (b) Schwefel

(c) Rosenbrock (d) Ackley

Fig. 1 Convergence using EA and CCEA for 50D problems (Fixed length partitions)

2.3 Shortcomings of Basic CCEA

Although it is clear from the previous subsection that CCEA is capable of achieving much better results than EA for certain problems, the performance is likely to deteriorate if the subsystems are not chosen appropriately. In particular, random partitioning as used in basic CCEA would perform well only if the interdependence between the variables is negligible, which allows each subsystem to be evolved towards the global optimum variable values independently of other subsystems. However, if the problems have a strong interdependence among the variables, it becomes imperative for good performance that the partitioning is intelligent enough to create subsystems such that: (a) The variables within any subsystems are strongly dependent on each other, and (b) interaction between different subsystems is minimal. The class of problems without substantial inter-variable dependence are commonly referred to as *separable* problems, whereas problems with strong variable interdependence are know as *non-separable* problems.

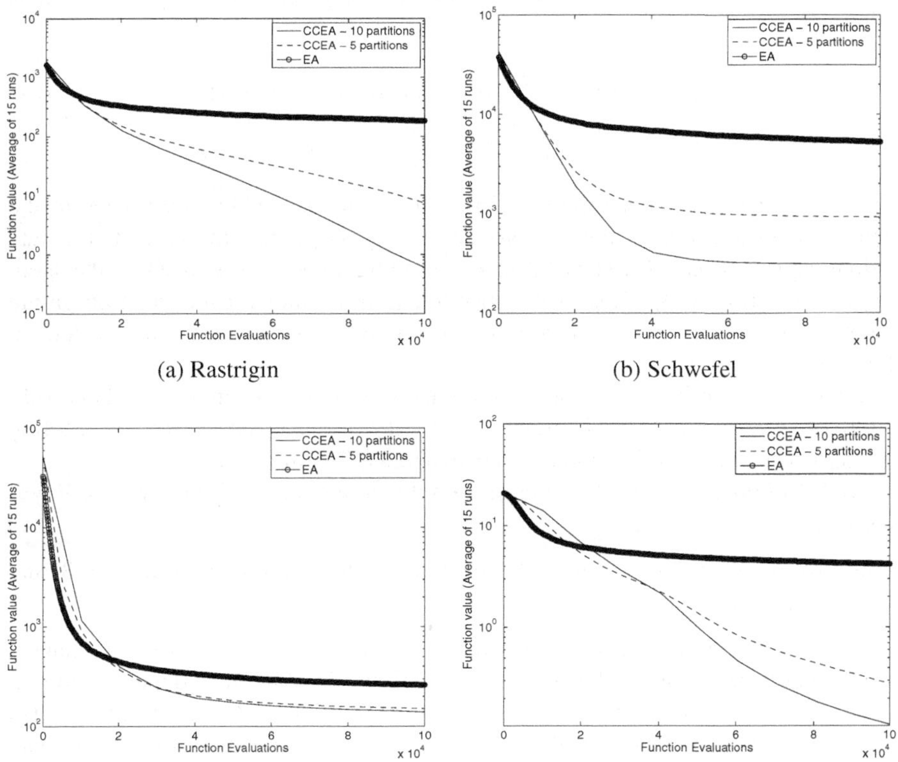

(a) Rastrigin (b) Schwefel

(c) Rosenbrock (d) Ackley

Fig. 2 Convergence using EA and CCEA for 100D problems (Fixed length partitions)

To illustrate this effect, two non-separable problems were chosen. First one is Shifted 50D Rotated Rastrigin function (abbreviated as Rastrigin-sr here) from IEEE CEC 2005 benchmarks [12], whereas the second one is 50D G2 [5]. The performance of CCEA and EA on these two problems, using 100,000 evaluations, is listed in Table 3. It is seen that EA outperforms CCEA for both these test problems just the opposite of what we observed for the earlier benchmark functions. Because of strong coupling of the variables, random partitioning is ineffective for evolving the solutions individually. The convergence plots for the problems are shown in Figure 3.

The above examples make it clear that for some problems partitioning variables into non-overlapping populations and coevolving may work well while for others keeping all variables together as a single population EA works better. In the next section, CCEA-AVP has been developed to which exploits the dependence between variables to adaptively create partitions, which results in good performance of the algorithm across wide variety of test problems.

Table 3 Results using CCEA (fixed partitions) and EA for Shifted Rotated Rastrigin and G2

	EA		CCEA			
			5 partitions		10 partitions	
	Mean	S.D.	Mean	S.D.	Mean	S.D.
Rastrigin-sr	118.282	29.700	224.228	43.033	278.469	79.055
G2	0.7489	0.0253	0.5833	0.0706	0.3997	0.0505

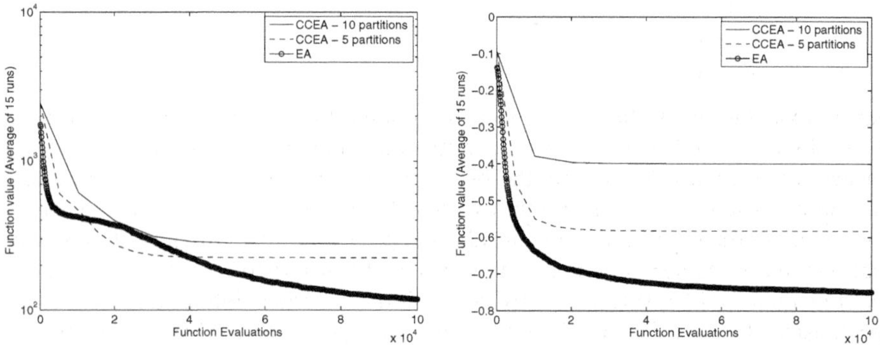

(a) Rastrigin-sr (b) G2

Fig. 3 Convergence using EA and CCEA for 50D problems (Fixed length partitions)

3 Proposed CCEA with Adaptive Variable Partitioning Based on Correlation

For problems having significant variable interactions, it is conceivable that design vectors with high global function fitness are likely to exhibit certain mathematical relationships among the variables. The proposed CCEA tries to derive benefit from this fact, and employs it for effective partitioning of variables during the search. The proposed CCEA algorithm is shown in Algorithm 2.

In the proposed CCEA algorithm, first of all the solutions are evolved as done in a conventional EA for a few generations, after which the partitioning is invoked. An archive (*arc*) of all the solutions explored during the run is maintained. To partition the design space into smaller subsets, the following strategy is employed. First of all, the solutions in *arc* are sorted based on the fitness function value. Thereafter, correlation between the variables is calculated based on the top 50% of the solutions in the sorted archive. For the presented studies, Pearson's correlation coefficient is used, which varies between 0 and 1, where 0 indicates no correlation, and 1 indicates very strong correlation in the multivariate data. Once the pairwise correlation values are obtained for each variable ($D \times D$ matrix, where D is the number of variables), the *absolute* correlation values exceeding a cutoff value are identified and collected to form a partition. The process identifies disjoint sets with strong interdependence of variables within the partition and weak interdependence across partitions. Such a

Algorithm 2. CCEA with Correlation based Adaptive Variable Partitioning

Require: N_{EA} (Number EA generations),N_G (Number of CCEA Generations), N_V (Number of variables for the problem), N_P (Population size), *CR* (Crossover Rate), *MR* (Mutation Rate), Distribution index of Crossover and Distribution index of Mutation.

Initialize Population (pop), Archive (arc)
Set $Soln_{best}$ as the best individual in the population
Run EA for a N_{EA} generations:
for $gen = 2$ to N_{EA} **do**
  $childpop_{gen} \leftarrow$ Evolve (pop_{gen-1})
  Evaluate $(childpop_{gen})$
  $P \leftarrow$ Sort $(childpop_{gen} + pop_{gen-1})$
  $pop_{gen} \leftarrow$ Reduce (P)
  Update arc, $Soln_{best}$
end for
Invoke CCEA:
for $cgen = 2$ to N_G **do**
  Determine number of partitions s based on correlation $(1 \le s \le S_{max})$ {Refer Algorithm 3}
  Split pop_{cgen} into s partitions $pop_{1,cgen}, pop_{2,cgen} \cdots pop_{s,cgen}$. Let the $\mathbf{V_j}$ denote the indices of design parameters contained in a partition j
  for $i = 1$ to s **do**
    Construct $spop_i$ by combining variable values in partition i with best/random variable values from other partitions
    for $subgen = 2$ to N_{SG} **do**
      $childpop_{i,subgen} \leftarrow$ Subevolve $(spop_{i,subgen-1})$
      Evaluate $(childpop_{i,subgen})$
      $CP \leftarrow$ Sort $(childpop_{i,subgen} + spop_{i,subgen-1})$
      $spop_{i,subgen} \leftarrow$ Reduce (CP)
    end for
    Set $pop_{cgen}(\mathbf{V_i}) = spop_i$
    Update $Soln_{best}$ {Update best solution}
  end for
  Update arc
end for

scheme for non-separable problems, would create large block(s) of interdependent variables, thereby behaving similar to EA. On the other hand, if the problem is separable, then the variable correlations would be considerably low, and the results would be closer to those obtained using uniform partitioning. Hence, the adaptive strategy tries to opt for the better option (whether to evolve the variables together or separately; and how to partition the variables), depending on the problem behavior. The following cases may arise while partitioning the variables:

1. Total number of correlation based partitions and uncorrelated variables amounts to exact number of maximum subsystems (S_{max}).
2. The total number of partitions based on correlation (N_{Sc}) are less than the S_{max}. In Such cases, if the remaining variables (N_{Su}) are greater than remaining N_r

Algorithm 3. Partitioning Strategy

Require: *arc* (Archive of explored solutions), S_{max} (Maximum number of subsystems), N_V (No. of variables for the problem), Correlation Threshold T

A = Sort (arc) {Sort the solutions in terms of function fitnesses}
$N_a = |A|$
$C \leftarrow$ Correlation $(A(1:N_a/2))$
for $i = 1$ to N_V **do**
  $S_i \leftarrow \{j | C_{i,j} \geq T\}$
end for
$Sc \leftarrow \{S_i \quad ; \quad |S_i| > 1\}$ (Sets of variables correlated with atleast one other variable)
$Su \leftarrow \{S_i \quad ; \quad |S_i| = 1\}$ (Variables not correlated with any other variable)
while $\{ S_i \cap S_j \neq \phi \forall \{S_i, S_j\} \in Sc, i \neq j \}$ **do**
  if $S_i \cap S_j \neq \phi$ (where $S_i, S_j \in Sc$) **then**
    $S_{new} = S_i \cup S_j$ (Merge sets with common elements)
    $Sc \leftarrow$ Insert S_{new}
    $Sc \leftarrow$ Delete S_i, S_j
  end if
end while
N_{Sc} = Number of sets in Sc
if $N_{S_c} \geq S_{max}$ **then**
  $Sc_{S_{max-1}} \leftarrow \{Sc_{S_{max-1}} \cup Sc_{S_{max}} \cup \ldots \cup Sc_{N_{Sc}}\} \cup Su$
else
  $N_r = S_{max} - N_{Sc}$ (Number of subsystems left after assigning correlation based partitions)
  $Su \leftarrow$ Distribute S_u into N_r partitions uniformly.
  $S = \{Sc, Su\}$
end if

subsystems ($N_r = S_{max} - N_{Sc}$), then they are distributed in those N_r systems as uniformly as possible. If on the other hand, the remaining variables are less than remaining subsets, then one variable each is assigned into corresponding number of subsets, and the leftover subsets are discarded.

3. If the number of correlation based partitions (N_{Sc}) are greater than S_{max}, then the excess partitions are merged into one partition, so that the total number of partitions do not exceed S_{max}. This limit is imposed on partitioning because too many partitions would create an overhead of function evaluations with less interaction frequency among the disjoint sets, which likely to result in inferior performance of CCEA.

The details of partitioning strategy are given in Algorithm 3. For CCEA-AVP, since different partitions can contain different number of variables, the number of *subevolve* generations for the partitions have not been kept the same (unlike [11]). Instead, each partition is allotted $N_G = 2 \times n_s$ generations for evolution, where n_s is the number of variables in that partition. An upper bound on this value is put as 10 generations, so that the interaction frequency (calls to the top level CCEA) are not reduced significantly.

4 Numerical Experiments

Six test functions have been used in the presented studies. The test functions are listed in Table 4. First four functions (Rastrigin, Schwefel, Rosenbrock, Ackley) have relatively lower level of variable interactions, where as the remaining two functions (Shifted Rotated Rastrigin and G2) are highly non-separable problems.

Table 4 Description of test functions used for studies

Test problem	Objective value
Rastrigin (Minimize)	$f(x) = \sum_{i=1}^{n} \left[x_i^2 - 10\cos(2\pi x_i) + 10\right]$ $-5.12 \leq x_i \leq 5.12$
Schwefel (Minimize)	$f(x) = 418.9829n - \sum_{i=1}^{n} \left(x_i \sin\left(\sqrt{\lvert x_i \rvert}\right)\right)$ $-500 \leq x_i \leq 500$
Rosenbrock (Minimize)	$f(x) = \sum_{i=1}^{n-1} \left[100(x_{i+1} - x_i^2)^2 + (x_i - 1)^2\right]$ $-30 \leq x_i \leq 30$
Ackley (Minimize)	$f(x) = -20\exp\left(-0.2\sqrt{\frac{1}{n}\sum_{i=1}^{n} x_i^2}\right) - \exp\left(\frac{1}{n}\sum_{i=1}^{n}\cos(2\pi x_i)\right) + 20 + e$ $-32 \leq x_i \leq 32$
Shifted Rotated Rastrigin [12] (Minimize)	$f(x) = \sum_{i=1}^{n} \left[z_i^2 - 10\cos(2\pi z_i) + 10\right]$ $-5.12 \leq x_i \leq 5.12, z = (x - o) \times M$, where o is the shifted global optimum, M is the linear Transformation Matrix (Rastrigin_M_D50).
G2 [5] (Maximize)	$f(x) = -\left\lvert \frac{\sum_{i=1}^{n}\cos^4(x_i) - 2\prod_{i=1}^{n}\cos^2(x_i)}{\sqrt{\sum_{i=1}^{n} i x_i^2}} \right\rvert$ Subject to $g_1(x) = 0.75 - \prod_{i=1}^{n} x_i \leq 0$ and $g_2(x) = \sum_{i=1}^{n} x_i - 7.5n \leq 0$ $0 \leq x_i \leq 10$ $(i = 1, \ldots, n)$

To study the performance of various algorithms on the aforementioned test functions, multiple (15) independent runs are made using each algorithm (for all combinations of parameters) on all the test functions. The parameters used for EA and CCEA are listed in Table 5. For Rastrigin, Schwefel, Rosenbrock and Ackley, 50 and 100 dimension problems have been studied, where as for Shifted Rotated Rastrigin and G2 function, 50 variable problem have been studied. Experiments were done with different values of partitions (number of fixed partitions for basic CCEA and maximum partitions in CCEA-AVP), and also with different values of threshold for correlation, as shown in Table 5. For CCEA-AVP, partitioning is invoked after 10000 function evaluations.

4.1 Results on 50D Test Problems

Table 6 lists the results obtained using basic CCEA with 10 partitions, CCEA-AVP with a maximum of 10 partitions, and EA. It is seen that for Rastrigin, Schwefel,

Table 5 Parameter used for the studies

Parameter	Value(s)
Population Size	100
Maximum function Evaluations	100000
Crossover Probability	1.0
Mutation Probability	0.1
Crossover Index	15
Mutation Index	20
Maximum partitions for CCEA-AVP	5,10
Correlation threshold T	0.3, 0.4, 0.5, 0.6

Rosenbrock and Ackley functions, best performance is achieved using basic CCEA (fixed 10) partitions, which is expected due to relatively low interactions between the variables for these test functions. The results are closely followed by CCEA-AVP. The results using both these CCEA strategies are much better than those obtained by evolving the solutions as single population in EA, which is expected. On the other hand, for the remaining two functions, *i.e.* Shifted Rotated Rastrigin and G2, the standings of the algorithms are reversed. EA gets the best solutions for this case, closely followed by CCEA-AVP. Basic CCEA performance for these two problems is inferior as compared to the other two algorithms, as expected.

The convergence plots for 50D test problems, are shown in Figure 5. For CCEA and EA, the average plots have been shown across all 15 runs. It is possible to show average plots for these two algorithms as the number of evaluations after each generation in the same for all runs. However, for CCEA-AVP, the number of evaluations after each generation may differ from run to run, depending on the partitioning of the variables (recall that number of *subevolve* generations used for CCEA-AVP is $2 \times n_s$, where n_s is the number of variables in the given partition p_s). Therefore, the convergence plot for *all* the runs are shown in the Figure(s). It can be visibly seen that for Rastrigin, Schwefel, Rosenbrock and Ackley, the CCEA-AVP runs resembles fixed partitioning whereas for Shifted Rotated Rastrigin and G2, it behaves similar to EA. Note that in the figure, only the convergence plots corresponding to CCEA-AVP with cutoff value of 0.6 have been shown. Other cases (cutoff 0.3, 0.4, 0.5) have been omitted for the sake of brevity.

Next, the same set of results are reproduced using 5 fixed partitions in basic CCEA, and 5 maximum partitions in CCEA-AVP, to observe the effect of partitioning size. The result are listed in Table 7 It is observed that while the relative standings of the algorithms remain unchanged for all the test problems, their converged values show difference from the 10 partition case. It is interesting to note that reducing the number of partitions for CCEA from 10 to 5 *worsens* the performance of CCEA for first four problems (Rastrigin, Schwefel, Rosenbrock, Ackley), whereas it *improves* the performance of CCEA for the remaining two cases. This is understandable as for separable functions, more the number of partitions, easier it is to evolve the subpopulations owing to their smaller size. On the other hand, the

Table 6 Results using EA and CCEA for 50 Variable problems (Maximum 10 partitions)

Test Problem	Algorithm	Best	Mean	Worst	S.D.
Rastrigin	Fixed 10	5.779232e-03	1.237750e-02	2.345529e-02	4.273683e-03
	AVP (Cutoff 0.3)	1.836837e-01	2.302523e+00	6.299632e+00	1.847724e+00
	AVP (Cutoff 0.4)	4.767614e-02	4.165830e-01	1.467531e+00	4.893863e-01
	AVP (Cutoff 0.5)	1.787012e-02	6.865380e-02	1.765419e-01	5.303837e-02
	AVP (Cutoff 0.6)	1.057469e-02	2.876147e-02	5.838951e-02	1.394009e-02
	EA	1.870910e+01	2.350803e+01	2.852830e+01	3.273037e+00
Schwefel	Fixed 10	4.569431e-02	7.977893e+00	1.185799e+02	3.059714e+01
	AVP (Cutoff 0.3)	5.926506e+02	1.209234e+03	1.778195e+03	3.805453e+02
	AVP (Cutoff 0.4)	4.737855e+02	1.192371e+03	1.776657e+03	3.965489e+02
	AVP (Cutoff 0.5)	4.737964e+02	1.200240e+03	1.776606e+03	3.950191e+02
	AVP (Cutoff 0.6)	4.738365e+02	1.200250e+03	1.776630e+03	3.950148e+02
	EA	5.291690e+02	1.254723e+03	1.829850e+03	3.858276e+02
Rosenbrock	Fixed 10	2.656768e+01	9.523261e+01	1.492941e+02	3.797645e+01
	AVP (Cutoff 0.3)	4.068637e+01	8.515012e+01	1.562630e+02	3.343325e+01
	AVP (Cutoff 0.4)	4.017283e+01	8.497414e+01	1.609554e+02	3.930656e+01
	AVP (Cutoff 0.5)	4.007405e+01	7.563297e+01	1.430863e+02	3.140305e+01
	AVP (Cutoff 0.6)	3.911556e+01	7.182199e+01	9.954174e+01	2.655364e+01
	EA	4.204690e+01	8.501223e+01	1.892910e+02	4.034408e+01
Ackley	Fixed 10	1.804900e-02	2.594491e-02	3.357565e-02	4.863999e-03
	AVP (Cutoff 0.3)	3.320793e-02	6.783809e-02	1.303181e-01	2.929305e-02
	AVP (Cutoff 0.4)	1.609523e-02	2.886849e-02	3.392610e-02	5.281810e-03
	AVP (Cutoff 0.5)	1.538815e-02	2.668428e-02	3.463329e-02	5.561409e-03
	AVP (Cutoff 0.6)	1.590089e-02	2.656794e-02	3.202351e-02	5.077989e-03
	EA	5.964560e-01	9.353366e-01	1.185250e+00	1.641141e-01
Rasrtigin-sr	Fixed 10	1.642592e+02	2.784686e+02	5.027227e+02	7.905534e+01
	AVP (Cutoff 0.3)	7.346480e+01	1.047515e+02	1.380228e+02	2.154041e+01
	AVP (Cutoff 0.4)	6.678959e+01	1.014526e+02	1.443156e+02	2.248234e+01
	AVP (Cutoff 0.5)	7.382558e+01	1.009139e+02	1.532730e+02	2.553048e+01
	AVP (Cutoff 0.6)	7.369251e+01	1.008083e+02	1.531503e+02	2.556444e+01
	EA	7.684920e+01	1.182818e+02	1.635920e+02	2.169965e+01
G2	Fixed 10	4.913381e-01	3.997472e-01	3.197929e-01	5.048437e-02
	AVP (Cutoff 0.3)	7.344912e-01	6.906985e-01	6.323944e-01	3.448017e-02
	AVP (Cutoff 0.4)	7.408774e-01	6.879742e-01	6.329717e-01	3.770002e-02
	AVP (Cutoff 0.5)	7.648944e-01	6.850655e-01	6.269927e-01	4.222448e-02
	AVP (Cutoff 0.6)	7.506297e-01	6.861708e-01	6.237227e-01	3.810408e-02
	EA	7.802640e-01	7.488963e-01	6.991770e-01	2.529587e-02

results using CCEA-AVP show very marginal variations between the two partitions size limits used. This is because the partitions are created adaptively and the maximum limit is imposed only if the number of adaptively created subsets exceeds the limit on maximum number of subsets (S_{max}). Hence, the proposed CCEA-AVP tends to be less sensitive to the number of maximum subsets allowed.

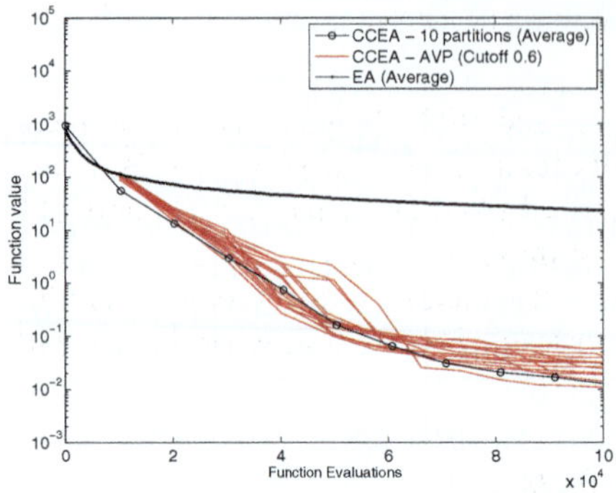

(a) Rastrigin

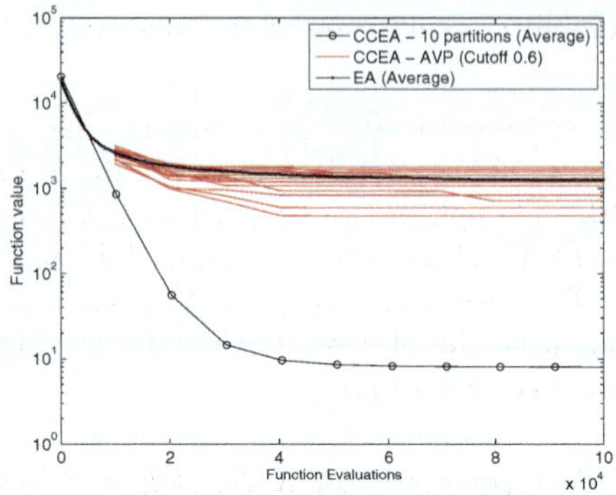

(b) Schwefel

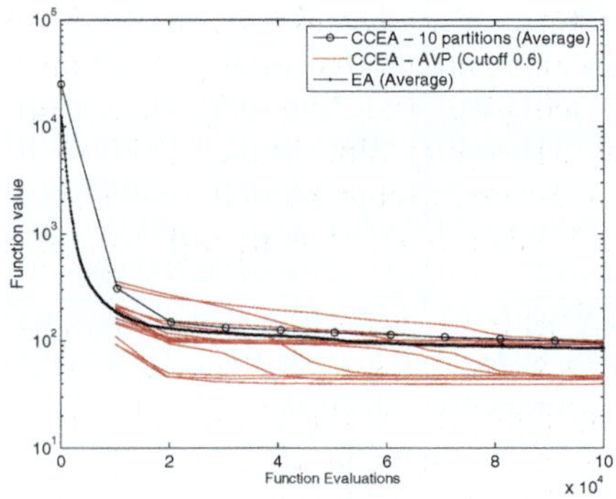

(c) Rosenbrock

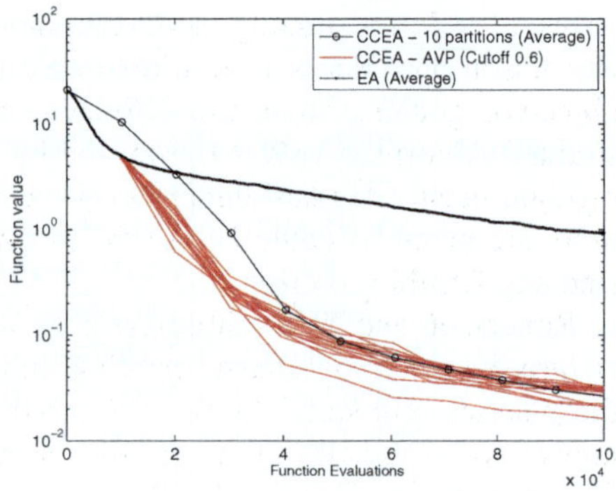

(d) Ackley

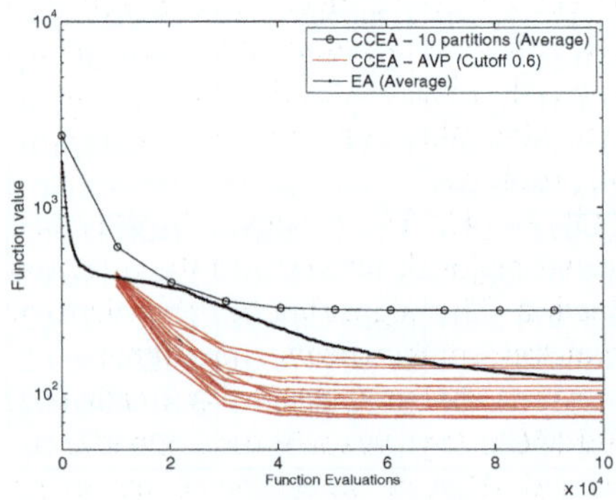

(e) Rastrigin-sr

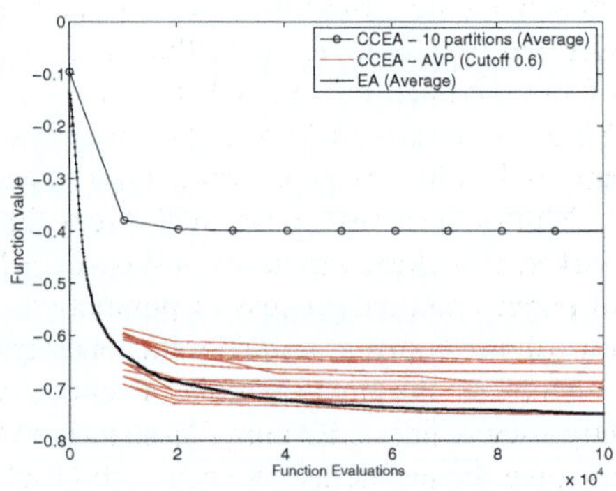

(f) G2

Fig. 4 Convergence plots for 50D problems; Comparison between CCEA with 10 Fixed partitions, CCEA-AVP with Maximum 10 partitions (cuttoff 0.6), and EA

The convergence plots for performance on 50D test problems using 5 partitions for basic CCEA, maximum 5 partitions for CCEA-AVP and EA are shown in Figure 5. The plots reflect the same trends as observed from Table 7.

Table 7 Results using EA and CCEA for 50 Variable problems (Maximum 5 partitions)

Test Problem	Algorithm	Best	Mean	Worst	S.D.
Rastrigin	Fixed 5	8.731096e-03	1.645049e-02	2.915661e-02	5.548972e-03
	AVP (Cutoff 0.3)	3.289065e-01	1.853667e+00	6.417341e+00	1.799776e+00
	AVP (Cutoff 0.4)	1.599303e-01	8.958731e-01	3.351182e+00	8.892952e-01
	AVP (Cutoff 0.5)	3.935962e-02	1.899336e-01	7.903401e-01	1.899213e-01
	AVP (Cutoff 0.6)	1.421517e-02	7.560980e-02	2.209411e-01	6.530767e-02
	EA	1.870910e+01	2.350803e+01	2.852830e+01	3.273037e+00
Schwefel	Fixed 5	4.353910e-02	1.501039e+02	4.738343e+02	1.377383e+02
	AVP (Cutoff 0.3)	5.923904e+02	1.215252e+03	1.776890e+03	3.696121e+02
	AVP (Cutoff 0.4)	4.738339e+02	1.192430e+03	1.776680e+03	3.965542e+02
	AVP (Cutoff 0.5)	4.738473e+02	1.200286e+03	1.776655e+03	3.950255e+02
	AVP (Cutoff 0.6)	5.922526e+02	1.208174e+03	1.776657e+03	3.803701e+02
	EA	5.291690e+02	1.254723e+03	1.829850e+03	3.858276e+02
Rosenbrock	Fixed 5	3.826361e+01	7.095662e+01	1.513366e+02	3.368268e+01
	AVP (Cutoff 0.3)	4.126986e+01	8.595546e+01	1.759625e+02	3.887288e+01
	AVP (Cutoff 0.4)	4.067930e+01	7.941118e+01	1.253789e+02	2.947897e+01
	AVP (Cutoff 0.5)	3.981255e+01	7.306836e+01	1.059848e+02	2.694808e+01
	AVP (Cutoff 0.6)	3.968770e+01	7.208999e+01	1.345994e+02	3.050495e+01
	EA	4.204690e+01	8.501223e+01	1.892910e+02	4.034408e+01
Ackley	Fixed 5	2.191173e-02	2.872615e-02	3.587124e-02	3.897202e-03
	AVP (Cutoff 0.3)	3.138552e-02	5.127192e-02	9.325807e-02	2.066917e-02
	AVP (Cutoff 0.4)	2.447292e-02	3.152120e-02	3.982890e-02	5.680858e-03
	AVP (Cutoff 0.5)	2.298943e-02	2.964303e-02	4.211521e-02	5.371868e-03
	AVP (Cutoff 0.6)	2.020320e-02	2.953798e-02	3.808804e-02	5.092418e-03
	EA	5.964560e-01	9.353366e-01	1.185250e+00	1.641141e-01
Rastrigin-sr	Fixed 5	1.632540e+02	2.242275e+02	2.918196e+02	4.303274e+01
	AVP (Cutoff 0.3)	7.019403e+01	1.009195e+02	1.379169e+02	1.989549e+01
	AVP (Cutoff 0.4)	7.253207e+01	1.033894e+02	1.373290e+02	2.099474e+01
	AVP (Cutoff 0.5)	6.474726e+01	9.785739e+01	1.440974e+02	2.138567e+01
	AVP (Cutoff 0.6)	6.474830e+01	9.747443e+01	1.438673e+02	2.142065e+01
	EA	7.684920e+01	1.182818e+02	1.635920e+02	2.169965e+01
G2	Fixed 5	7.219800e-01	5.833207e-01	4.526570e-01	7.063032e-02
	AVP (Cutoff 0.3)	7.760420e-01	7.082753e-01	6.425938e-01	3.290058e-02
	AVP (Cutoff 0.4)	7.694752e-01	7.073772e-01	6.459791e-01	3.955726e-02
	AVP (Cutoff 0.5)	7.718898e-01	6.994713e-01	6.478198e-01	3.797481e-02
	AVP (Cutoff 0.6)	7.741146e-01	6.975395e-01	6.432024e-01	4.067741e-02
	EA	7.802640e-01	7.488963e-01	6.991770e-01	2.529587e-02

4.2 *Results for 100D Problems*

In extension to the studies on 50D problems described in the previous subsection, additional studies were done on 100D test problems using the proposed CCEA-AVP algorithm. This was done in order to ascertain that increasing the problem dimensionality does not effect on the performance of the proposed CCEA-AVP adversely. 100 variable Rastrigin, Schwefel, Rosenbrock and Ackley functions were used for

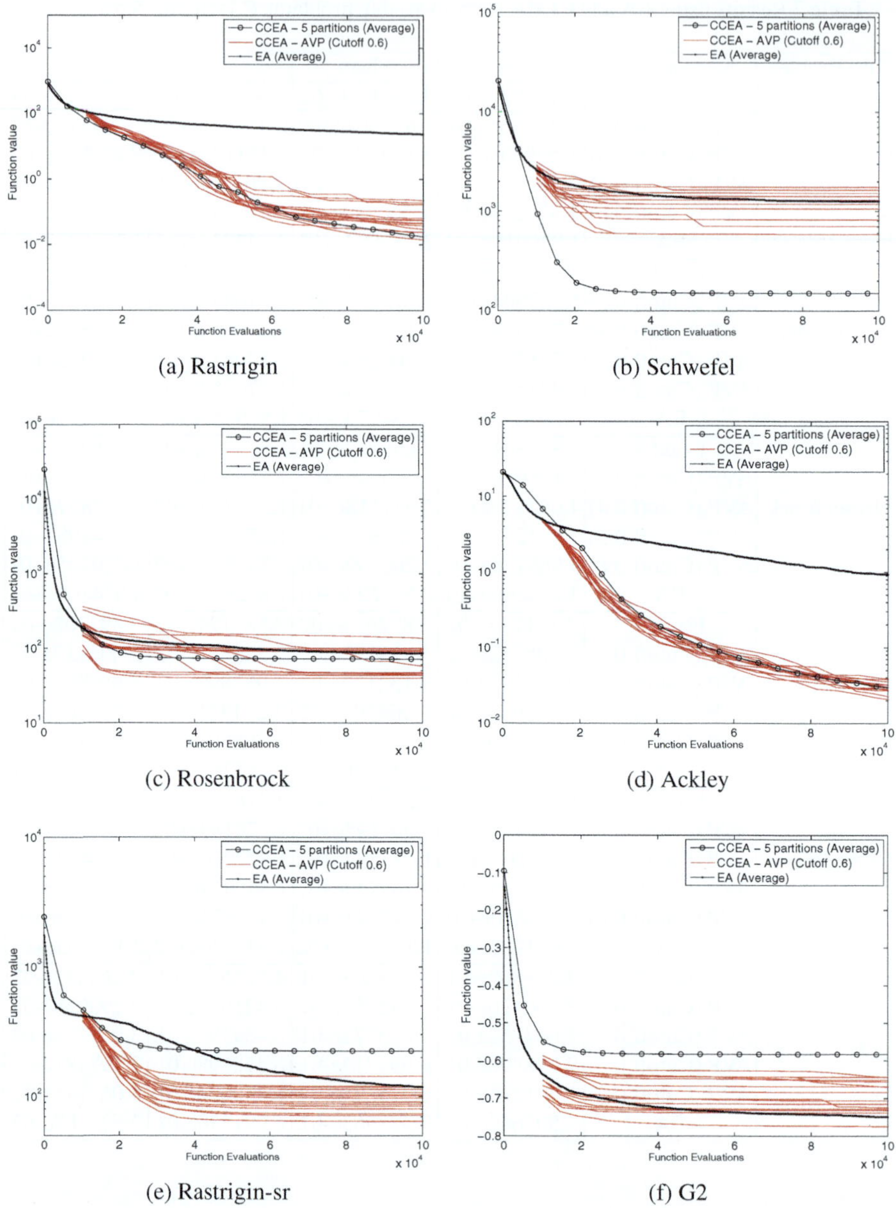

(a) Rastrigin (b) Schwefel

(c) Rosenbrock (d) Ackley

(e) Rastrigin-sr (f) G2

Fig. 5 Convergence plots for 50D problems; Comparison between CCEA with 5 Fixed partitions, CCEA-AVP with Maximum 5 partitions (cuttoff 0.6), and EA

this study. As for 50D case, 15 independent runs were made using 5 and 10 partition size for basic CCEA and maximum partition size 5 and 10 for CCEA-AVP. The parameters used for the studies are same as those used for 50D studies, as listed in Table 5.

Table 8 shows the mean results obtained using basic CCEA (10 partitions), CCEA-AVP (maximum 10 partitions), and EA for the 100D test problems. The observations are consistent with the 50D case, i.e. basic CCEA performs best for these problems, closely followed by CCEA-AVP, whereas EA performance is significantly inferior to these two algorithms. The corresponding convergence plots are shown in 6.

Finally, the same set of studies for 100D problems were repeated with 5 partitions for CCEA and maximum 5 partitions for CCEA-AVP. The average converged values after 100000 evaluations are shown in Table 9, and corresponding convergence plots in Figure 7. Once again, it is seen that while CCEA performance consistently degrades for all four test problems when the number of partitions are reduced, the CCEA-AVP performance is not significantly different from the maximum 10 partition case.

Table 8 Results using EA and CCEA for 100 Variable problems (Maximum 10 partitions)

Test Problem	Algorithm	Best	Mean	Worst	S.D.
Rastrigin	Fixed 10	2.826129e-01	5.561476e-01	2.433088e+00	5.257507e-01
	AVP (Cutoff 0.3)	1.681521e+01	3.910664e+01	6.670020e+01	1.630205e+01
	AVP (Cutoff 0.4)	8.706855e-01	1.373383e+01	3.341113e+01	9.444110e+00
	AVP (Cutoff 0.5)	1.483221e+00	4.900714e+00	1.321691e+01	3.328606e+00
	AVP (Cutoff 0.6)	3.788492e-01	2.548752e+00	7.478845e+00	2.183728e+00
	EA	1.664640e+02	1.833663e+02	2.011990e+02	1.323145e+01
Schwefel	Fixed 10	1.507234e+00	3.099444e+02	8.309508e+02	2.136654e+02
	AVP (Cutoff 0.3)	3.438069e+03	4.092864e+03	4.802564e+03	3.511965e+02
	AVP (Cutoff 0.4)	3.096133e+03	3.482946e+03	4.094800e+03	2.999574e+02
	AVP (Cutoff 0.5)	2.964297e+03	3.332869e+03	4.030107e+03	2.959289e+02
	AVP (Cutoff 0.6)	2.844582e+03	3.358064e+03	4.028548e+03	3.027776e+02
	EA	4.624850e+03	5.296503e+03	6.093950e+03	4.336429e+02
Rosenbrock	Fixed 10	9.457729e+01	1.382056e+02	3.167990e+02	5.804764e+01
	AVP (Cutoff 0.3)	1.645619e+02	2.696512e+02	3.439314e+02	5.965796e+01
	AVP (Cutoff 0.4)	1.430433e+02	2.222614e+02	2.792287e+02	4.110667e+01
	AVP (Cutoff 0.5)	1.066705e+02	2.017872e+02	2.897350e+02	4.897000e+01
	AVP (Cutoff 0.6)	9.615495e+01	2.006843e+02	2.689460e+02	5.037139e+01
	EA	1.792070e+02	2.600944e+02	3.612080e+02	5.108574e+01
Ackley	Fixed 10	8.540569e-02	1.085568e-01	1.413097e-01	1.707687e-02
	AVP (Cutoff 0.3)	9.503202e-01	1.676936e+00	2.208593e+00	4.268166e-01
	AVP (Cutoff 0.4)	1.032025e-01	4.013324e-01	1.544246e+00	3.858523e-01
	AVP (Cutoff 0.5)	6.707651e-02	9.702747e-02	1.907402e-01	2.954190e-02
	AVP (Cutoff 0.6)	6.798911e-02	8.992145e-02	1.220725e-01	1.397687e-02
	EA	3.979840e+00	4.184789e+00	4.423220e+00	1.017640e-01

The study clearly shows that a fixed partitioning scheme as in CCEA can lead to significantly poor performance for certain classes of problems while it may be useful for others. CCEA-AVP is designed to be applicable as a generic optimizer

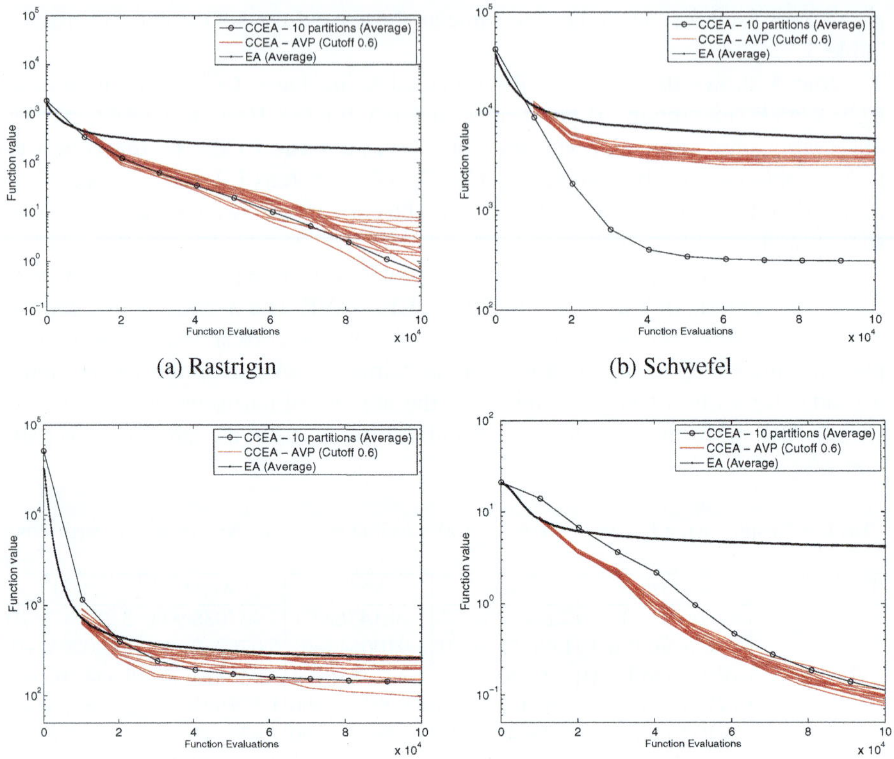

(a) Rastrigin (b) Schwefel

(c) Rosenbrock (d) Ackley

Fig. 6 Convergence plots for 100D problems; Comparison between CCEA with 10 Fixed partitions, CCEA-AVP with Maximum 10 partitions (cuttoff 0.6), and EA

with good mean performance when compared to EA and CCEA over the wide range of problems.

4.3 Variation in Performance of CCEA-AVP with Different Values of Correlation Threshold

In the preceding subsections, studies have been presented using different values of correlation thresholds in order to combine the variables together into one partition. Although all four values (0.3, 0.4, 0.5 and 0.6) have worked well for the given problems, there is slight variation in the performance introduced by the cutoff value. The relative standings of CCEA-AVP using the above four cutoff values, for 50D maximum 10 partitions case, are shown in Figure 8. It is observed that 0.6 correlation value works best for most of the problems (although the for Schwefel and G2 its performance is worse compared to other case, but a look into the mean values reveal that the difference in performance was negligible for these two cases across all cutoff values).

Table 9 Results using EA and CCEA for 100 Variable problems (Maximum 5 partitions)

Test Problem	Algorithm	Best	Mean	Worst	S.D.
Rastrigin	Fixed 5	4.278939e+00	6.720367e+00	1.060458e+01	1.763830e+00
	AVP (Cutoff 0.3)	2.670488e+01	4.598337e+01	6.578003e+01	1.180991e+01
	AVP (Cutoff 0.4)	1.396628e+01	2.387369e+01	3.822304e+01	7.503289e+00
	AVP (Cutoff 0.5)	5.136025e+00	1.283778e+01	2.377627e+01	5.027574e+00
	AVP (Cutoff 0.6)	3.562711e+00	8.710728e+00	1.334070e+01	2.833059e+00
	EA	1.664640e+02	1.833663e+02	2.011990e+02	1.323145e+01
Schwefel	Fixed 5	4.820491e+02	9.238445e+02	1.428535e+03	2.797439e+02
	AVP (Cutoff 0.3)	2.801671e+03	3.796555e+03	4.945437e+03	6.523428e+02
	AVP (Cutoff 0.4)	2.568155e+03	3.426769e+03	4.057032e+03	4.537557e+02
	AVP (Cutoff 0.5)	2.630590e+03	3.365949e+03	4.038401e+03	4.402954e+02
	AVP (Cutoff 0.6)	2.624483e+03	3.324425e+03	4.039300e+03	4.575261e+02
	EA	4.624850e+03	5.296503e+03	6.093950e+03	4.336429e+02
Rosenbrock	Fixed 5	8.440033e+01	1.499516e+02	2.613990e+02	4.467410e+01
	AVP (Cutoff 0.3)	1.217372e+02	2.279518e+02	3.412558e+02	5.395131e+01
	AVP (Cutoff 0.4)	1.107537e+02	1.904279e+02	2.588505e+02	3.607417e+01
	AVP (Cutoff 0.5)	1.021898e+02	1.665388e+02	2.169891e+02	3.543739e+01
	AVP (Cutoff 0.6)	8.913882e+01	1.544552e+02	2.128978e+02	4.247899e+01
	EA	1.792070e+02	2.600944e+02	3.612080e+02	5.108574e+01
Ackley	Fixed 5	2.285906e-01	2.692870e-01	3.093061e-01	2.164270e-02
	AVP (Cutoff 0.3)	5.068778e-01	1.294581e+00	2.170572e+00	4.468844e-01
	AVP (Cutoff 0.4)	1.914989e-01	3.442524e-01	8.454164e-01	1.725802e-01
	AVP (Cutoff 0.5)	2.387114e-01	2.962445e-01	4.632740e-01	5.550161e-02
	AVP (Cutoff 0.6)	2.657690e-01	3.346279e-01	4.567908e-01	5.411669e-02
	EA	3.979840e+00	4.184789e+00	4.423220e+00	1.017640e-01

Although it may not be as straightforward choice to determine the optimum cutoff value for any problem, empirically it can be said that if too low a value of cutoff ($\approx$ below 0.3) is used, then even the weekly correlated variables would be evolved in the same partition, and the algorithm's performance would be roughly equivalent to an EA. On the other extreme, if too high a value of cutoff is used, then even strongly correlated variables might end up in different partitions, which would result in the performance of the algorithm being like basic CCEA. Therefore, the cutoff value should be reasonably chosen in order to avoid both extremes and achieve a balance between interacting variable partitions. Based on our studies, it is expected that a value of 0.4-0.6 should work reasonably well for a wide variety of problems.

5 Conclusions and Further Studies

This paper investigates the performance of a Single Best Collaboration Strategy Coevolutionary Algorithm for a set of benchmark test functions The results suggest that the while CCEA shows a definite promise in solving large scale optimization problems, it is sensitive the choice of a number of parameters and the underlying partitioning scheme. Studies suggest that basic CCEA, with random partitioning

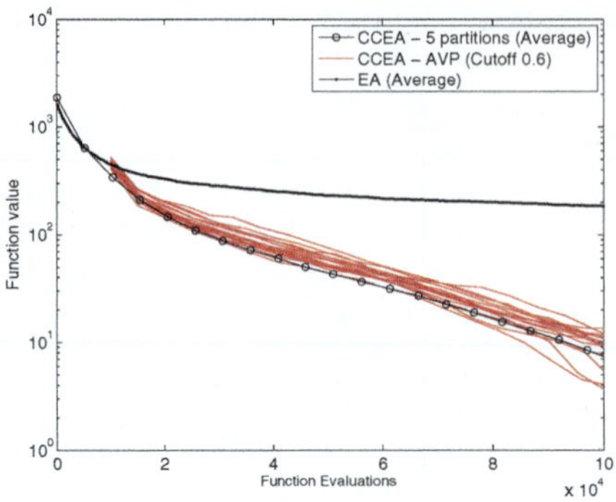

(a) Rastrigin

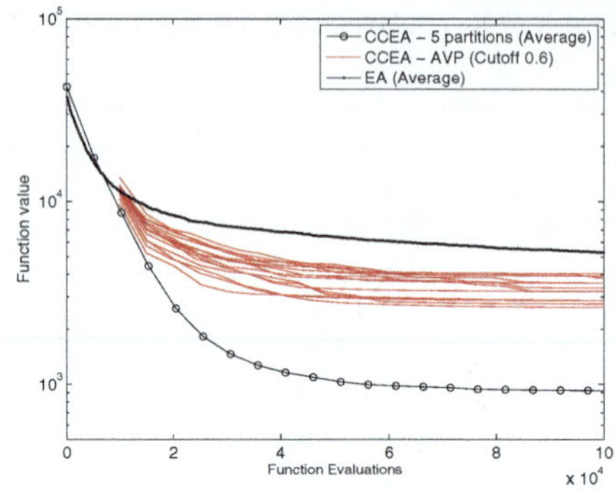

(b) Schwefel

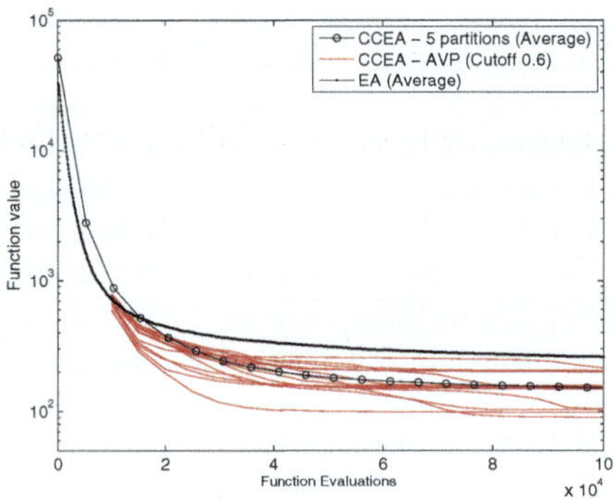

(c) Rosenbrock

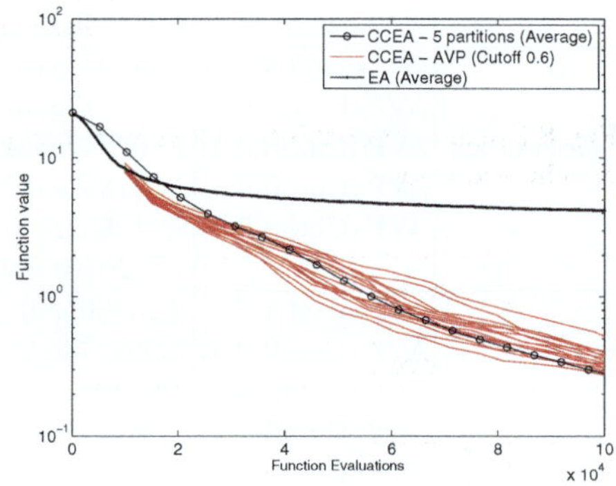

(d) Ackley

Fig. 7 Convergence plots for 100D problems; Comparison between CCEA with 5 Fixed partitions, CCEA-AVP with Maximum 5 partitions (cuttoff 0.6), and EA

may not be suitable for nonseparable problems, and in such cases, evolving the variables together as in EA, might be more beneficial. It is also important to highlight that the performance of CCEA is also dependent on the underlying recombination scheme. Previous studies with (1+1) CCEA and (1+1)EA with binary representation have highlighted that both exhibit quite different exploration capabilities.

Drawing motivation from the aforementioned observations, a novel CCEA algorithm with adaptive variable partitioning (CCEA-AVP) has been proposed as extension to Ray and Yao's earlier work [11]. The algorithm identifies correlations between the variables in each generation, and partitions the strongly correlated variables together. Such a scheme preserves the advantages of both, the basic CCEA as well as EA. This is because modularity of the problem is not compromised while choosing partitions. If the problem exhibits sufficient separability, the partitioning is similar to basic CCEA, whereas if the variables are highly interdependent, the algorithm behaves similar to EA. This adaption ensures that the proposed algorithm performs well for both separable and non-separable problems.

Studies were conducted on six benchmark functions to test the efficacy of the proposed algorithm and the results obtained were encouraging. On an average,

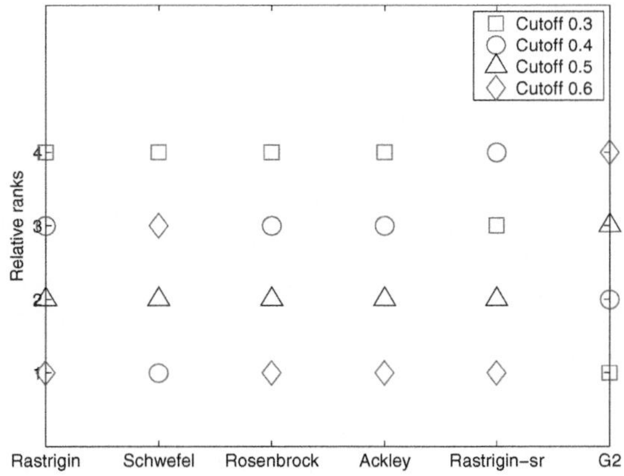

Fig. 8 The performance of CCEA-AVP with different Correlation Threshold values used in order to assign the variables into the same partition

for separable problems, the algorithm obtains results comparable to basic CCEA, whereas for non-separable problems, it obtains results comparable to EA. Since the separability for a problem may often not be know apriori, the proposed algorithm demonstrates a wider applicability as a generic optimization tool. Additionally, it was also observed that CCEA-AVP tends to be less sensitive to changes in the number of (maximum) partitions, as it only uses them as upper bound, and the partitions are otherwise created adaptively.

Although the presented studies have highlighted the advantages of the proposed method, there remain a number of related fronts where improvements are necessary. The first and foremost is the fact that the linear correlations have been used in the studies in order to capture the dependence of the variables. There is a need to capture non-linear correlation among variables in order to improve the efficiency of the method. The next improvement would be to design specific operators to sample along the directions of interest for a set of positively or negatively correlated variables. The studies are currently underway.

Acknowledgements. The authors would like to acknowledge the support of Defence and Security Applications Research Center (DSARC) at University of New South Wales at Australian Defence Force Academy (UNSW@ADFA), for the presented research work.

The authors would also like to thank Mr. Amitay Isaacs from School of Engineering and Information Technology (SEIT), UNSW@ADFA, for useful discussions and implementation of a part of the algorithm.

References

1. Deb, K., Pratap, A., Agarwal, S., Meyarivan, T.: A fast and elitist multiobjective genetic algorithm: NSGA-II. IEEE Transactions on Evolutionary Computation 6, 182–197 (2002)

2. Goh, C.K., Tan, K.C.: A competitive-cooperative coevolutionary paradigm for dynamic multiobjective optimization. IEEE Transactions on Evolutionary Computation 13(1) (2009)
3. Liu, B., Ma, H., Zhang, X.: A coevolutionary differential evolution algorithm for constrained optimization. In: Proc. of Third International Conference on Natural Computation, pp. 51–57 (2007)
4. McNamara, J.M., Barta, Z., Fromhage, L., Houston, A.I.: The coevolution of choosiness and cooperation. Nature 451(7175), 189–192 (2008)
5. Michalewicz, Z., Schoenauer, M.: Evolutionary Algorithms for Constrained Parameter Optimization Problems. Evolutionary Computation 4(1), 1–32 (1996)
6. Popovici, E., De Jong, K.: Relationships between internal and external metrics in coevolution. In: IEEE Congress on Evolutionary Computation CEC 2005, vol. 3, pp. 2800–2807 (2005), doi:10.1109/CEC.2005.1555046
7. Popovici, E., De Jong, K.: Understanding cooperative co-evolutionary dynamics via simple fitness landscapes. In: GECCO 2005: Proceedings of the, conference on Genetic and evolutionary computation, pp. 507–514. ACM, New York (2005), `http://doi.acm.org/10.1145/1068009.1068094`
8. Popovici, E., De Jong, K.: The effects of interaction frequency on the optimization performance of cooperative coevolution. In: GECCO 2006: Proceedings of the 8th annual conference on Genetic and evolutionary computation, pp. 353–360. ACM, New York (2006), `http://doi.acm.org/10.1145/1143997.1144061`
9. Popovici, E., De Jong, K.: Sequential versus parallel cooperative coevolutionary algorithms for optimization. In: IEEE Congress on Evolutionary Computation CEC 2006, pp. 1610–1617 (2006), doi:10.1109/CEC.2006.1688501
10. Potter, M.A., Jong, K.A.D.: A cooperative coevolutionary approach to function optimization. In: Davidor, Y., Männer, R., Schwefel, H.-P. (eds.) PPSN 1994. LNCS, vol. 866, pp. 249–257. Springer, Heidelberg (1994)
11. Ray, T., Yao, X.: A cooperative coevolutionary algorithm with correlation based adaptive variable partitioning. In: Proceedings of 2009 IEEE Congress on Evolutionary Computation CEC 2009, pp. 983–989 (2009)
12. Suganthan, P.N., Hansen, N., Liang, J.J., Deb, K., Chen, Y.P., Auger, A., Tiwari, S.: Problem Definitions and Evaluation Criteria for the CEC 2005 Special Session on Real-Parameter Optimization. Tech. rep., Nanyang Technological University, Singapore and Kanpur Genetic Algorithms Laboratory, IIT Kanpur, India (2005)
13. Tan, C., Goh, C., Tan, K., Tay, A.: A cooperative coevolutionary algorithm for multi-objective particle swarm optimization. In: IEEE Congress on Evolutionary Computation CEC 2007, pp. 3180–3186 (2007), doi:10.1109/CEC.2007.4424878
14. Van den Bergh, F., Engelbretch, A.P.: A cooperative approach to particle swarm optimization. IEEE Transactions on Evolutionary Computation 8(3) (2004)
15. Wiegand, R.P., Liles, W.C., Jong, K.A.D.: An empirical analysis of collaboration methods in cooperative coevolutionary algorithms. In: Proceedings from the Genetic and Evolutionary Computation Conference, pp. 1235–1242. Morgan Kaufmann, San Francisco (2001)
16. Yang, Z., Tang, K., Yao, X.: Large scale evolutionary optimization using cooperative coevolution. Information Sciences 178(15), 2985–2999 (2008), doi:10.1016/j.ins.2008.02.017

Complex Emergent Behaviour from Evolutionary Spatial Animat Agents

K.A. Hawick and C.J. Scogings

Abstract. Spatial multi-agent systems provide a powerful model framework for investigating evolutionary behaviour amongst animat agents. We have developed a microscopic animat-based model in which autonomous agents are microscopically controlled by a rule-set that can be evolved using suitable operators. Our system can support over a million animats co-existing over many generations and has already been used to explore several collective phenomena including clustering; segregation; tribal warfare and battlefront formation. We incorporate simple microscopic behaviour rules based on local view information that determine animat feeding; breeding; movement; seeking; and avoidance. We use a simple agent state model consisting of position, current health and age and a genetic code for each animat. We find a number of spatially-rich and complex, emergent patterns from the microscopic model and discuss how the model's convergence to stable macroscopic behaviour cycles is related to the localized rule parameters. We illustrate how an animat agent population of predators and prey can evolve more effective individuals by applying genetic algorithms to the species rule-sets and how our model framework and approach can be applied to sociological and predatory phenomena.

1 Introduction

Understanding the emergent behaviour in highly non-linear complex systems in terms of some sort of descriptive model remains one of the on-going goals of reductionist science [25, 15]. Many complex systems such as biological [19] and social phenomena [8] involving a number of individual components exhibit a richness of behaviour and a set of emergent properties that are not obvious from even a detailed knowledge of the workings of the individual component parts. Simulation models provide a means of exploring these emergent phenomena, whereby a system can

K.A. Hawick · C.J. Scogings
Massey University, Albany, Auckland, New Zealand
e-mail: K.A.Hawick@massey.ac.nz, C.Scogings@massey.ac.nz

R.A. Sarker and T. Ray (Eds.): Agent-Based Evolutionary Search, ALO 5, pp. 139–159.
springerlink.com

be set up in a number of detailed starting conditions and its evolution carefully observed during a computational experiment.

One of the main problems in modeling and exploring evolutionary behaviour through simulations is the sheer size of the parameter space or fitness landscape that is usually encountered. It is extraordinarily difficult to apply brute-force search methods to the phase spaces of many biologically-inspired computing models. The notion of genetic algorithms makes use of biologically-inspired mechanisms to combine and adapt existing "solutions" to find even better ones. This approach can still take very large computational resources to run simulated evolutionary models for long enough to see statistically significant changes or dramatically new solutions. It also presupposes that the starting conditions are sensible solutions and that in effect the experimenters are looking in roughly the right place or at least a plausible position in solution space [12, 14]. While real biological systems proceed using thermodynamically scaled system sizes and very long time scales, even for simple computational models, it is not sufficient to start somewhere completely random and "hope" to evolve a solution somewhere in phase space. We discuss an approach and a model that we have developed, whereby we can employ evolutionary inspired methods in a very controlled manner, starting from well defined areas of model space and with a parameter space that is heavily constrained.

Multi-agent systems have proved a very interesting tool for exploring evolutionary behaviour since it is possible to encode relatively simple rules into individual agents and to perform computational experiments involving many interacting individual agents. However even this approach has its limitations, not least due to the need for large numbers of individual agents in the system to have any hope of duplicating the phenomena thought to occur in real biological or sociological systems.

Popular science accounts of genetics often fail to draw attention to the large number of components and time-scales involved in the evolution of real genetic systems [31]. In this article we describe how our software agent-based model can be used to explore complex emergent behaviours based on a relatively small and manageable space size of microscopic parameters that control individual spatial agents or animats [36, 35]. Animat simulations [3, 18] provide a powerful framework for studying the collective behaviour of many individual intelligent agents [7]. Complex systems using spatial agents or animats have existed for some time—see for example [32, 1, 13]—and have yielded rich insights into emergent group behaviour in physical, biological and sociological simulation settings.

A useful starting point for us was to identify a fundamental behaviour such as predation and build up a microscopic model around it. We have refined our predator prey animat model over a number of years and it has been introduced and discussed in several previous publications including [26, 10]. Unlike other models which focus on the evolution of animats and the emergence of new species, we concentrate on making explicit, well-defined changes to the microscopic control variables of the model and then analyzing any new (emergent) animat collective behaviours. In particular we have documented fascinating emergent macro-behaviours such as the defensive spirals and other features discussed in [10]. Figure 1 shows a typical example of the rich emergent spatial patterns formed in animat models. In this example the predators and prey have formed a spiral battlefront which will slowly rotate around as animats are born, eat, die and move.

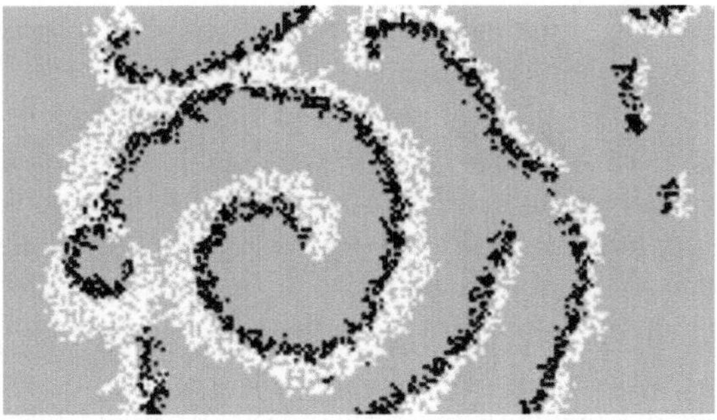

Fig. 1 A complex emergent pattern showing a spiral battlefront of predators (black) pursuing prey (white). The spiral has appeared alongside various other clustering formations

Our model exhibits some typical (and highly robust) wave front and proto-spiral pattern generation behaviours. We have been able to study these in a quantifiable manner by applying automatic feature detection techniques to the spatial animat patterns [10]. In this article we focus on evolutionary aspects of the model. We review some key animat based models and associated research work in Sect. 2 and describe the detailed working of our own animat model in Sect. 3. We include discussion of the features we incorporated to support evolutionary behaviour in Sect. 3.2 and the parameter tuning that was necessary to attain stable systems in Sect. 3.3. We present a number of observations on the static and dynamic properties of the animat model in Sect. 4. We review some typical properties and issues associated with genetic algorithms and other mechanisms for exploring fitness landscapes in Sect. 5. We present some results of animat evolution in Sect. 6, including control systems (Sect. 6.1), genetic crossover results (Sect. 6.2) and results when mutation was incorporated (Sect. 6.3). We offer a discussion of ideas on the role that variable spatial conditions and environments cause in Sect. 6.4 and some concluding remarks in Sect. 7.

2 A Review of Animat Models

Wilson [36] coined the word "animat" to refer to artificial creatures living in a virtual environment. However the history of what may be termed animat-based models stretches back for some time and a brief time line and review of some of the better-known models is provided here.

Cellular automata were developed at the Los Alamos National Laboratory in the 1940s by John von Neumann after a suggestion by his colleague Stanisław Ulam. A cellular automaton is a discrete model consisting of a regular grid of cells, each of which is in one of a finite number of states. Time is also discrete and the state of a cell at time t is a function of the states of a finite number of cells (called the neighbourhood) at time $t-1$. These neighbours are a selection of cells relative to the specified cell and do not change. Every cell has the same rule for updating, based on the states of the cells in the neighbourhood. Each time the rules are applied to the whole grid, a new generation is created.

Evolutionary Algorithms were introduced in Lawrence Fogel's PhD thesis [6] in 1964. An Evolutionary Algorithm applies reproduction, mutation, recombination and selection to a candidate solution in order to produce successive generations of "better" solutions. A fitness function, and possibly an entire fitness landscape, is used to assess the usefulness of each solution. Within the broader area of Evolutionary Algorithms the most popular are *Genetic Algorithms* introduced by John Holland [12].

The Game of Life is a cellular automaton devised by John Conway in 1970. The Game consists of an infinite 2D grid of square cells and each cell has two possible states, *live* or *dead*. At each step, a cell can change its state based on the number of live neighbours. The Game is initialized with a pattern of live and dead cells. The first generation is created by applying the rules simultaneously to every cell and then the rules are applied repeatedly to subsequent generations. The Game has attracted much interest because of the surprising ways in which patterns can evolve. An amazing number of initial patterns and various extensions, along with much animated discussion, can be found on many web sites.

Core Wars (1984) is a programming game in which computer programs compete for control of the Memory Array Redcode Simulator (MARS) which is a virtual computer. The programs are written in an assembler-type language called Redcode. The object of the game is to cause all processes of the opposing programs to terminate leaving your program as the winner. The International Core Wars Society was established in 1985 and still produces articles and newsletters for Core Wars fans. Steen Rasmussen evolved *Coreworld* in 1990 from Core Wars by introducing a genetic algorithm that would automatically write programs [20]. However the programming language in which Coreworld programs were written was very brittle, and most mutations completely destroyed the functionality of a program.

Tierra (1991) is a simulation derived from Core Wars and Coreworld developed by Thomas S. Ray [21]. Ray made key changes to the Coreworld programming language to ensure that mutations were much less likely to destroy a program and was thus able to observe for the first time computer programs that did evolve in a meaningful way. In Tierra, small computer programs (called "digital organisms") compete for CPU time and access to main memory (called "the soup"). The exact instructions making up each program are known as "the genome". These computer programs can evolve through mutation, replication and recombination. One interesting difference from the typical genetic algorithm approach is the absence of an explicit "fitness function". It was hoped that this might allow for more "open-ended" evolution in which the dynamics of the evolutionary process can themselves change over time. However, this was not achieved and true open-ended evolution has yet to be implemented in an artificial system.

Inspired by Tierra, Chris Adami and others constructed *Avida* [1] in 1993. In Tierra computer programs compete for CPU time and access to main memory. In Avida, every computer program lives in its own protected region of memory and is executed by its own virtual CPU. Thus Avida simulates a parallel computer whereas Tierra simulated a sequential computer. These virtual CPUs can run at different speeds, such that one organism may, for example, execute twice as many instructions in the same time as another organism. Organisms perform tasks (logical computations) and successful completion of a task may increase the speed of an organism's CPU. Thus "successful" organisms have an increased chance of evolving further. Avida is still under active development.

Echo [13] (1994) was created as a complex adaptive system in which agents interact to develop survival strategies in environments with limited resources. Interaction can take the form of combat, trade or mating. Individual genotypes encode rules for interactions and usually populations of these genomes evolve networks which regulate the flow of resources. Resulting networks can resemble species communities in ecological systems.

Ant Colony optimization [5] and the related area of *Swarm Intelligence* [11] have also received much recent attention in the research literature. Both techniques make use of collective phenomena based on individual microscopic behaviours and therefore also have much in common with the animat agent approach.

Self-organizing and *nature-inspired* computational behaviour are two further topics of recent interest that relate to animat models. Reviews of collective emergent phenomena and self-organizing behaviours are provided in [30] and [37].

3 The Animat Model

Our animat model is a predator-prey simulation that combines the replication rules of Conway's Game of Life with the behavioural rules of Boids [22] and the evolutionary processes of Tierra [21] and Avida [1] to produce a unique and exciting world of animats that eat, breed and interact to create fascinating emergent macro-behaviours. Two species of animat exist in an unbounded 2-dimensional plain—the predators that must eat prey to survive; and the prey that must eat "grass" to survive. Predator-prey models have been studied since the work of Lotka [17] and Volterra [33] and are known to exhibit periodic oscillations in species populations. Our model exhibits this behaviour with predator populations lagging behind those of prey in the well known boom-bust cycles. A mean population value can still be taken as an indicative measure of the success or sustainability of the animats and this can be used as a metric to compare different microscopic parameter sets of the model.

3.1 Model Basics

The state of each animat in the model is defined at all times by the following variables: species (predator or prey); gender (male or female); location (xy-coordinates);

current age; current health; and a set of rules. The model also maintains the following global constants for each species: maximum age; maximum health; vision range (used to locate other animats); and a crowding number (if more than this number of animats are adjacent then they are crowded).

When an animat is created the current age is initialized to zero and is then incremented every time step. If the maximum age is reached the animat "dies of old age" and is removed. Similarly, an animat starts with its current health set to the average of its parents'. In early versions of the model, the current health was initialized to the maximum but it was found that this allowed populations on the brink of starvation to continue for prolonged periods. Each time step the current health is reduced and if it reaches zero the animat "starves to death". Current health can be increased by eating but can never be greater than the maximum health for that species. It is useful to define "hungry" to mean that current health is less than half the maximum and "well-fed" to mean that current health is equal to or greater than half the maximum.

At every time step, each animat executes one of a small set of simple rules such as "eat prey". Rules are provided as a sequence of letter codes—which makes them easy to evolve and mutate when required. A new rule could be invented at any time and the rules currently in use are summarized in Table 1. Rules have conditions; for example a predator can only eat prey if (a) the predator is "hungry" and (b) the prey is adjacent. Breeding has several conditions including being well-fed and having an adjacent mate. In addition, there is a "birth rate" for each species which is the chance of breeding successfully. This provides a simple variable to control an otherwise complex mix of factors such as availability of food and shelter, possible birth complications, infant mortality rates, etc. Birth rates can be adjusted to change population levels in order to create a stable environment. Typical birth rates used in the model are 15% for predators and 55% for prey. Note that breeding can only be performed by females. Males can carry the breed rule in their rule set as a useless rule that will never be executed, but which can be passed on to female offspring that can execute it.

Table 1 lists the principle rules available to the animats. Each animat agent incorporates a list of rules which it tries to execute in order, trying each in turn until one succeeds. The last rule in the list may be the "do nothing" rule, which is independent of environment and therefore always succeeds. Adaptations of the rule list form the basis of our evolutionary experiments, described below.

Normally the model is run with one set of rules for each species. This set of rules never varies and any new animats are exact clones of their parents. This is in line with our policy of carefully introducing small controlled changes to the model in order to more accurately measure the effects that such changes have on the system. Rules are placed in a priority order and each animat always executes the first rule in its list for which the conditions are satisfied. We have experimented with changing the order of priority of the rules [9] (but not the rules themselves) and thus produced different "tribes" of animats where each tribe has a different rule set (genotype) that consists of the same rules but in a different priority order. The "move randomly" rule has an artificial condition in that it is only successful 50% of the time. If a rule has no conditions then any following rules in the rule set would never be executed.

Table 1 Rules currently used in the model. A rule is only executed if the conditions are met; otherwise that rule is ignored. The birth conditions are discussed in the text

Code	Action	Predator	Prey	Conditions
A	move Away	No	Yes	crowded; adjacent prey
B	Breed	Yes	Yes	well-fed; adjacent mate; birth conditions
E	Eat prey	Yes	No	hungry; adjacent prey
F	Flee predator	No	Yes	adjacent predator
G	eat Grass	No	Yes	hungry; not crowded; requires grass
L	aLtruistic	Yes	No	adjacent needy predator
M	seek Mate	Yes	Yes	well-fed; no adjacent mate
P	seek Prey	Yes	No	hungry; no adjacent prey
R	move Randomly	Yes	Yes	50% chance of success

Table 2 presents the model parameters and the typical values that we used in the reported experiments.

Table 2 Animat model parameters showing typical values, valid ranges and a brief description

Parameter	Typical Value	Limits	Description
Prey-vision	20 cells	> 10	Prey vision range
Prey-health	100	> 50	Prey maximum health
Prey-age	20 steps	> 10	Prey maximum age
Prey-breed-crowd	10	4 to 10	too many neighbours to breed
Prey-graze-crowd	10	4 to 10	too many neighbours to graze
Prey-birthrate	55%	30% to 80%	Prey chance of succesful birth
Predator-vision	60 cells	> prey vision	Predator vision range
Predator-health	200	> prey health	Predator maximum health
Predator-age	50 steps	> prey age	Predator maximum age
Predator-crowd	6	4 to 10	too many neighbours to breed
Predator-birthrate	15%	10% to 40%	Predator chance of succesful birth

3.2 Changes to the Model

In early versions of the model it was common for prey animats to breed rapidly and form dense clusters, particularly in areas where predators were temporarily absent. This often led to extremely large prey populations which prevented the formation of interesting emergent clusters and also caused a dramatic slowdown in the execution time for the model. (Note that this was prior to the introduction of "grass" as the first versions of the model assumed that prey could always find food and hence were implicitly feeding at all times.) The concept of a crowd was introduced in order to solve this problem by postulating that if prey are too closely crowded together their chances of feeding and breeding should be reduced. The rule "(A) move away from other prey animats" was introduced and this caused prey to spread out. This in turn prevented very dense clusters from forming and also made it more likely that a large formation of prey would spread and be discovered by predators. The crowding rules are still retained in the model but have had less impact since the introduction of grass.

Inspired by Ronkko's model [24], grass was introduced into the system along with the rule "(G) eat grass". Grass can be placed at specific locations on the map and can thus be used to contain the animat populations and prevent endless expansion in both space and animat numbers. A restricted population enables longer simulation runs that can then exhibit further emergent behaviour. Previous work [27] has shown that animat behaviour is not affected by the edges of the grassed area.

Grass is assigned a "nutritional" value and this is the amount that the current health of an animat will increase when it eats the grass. Figure 2 demonstrates the effect of bands of different grass values. Animats have not survived on the inner grass values which are low (20 to 30) but are thriving on the outer sections (grass values 70 to 80) where the usual emergent spiral formations are clearly visible. Grass with a value of 70 will increase current health by 70 although it may not exceed the maximum health which is usually 100 for prey. However, grass with a value of 30 will only increase current health by 30 when eaten. The grass value allows the simulation of plentiful or scarce resources and any consequent changes in animat behaviour. Grass is never diminished and is assumed to retain the same nutritional value for the duration of the simulation. Future work may introduce grass that changes its value over time.

3.3 Fine Tuning

One of the significant problems of this type of model is execution speed. In particular, when every animat has to locate all its neighbours in every time-step, the increase in execution time becomes exponential. It is partly for this reason that many early models (for example Avida and Tierra) contain organisms that act in isolation with no knowledge of their neighbours.

We have put significant effort into fine tuning our model and optimizing execution speed [26] and the speedup obtained is shown in Fig. 3. Initially n animats were placed on an unbounded (effectively infinite) plain but the $O(n^2)$ search for

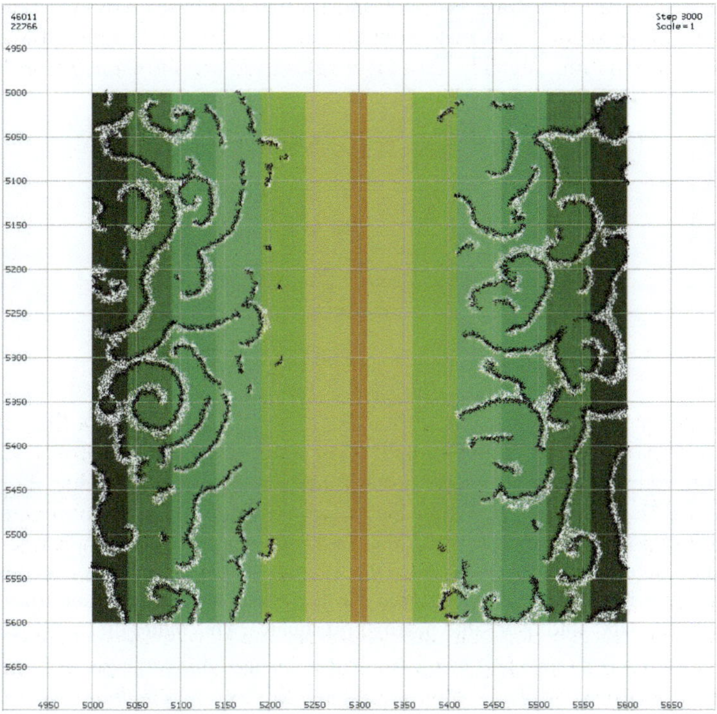

Fig. 2 Clusters of animats at step 3000 on a background of varying grass values with 80 on the outer edges and 20 in the centre. Predators are black and prey are white. Animats can not exist on the lowest grass values in the centre but are thriving on the higher values on the outer edges. Note the emergent spiral clusters

neighbours soon rendered this impractical. Then the plain was divided into coarse-grained squares ("the grid") with several animats in each square. Animats could then search for neighbours only within the square they occupied—except for those that were positioned near the border of a square. Finally a fine-grained matrix was introduced with each animat occupying a cell and searching only for neighbours within the vision range.

The animats tend to form quite dense localized clusters. Other space partitioning techniques such as oct-trees have been successfully used to lower the computational complexity of neighbour-interactions in particle and astronomical simulation codes [34]. The overheads involved in using the two-dimensional equivalent (quad-tree partitioning) were found to obviate any advantage for our simulation sizes and therefore we have continued to use a grid partition. In order to reduce speed further the need for cumbersome distance calculations was removed by the introduction of a table of offsets. In order to find a neighbour the animat simply has to check all cells in the offset list as shown in the example in Fig. 4.

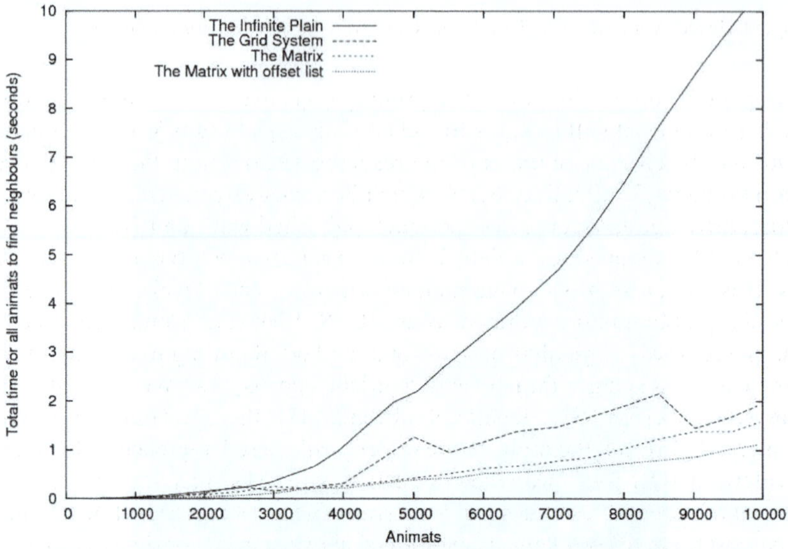

Fig. 3 Run time of the model using the four different spatial approaches. The algorithms show continuous improvement although the "Grid System" shows execution time peaks at points where animats "spill over" into nearby grid squares. This figure first appeared in [26]

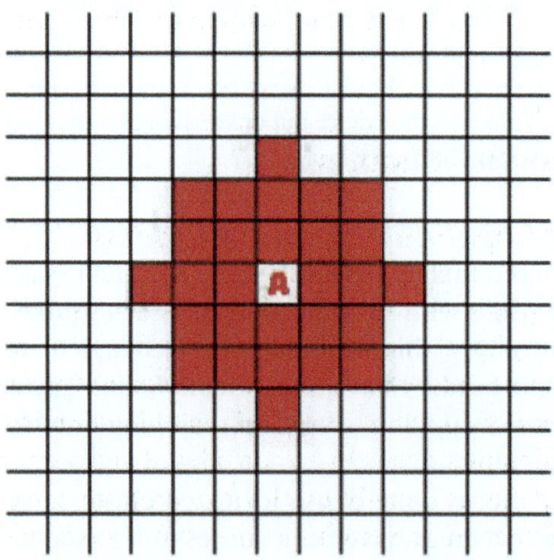

Fig. 4 The radial lookup table prescribes those cells that might contain another animat. The cells are sorted by distance from the centre so that only the minimal list of possible locations necessary to find the nearest neighbours will be traversed. In this example the vision range is assumed to be 3 (it is normally 20 to 60) and the offset list would contain (0,0)(-1,0)(0,-1)(0,1)(1,0)(-1,-1)(1,-1)(1,1)(-2,0)...(0,3)(3,0). Animat A would then search only through the shaded cells in order from nearest to furthest

4 Animat Model Observations

The system can manage significant numbers of animats. A typical run contains tens of thousands and the current record is over a million animats. These high population figures enable emergent macro-behaviours that would not be possible with lower numbers. The interaction of thousands of animats of both species as they follow their rules has produced fascinating emergent features in the form of macro-clusters. We have analyzed and documented these emergent clusters in [10]. The most significant of these are distinct spiral patterns, an emergent behaviour which can not be traced to any of the simple rules guiding the lives of individual animats. Several spirals are visible in the Figs. in this chapter.

One of the biggest problems facing this type of simulation is what we call the "Hand of God" effect in which a model contains numerous global parameters that are imposed on all animats equally, irrespective of any local situation. For example, many genetic algorithms use globally imposed fitness functions to decide which organisms live or die. Such global effects may constrain emergent behaviours. We have refined our model to ensure that animat behaviour is governed almost entirely by local criteria. Each animat eats, breeds or moves depending only on the proximity (and actions) of its neighbours. In our model there is no global fitness function—there is only life or death.

This system enables research into far more complex emergent behaviours than many previous models of this type because it simulates higher-order animals. For instance we have demonstrated how altruism can benefit both groups and individuals when resources are scarce but that selfishness tends to dominate when resources are plentiful [28]. In order to introduce altruism into the model we introduced the rule "(L) Altruistic". When an animat executes this rule it finds an adjacent neighbour of its own species which has a lower current health level than its own. Both animats then receive the average of the two health levels. For example, an animat with a current health level of 60 has an adjacent neighbour with a current health level of 40. The first animat executes the altruistic rule and both animats receive a current health level of 50. It is interesting to note that this rule is very localized as it is only executed if an immediate neighbour has a lower current health level and no attempt is made to seek out other "needy" animats that may not be adjacent. And yet this localized rule made a significant difference to the overall population when resources (grass value) were low.

Another study of higher-order behaviours involved a study of the effects of predators working in teams (or "wolf packs") when hunting. The model was altered such that the "eat prey" rule was not always successful and there was a chance that predators might fail to catch and eat prey. Various techniques of cooperating hunters were then introduced and analyzed. These experiments are discussed in [29]. This type of experiment demonstrates the versatility of our model which can easily be adapted to investigate a multitude of behaviours. Further work is planned to study the interactions of different types of animats within the same community, such as "workers" and "soldiers".

5 Evolution Algorithms

Genetic algorithms provide a mechanism for exploring the phase space of a model in search of optimal solutions [12]. There are two biologically inspired mechanisms that are commonly used: combination and mutation. Genetic crossover is a specific means of combining genetic information from the two parents to produce a new offspring that hopefully takes the best characteristics from its parents. Mutation is a simple random mechanism to introduce completely new information and avoid some of the problems of solutions being stuck in local minima of their fitness landscape. The key problem in applying genetic algorithms is in setting up an appropriate genetic representation of the properties of individual life forms (solutions) so that a sensible interpretation can be made of "mutation" and "crossover." It is by no means trivial to follow the biological metaphor exactly and establish a genetic strand of pseudo DNA to which these notions can be directly applied.

Our model has a physical landscape which is essentially featureless. Individual animats do encounter a physical environment but its properties are solely a consequence of animats' neighbours. Predators need prey, and need other predators to breed with. Prey are assumed to eat "grass" which is replenished at a uniform rate in our model. This is analogous to a complex entropic system through which a source of energy (sunshine) continually flows. Energy is not conserved in our model system—it is not intended to be as it is not a closed system [2]. Through careful experimentation with our microscopic parameters we have arrived at a set of feasible prey and predator parameters. We can initialize our system with a viable animat population and not have to wait in hope for artificial life to emerge from the "primordial soup." In this sense we have vastly reduced the area of fitness landscape we need the evolutionary model to explore. We are able to launch the model into a regime whereby the population eventually and typically stabilizes into periodic boom-bust oscillations around either a flat mean value or sometimes around a slowly-changing envelope function. This avoids many abortive experiments with non-viable populations that immediately crash into extinction, and thus lets us explore perturbative changes to the model starting from a viable population.

While we might have introduced a satisfactory predator type and a satisfactory prey type, by applying evolutionary operations we hope to find the "perfect predator" and the "perfect prey." However as both species are evolving together this means we have a situation of changing goals or in the language of genetic algorithms, the niches in our fitness landscape are varying too. The spatial aspects of our landscape provide an interesting mechanism for mixing species but this also means that different niches might arise in different physical parts of the model system. We proceed on the assumptions that we can in fact model a big enough system of animats, and that these effects are not significant when averaged over many independent starting conditions and run times. We are able to model systems of around 10^6 animats over several thousand time steps on a typical desktop computer overnight. Parallel computing techniques make it relatively simple to average over around 100 separate independent starting configurations and runs.

6 Evolution Experiments

Our model has been designed to use genetic algorithms to evolve animat rules and rule sets. Each rule is supplied as a string of characters in order to maximize the scope for mutation and crossover. However, we have adopted a cautious approach, preferring to edit one small aspect of the model at a time and then measure the effect that each change has on the model. As a first step we arranged the existence of several "tribes" or sub-groups of each species. All animats of the same species used the same rules but each tribe used the rules in a different priority order. Although no evolution was taking place, this "survival of the fittest" process ensured that successful tribes dominated and tribes with inefficient rule sets became extinct. This work [9] enabled us to identify efficient rule sets which led to robust and stable animat populations and these rule sets could then be used as the "standard" rule sets in future work. The standard rule set for predators was defined to be:

1. Breed (B)
2. Eat prey (E)
3. Seek mate (M)
4. Seek prey (P)
5. Altruistic (L)
6. Move randomly (R)

Using the codes given in Table 1 this rule set is written as BEMPLR. The standard rule set for prey was defined to be:

1. Breed (B)
2. Eat grass (G)
3. Move randomly (R)
4. Seek mate (M)
5. Move away from other prey (A)
6. Flee from predator (F)

Using the codes given in Table 1 this rule set is written as BGRMAF. The position of the rule "Flee from predator" was a surprise. We had initially assumed that it would be an important rule for prey and had placed it high up in the priority list. However in the "tribal" experiments described above, all successful tribes placed the rule very far down the list. This illuminates one of the many small but interesting facets of working with emergence.

We describe here the second stage of the evolutionary process in which we allow rule sets to evolve and mutate over time. This is not yet what some would regard as a full genetic algorithm because we are only shifting existing rules around and not changing the rules themselves. For example we do not allow prey to evolve the rule "Eat prey". The system is capable of implementing this type of "full" evolution but we prefer to introduce one small change at a time and attempt to measure the effects of that change. We prefer also to analyze higher-order behaviours such as hunting, trading or altruism within a predator-prey context whereas a model in which prey could eat other prey would no longer be a predator-prey model. All the experiments described below were performed with a predator birth rate of 15% and a prey birth rate of 55%.

The model can be initialized with a random mix of animats spatially scattered uniformly. It will typically undergo an initial period of under or over shoot before the populations adjust to the grass mass and predatory-prey equilibrium balance. We typically start an experiment using a stable and viable distribution of animat positions from a prior run that has already stabilized around mean population values. In the work reported below, this involves around 10^4 animats spatially mixed uniformly or from a prior stable mixed pattern.

6.1 Simulation 1 – The Control

The first experiment acted as the control. Animats used the standard rule sets and every animat of each species was therefore identical to all the others. The grass value was set to a standardized level of 60. The control simulation produced the expected results with the usual emergent spiral and other clustering formations.

6.2 Simulation 2 – Evolution by Crossover Only

The grass value was again set to 60 but animats were allowed to evolve rule sets. This was achieved by allocating rules randomly to the initial rule sets for every animat. When a new animat was produced during the simulation, its rule set was created by applying a single-point crossover [16] to the rule sets of its parents. For example if parent predators had rule sets BEMPLR and PLEMRB then their offspring would possess the rule set BEMMRB where the first three "chromosomes" are drawn from the rule set of the main parent (defined to be the parent with the same gender as the offspring) and the last three are drawn from the rule set of the other parent.

Note that this means that some rules may appear several times in a rule set and others may not appear at all. A repeated rule may increase the probability that it will be executed by the animat (for example: breed). In many cases repeating the rule will not assist the animat and the later repetitions become in effect redundant rules. However, carrying a repeat of a "useful" rule at various points in the rule set will ensure that it is passed on to offspring. By step 3000, the following rule sets were found to be dominating for the 98,636 prey:

10,256 male prey were using the rule set BGRGBR
9,823 male prey were using the rule set GAMBGR
9,055 male prey were using the rule set GRMBGR
6,853 male prey were using the rule set MGBFGR
15,802 female prey were using the rule set BGRBGR
9,249 female prey were using the rule set BGRGBR
8,370 female prey were using the rule set BGRFGR

This clearly shows that the most important rules for prey are B(breed) and G(graze) and both of these are repeated and placed near the top of the list (i.e. high priority).

Rule M(seek mate) also features prominently but other rules, including A(move away to relieve overcrowding) have vanished completely. Note that the B(breed) rule is only executed by females and thus is the leading rule in all successful female rule sets. The B(breed) rule is not important to males and thus appears lower down the list in successful male rule sets. The F(flee) rule has almost entirely disappeared and is clearly not important for successful rule sets. By step 3000, the following rule sets were dominating for the 33,708 predators:

2,359 male predators were using the rule set LPMBER
1,516 male predators were using the rule set EPMBLR
1,348 male predators were using the rule set LEPPRM
1,293 male predators were using the rule set PMBBER
1,853 female predators were using the rule set BPMBER
1,244 female predators were using the rule set PEBBLR
1,120 female predators were using the rule set EPBPRM
1,011 female predators were using the rule set BEPRLM

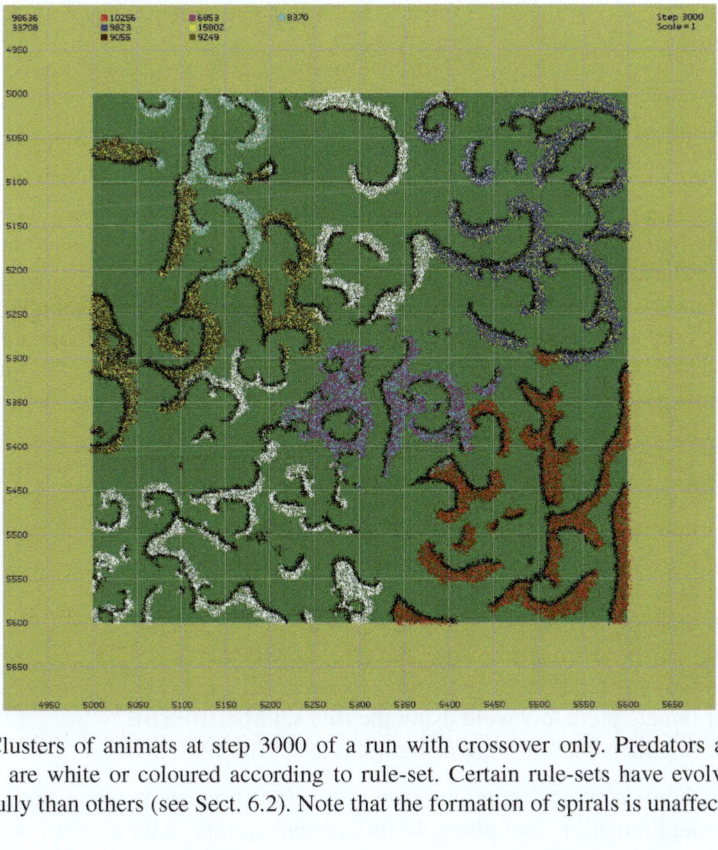

Fig. 5 Clusters of animats at step 3000 of a run with crossover only. Predators are black and prey are white or coloured according to rule-set. Certain rule-sets have evolved more successfully than others (see Sect. 6.2). Note that the formation of spirals is unaffected

Figure 5 shows animats clustering when a crossover mechanism has been applied. The colours show the different subspecies or tribes amongst the prey. Spatial patterns such as spirals still form, as found in the original (non-evolutionary) model, but these are now superimposed on the tribal segregatory patterns.

6.3 Simulation 3 – Evolution by Crossover and Mutation

This experiment retained the grass value of 60 but added mutation to crossover. When a new animat is created and a rule set is constructed for it (using crossover) there is a 5% chance that a mutation may take place. A mutation consists of a random rule being placed in a random position in the rule set. By step 3000, the following rule sets were dominating for prey:

897 of male prey were using the rule set BMGAGRBGGR
658 of male prey were using the rule set BMGAGRFGFR
629 of male prey were using the rule set MGAMGRFFRR
1,166 of female prey were using the rule set GBGBGRFGFR
912 of female prey were using the rule set BBGGRRBGGR
676 of female prey were using the rule set GBABGRFFRR

When comparing the results for crossover only (see Sect. 6.2) and the results for crossover and mutation together (shown here) it is clear that the same rules are favoured, namely B(breed) and G(graze). However there are also significant differences as listed below:

- Mutation allows the list of rules to grow and/or shrink. In this case it is apparent that the replication of certain rules, such as G(graze), was advantageous and the rule sets have lengthened.
- The number of animats with the same rule set is lower. For example, 15,802 female prey animats carried the same rule set with crossover only but only 1,166 female prey animats carried the same rule set after the introduction of mutation.
- There are more rule sets but some are very similar. For example the first two male prey rule sets only differ by two characters.

By step 3000, the following rule sets were dominating for predators:

416 of male predators were using the rule set MEBBPBBR
350 of male predators were using the rule set EPMRBRB
292 of male predators were using the rule set LPMEPRPR
267 of male predators were using the rule set BMLLMERBP
215 of male predators were using the rule set MPEPPRB
598 of female predators were using the rule set BBEEPRPR
432 of female predators were using the rule set MBEBPBBR
271 of female predators were using the rule set BEEPMBRB

Figure 6 shows a typical configuration after 3000 time-steps when both genetic crossover and mutation take place. In this system spirals again persist but the segregatory effect is much less pronounced. Many more different tribes survive but are

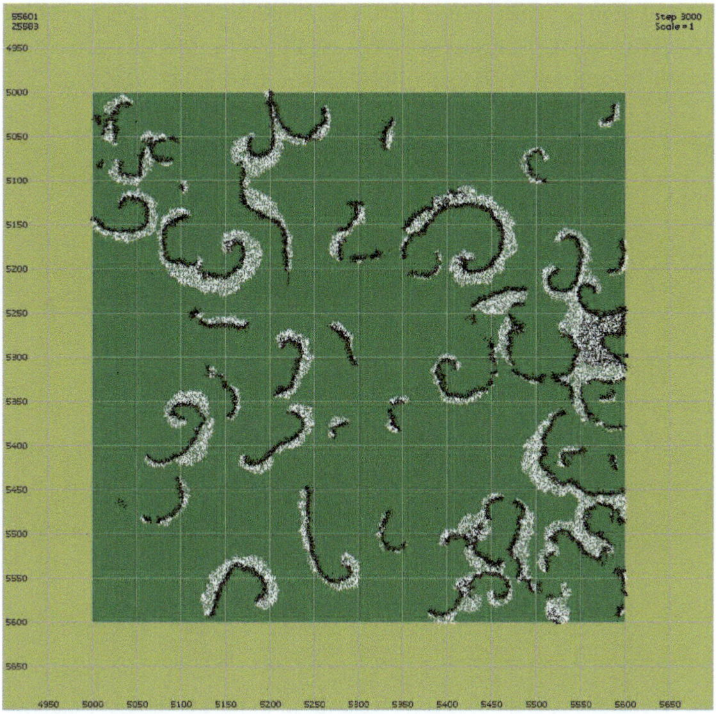

Fig. 6 Clusters of animats at step 3000 of a run with both crossover and mutation. Predators are black and prey are white. Individual rule sets are not shown as there are more of them and each set is much smaller. The total population is smaller than that with crossover alone. Spirals and other cluster formations continue to emerge

smaller in number. As a result the overall population is less successful and there are fewer animats.

6.4 Simulation 4 – Evolution with Scarce Resorces

Once the initial process of evolution had been established, a series of experiments were performed in order to analyze how well evolution assisted animats to adapt to a harsher environment. The nutritional value of the underlying grass area was modified in order to make it harder for prey (and hence predators) to survive. The non-evolutionary model case defaults to newborns being exact clones of the main parent. The evolutionary experiments incorporated both crossover and mutation as defined above.

Figure 7 shows the typical boom-bust cycles in the population of predators superposed on the time varying envelop functions that differentiate typical population trends with and without evolutionary effects being incorporated. In each case enabling animats to evolve over time increases the overall predator population. When resources are scarce (a grass value of 30) and animats are not able to evolve they die out completely. When evolution is introduced the predators succeed almost as well as in the control situation with no evolution and a grass value of 60.

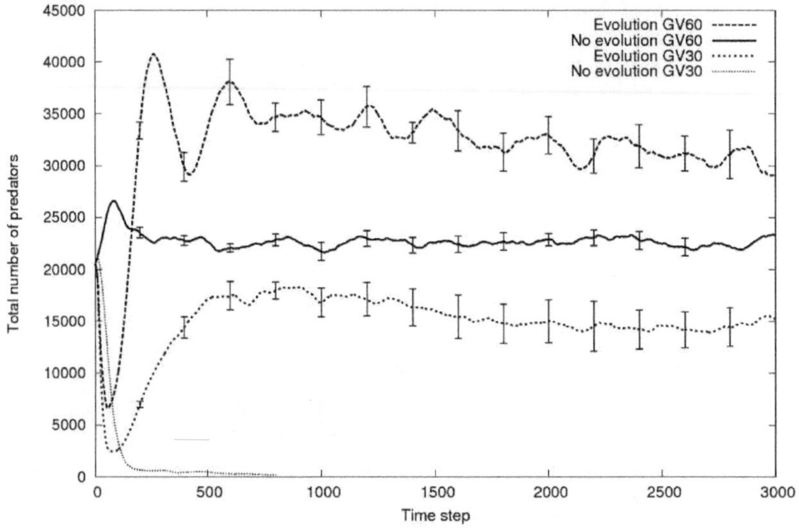

Fig. 7 Total predator population as it varies with time for the different algorithms—with and without evolution and with different grass richness values. These population figures are obtained from an average of ten runs

7 Discussion and Conclusions

We have described our spatial animat model based on a predator-prey intermix of animat agents that co-exist on a simulated flat world. We have related how this model has lead to study and worthwhile interpretation of a number of interesting emergent phenomena, including: spiral battlefront patterns; tribal segregation; resource scarcity effects; pack formation and a limited degree of altruism. Our model shows a remarkable robustness against its initial conditions and a scattered spatial mix of animats will typically arrive at a stable mean population balance between predators and prey with periodic boom-bust oscillations around them. We find that 1000 initial time steps is enough to bring the system to this stable oscillatory phase, and certainly to recover from initial shock effects. We have found that 3000 time steps is usually enough to find the long term statistical behaviour as not much subsequently changes in terms of the model's gross statistical properties, even though we might run to 10,000 steps and beyond.

Our microscopic model can usefully be compared with the Lotka-Volterra predator-prey model [17, 33]. Let $R(t)$ be the number of prey ("rabbits") at time t and $F(t)$ the number of predators ("foxes"). The uncoupled equations for predators and prey in a non-interacting world are then:

$$\frac{dR}{dt} = |a|R \tag{1}$$

$$\frac{dF}{dt} = -|g|F \tag{2}$$

so that unmolested by predators, the prey grow exponentially in number, and the predators starve through lack of prey and die off exponentially. The two controlling rate constants are positive numbers shown by the absolute value symbols in Eqs. 1 and 2. We drop this hereafter and assume that $a >= 0$ and $g >= 0$. It is of course interesting to consider what happens when the two populations do interact, and ignoring spatial distribution effects, we model this through a coupling term assumed to be proportional to the product RF which is related to the probability of a predator-prey encounter. We then obtain:

$$\frac{dR}{dt} = aR - bRF \tag{3}$$

$$\frac{dF}{dt} = hRF - mR \tag{4}$$

Solving these equations numerically, one can obtain equilibrium periodic solutions $(a = 2; b = c = h; m = 1.1)$ where the prey population reaches a high mean value with boom-bust periodic oscillations superposed on it. The predator population attains a lower mean oscillating value that lags behind the prey booms. Our spatial model does indeed show this *average* effect after some long term spatial fluctuations. It remains to link the equation parameters to the microscopic properties of our animat model, using diffusion constants. We have found our 2-species systems relatively forgiving in that local fluctuations of prey diminish the predators but they can typically recover as animats drift in from other spatial regions. The effects we have found in our model have been found in a number of real systems such as flocking and herding of birds or animals [23, 11] and ant colonies [4].

We have focused on the effects of introducing a limited form of evolutionary adaptation. Individual animats can give birth to new offspring that either combine features from their parents or can have small random mutations added. In our system we found that a form of crossover had the most beneficial effect with some subspecies tribes forming and being generally more successful than the random mutated species. We observe the interesting effect that evolution can help the model population to greater success in terms of animats supported per unit area of space. In particular in a situation of scarce resources and fewer prey available on average to predators, the continuing existence of the species depends on evolution. We believe models such as ours provide a valuable platform to experiment with evolutionary agent algorithms and effects, and that as well as giving general insights into emergent properties of a complex statistical system, there are also parallels with ecological and sociological systems in the real world.

References

1. Adami, C.: On modeling life. In: Brooks, R., Maes, P. (eds.) Proc. Artificial Life IV, pp. 269–274. MIT Press, Cambridge (1994)
2. Adami, C.: Introduction to Artificial Life. Springer, Heidelberg (1998)
3. Bajec, I.L., Zimic, N., Mraz, M.: Fuzzifying the thoughts of animats. In: De Baets, B., Kaynak, O., Bilgiç, T. (eds.) IFSA 2003. LNCS, vol. 2715, pp. 1103–1165. Springer, Heidelberg (2003)
4. Bonabeau, E., Theraulaz, G., Fourcassie, V., Deneubourg, J.L.: Phase-ordering kinetics of cemetery organization in ants. Phys. Rev. E 57, 4568–4571 (1998)
5. Dorigo, M., Maniezzo, V., Colomi, A.: Ant system: optimization by a colony of cooperating agents. IEEE Transactions on Systems, Man and Cybernetics – Part B 26, 29–41 (1996)
6. Fogel, L.J.: On the organization of intellect. PhD thesis. University of California, Los Angeles (1964)
7. Franklin, S.: Artificial Minds. MIT Press, Cambridge (1997)
8. Guye-Vuilleme, A., Thalmann, D.: A high-level architecture for believable social agents. Virtual Reality 5, 95–106 (2000)
9. Hawick, K.A., Scogings, C.J., James, H.A.: Spatial emergence of genotypical tribes in an animat simulation model. In: Henderson, S.G., Biller, B., Hsieh, M.H., Tew, J.D., Barton, R.R. (eds.) Proc. 2007 Winter Simulation Conference 2007 (WSC 2007), Washington DC, USA, pp. 1216–1222 (2007)
10. Hawick, K.A., Scogings, C.J., James, H.A.: Defensive spiral emergence in a predator-prey model. Complexity International, 1–10 (2008)
11. Hinchey, M.G., Sterritt, R., Rouff, C.: Swarms and swarm intelligence. Computer, 111–113 (2007)
12. Holland, J.H.: Adaptation in natural and artificial systems, 2nd edn. MIT Press, Cambridge (1992)
13. Holland, J.H.: Echoing emergence: Objectives, rough definitions, and speculations for echo-class models. In: Cowan, G.A., Pines, D., Meltzer, D. (eds.) Complexity: Metaphors, Models and Reality, pp. 309–342. Addison-Wesley, Reading (1994)
14. Holland, J.H.: Hidden order: How adaptation builds complexity. Addison-Wesley, Reading (1995)
15. Kauffman, S.A.: Investigations. Oxford University Press, Oxford (2000)
16. Lenski, R.E., Ofira, C., Collier, T.C., Adami, C.: Genome complexity, robustness, and genetic interactions in digital organisms. Nature, 661–664 (1999)
17. Lotka, A.J.: Elements of Physical Biology. Williams and Wilkins, Baltimore (1925)
18. Mouret, J.B., Doncieux, S.: Incremental evolution of animats behaviors as a multi-objective optimization. In: Asada, M., Hallam, J.C.T., Meyer, J.-A., Tani, J. (eds.) SAB 2008. LNCS (LNAI), vol. 5040, pp. 210–219. Springer, Heidelberg (2008)
19. Murray, J.D.: Mathematical Biology. Springer, Heidelberg (2001)
20. Rasmussen, S., Knudsen, C., Feldberg, R., Hindsholm, M.: The coreworld: Emergence and evolution of cooperative structures in a computational chemistry. Physica D 42, 11 (1990)
21. Ray, T.: An approach to the synthesis of life. In: Artificial Life II. Santa Fe Institute Studies in the Sciences of Complexity, vol. xi, pp. 371–408 (1991)
22. Reynolds, C.: Flocks, herds and schools: A distributed behavioral model. In: SIGRAPH 1987: Proc. 14th Annual Conf. on Computer Graphics and Interactive Techniques, pp. 25–34. ACM, New York (1987)

23. Romanczuk, P., Couzin, I.D., Schimansky-Geier, L.: Collective motion due to individual escape and pursuit response. Phys. Rev. Lett. 102, 010602–1–4 (2009)
24. Ronkko, M.: An artificial ecosystem: Emergent dynamics and lifelike properties. J. ALife 13, 159–187 (2007)
25. Ross, D., Brook, A., Thompson, D. (eds.): Dennett's Philosophy: A Comprehensive Assessment. MIT Press, Cambridge (2000)
26. Scogings, C.J., Hawick, K.A., James, H.A.: Tools and techniques for optimisation of microscopic artificial life simulation models. In: Nyongesa, H. (ed.) Proceedings of the Sixth IASTED International Conference on Modelling, Simulation, and Optimization, Gabarone, Botswana, pp. 90–95 (2006)
27. Scogings, C.J., Hawick, K.A.: Global constraints and diffusion in a localised animat agent model. In: Proc. IASTED Int. Conf. on Applied Simulation and Modelling, Corfu, Greece, pp. 14–19 (2008)
28. Scogings, C., Hawick, K.: Altruism amongst spatial predator-prey animats. In: Bullock, S., Noble, J., Watson, R., Bedau, M. (eds.) Proc. 11th Int. Conf. on the Simulation and Synthesis of Living Systems (ALife XI), pp. 537–544. MIT Press, Winchester (2008)
29. Scogings, C., Hawick, K.: Pack-hunting multi-agent animats. In: Proc. IASTED Int. Symp. on Artificial Intelligence and Applications, Innsbruck, Austria, pp. 220–224 (2009)
30. Serugendo, G.D.M., Karageorgos, A., Rana, O.F., Zambonelli, F. (eds.): Engineering Self-Organising Systems. Springer, Heidelberg (2004)
31. Turney, J.: Frankenstein's Footsteps - Science, Genetics and Popular Culture. Yale University Press, New Haven (1998)
32. Tyrrell, T., Mayhew, J.E.W.: Computer simulation of an animal environment. In: Meyer, J.A., Wilson, S.W. (eds.) From Animals to Animats 1: Proceedings of the First International Conference on Simulation of Adaptive Behavior, pp. 263–272. MIT Press, Cambridge (1991)
33. Volterra, V.: Variazioni e fluttuazioni del numero d'individui in specie animali conviventi. Mem. R. Accad. Naz. dei Lincei, Ser VI 2 (1926)
34. Warren, M.S., Salmon, J.K.: A parallel hashed oct-tree n-body algorithm. In: Supercomputing, 12–21 (1993)
35. Watts, J.M.: Animats: computer-simulated animals in behavioural research. J. Anim. Sci. 76, 2596–2604 (1998)
36. Wilson, S.W.: The animat path to AI. In: Meyer, J.A., Wilson, S.W. (eds.) From Animals to Animats 1: Proceedings of The First International Conference on Simulation of Adaptive Behavior, pp. 15–21. MIT Press, Cambridge (1991)
37. Wurtz, R.P. (ed.): Organic Computing. Springer, Heidelberg (2008)

An Agent-Based Parallel Ant Algorithm with an Adaptive Migration Controller

Ying Lin and Jun Zhang*

Abstract. This paper presents an agent-based parallel ant algorithm (APAA) for numerical optimization. By utilizing two cooperating agents to optimize different components of the solution vector, APAA reduces the scale of the problem handled by each agent and thus achieves improvement in algorithm performance. In APAA, each agent has tunable and untunable vectors. Tunable vectors are optimized through a novel continuous ant algorithm with untuable vector fixed. Excellent tunable vectors in one agent are migrated into the other agent as new untunable vectors through a stagnation-based asynchronous migration controller (SAMC), in which the migration strategy is adaptively adjusted according to the stagnation degree in the optimization process of the agent. The proposed APAA is especially suitable for large-scale problems. Experimental studies on a set of benchmark functions show that APAA can obtain better results at a faster speed for functions in high dimensional space.

1 Introduction

Ant colony optimization (ACO) is a new kind of metaheuristic inspired from the foraging behavior of ants in nature [1]. It has been proven to be an efficient way for solving discrete problems, such as graph coloring [1], cash flow scheduling [2], protein folding [3], and multicast routing [4], etc. ACO has also been extended to continuous domain and encouraging results have been reported [5, 6]. In the research of discrete and continuous ACO, an important and promising area is their parallel implementation. It has been shown that a proper parallel strategy can improve the performance of ACO effectively [7,8].

There is no golden rule to parallelize ACO [9]. This paper classifies existing parallel ant colony optimization (PACO) in literature into three categories: parallel independent runs of ACO algorithms, parallel evaluation, and parallel optimization through cooperative colonies.

Ying Lin · Jun Zhang
Department of Computer Science, Sun Yat-Sen University, Guangzhou, 510275, China
e-mail: junzhang@ieee.org

* Corresponding author.

R.A. Sarker and T. Ray (Eds.): Agent-Based Evolutionary Search, ALO 5, pp. 161–177.
springerlink.com

The parallel max–min ant system (MMAS) proposed by Stützle [9] is a typical instance of the first type of PACO. In literature [9], the efficiency of the parallel MMAS, including the situations that MMAS were executed under similar or different settings, was discussed in detail. However, as the parallel algorithms are generally isolated by little communication, the advantage of cooperation is weakened. The second type of PACO can be viewed as a distributed computation model because only the evaluation component is executed in parallel. Its effectiveness is limited to problems with costly or complex objective functions [10]. The third type of PACO comprises parallel strategies using multiple ant colonies. According to the proportion of sub-colonies, the third type of PACO can be further classified into coarse-grained [7, 8, 11] and fine-grained models [10]. Randall and Lewis [10] applied a fine-grained PACO named the parallel ant to solve the traveling salesman problem (TSP). Michel and Middendorf [12] embedded the idea of island model in ACO and set up a coarse-grained ant system with the best solution exchanged at a fixed frequency. Bullnheimer et al. [13] proposed a coarse-grained and a fine-grained parallel ant system and concluded that the coarse-grained one is better because of its proper communication cost.

In recent days, multi-colony PACO has received more and more consideration because of its more appropriate communication manner and better adaptability to different problems. However, the ants' migration among different colonies remains a challenging issue in multi-colony ACO [14]. As a matter of fact, this problem widely exists in all kinds of parallel algorithms with multiple populations [15, 16, 17]. Many research efforts have been conducted on adjusting the timing and the contents of migration. The timing of migration is critical because either over-frequent or scarce migration will reduce the efficiency of parallelization. Frequent migration makes information disseminated quickly but it also increases the communication cost. Sparse migration isolates the colonies and weakens the benefits of cooperation. Besides the timing of migration, the size of migration contents also influences the communication cost. Moreover, migration contents can help adjust balance between exploration and exploitation in each colony.

According to the characteristics of ant colonies, a sub-colony in multi-colony PACO can be considered as an agent [18, 19, 20]. This paper proposes an agent-based parallel ant algorithm (APAA) for continuous problems. The existing multi-colony PACO utilizes different colonies to optimize the same object. When solving large-scale problems, each colony faces the same difficulty with the colony of sequential ACO (SACO). However, the proposed APAA has different agents optimize different components of the solution vector of the optimization problem, which reduces the scale of the optimization problem for each agent. Therefore, the proposed APAA is especially suitable for large-scale problems. In order to handle the migration issue in APAA, a stagnation-based asynchronous migration controller is also proposed. It is based on analyzing the stagnation degree in the optimization process of each agent to adjust the timing and contents of migration. The proposed APAA with SAMC has been applied to a set of benchmark functions with unimodal and multimodal features. Experimental results demonstrate the efficiency of the proposed method in solving problems in high dimensional space.

The rest of this paper is organized as follows: Section 2 introduces a continuous ant algorithm as the basis of APAA; Section 3 details the implementation of the proposed APAA with SAMC; Section 4 displays the numerical experiments and related discussions; Section 5 draws a conclusion to the whole paper.

2 A Brief Introduction to a Continuous Ant Algorithm

To implement the proposed parallel strategy, this paper employs an improved version of the continuous ant algorithm from literature [21]. The detail process of the employed algorithm is described as follows.

2.1 *Initialization*

As for a continuous problem of n dimensions, each ant in the algorithm is initialized with a randomly generated solution

$$X^{(i)} = (x_1^{(i)}, x_2^{(i)}, \ldots, x_n^{(i)}) \tag{1}$$

where $x_j^{(i)}$ represents the value of the j-th variable in the solution of the i-th ant, i=1,2,…,N, and N is the total of ants in the algorithm.
After evaluating the ants and sorting them from the best to the worst, the pheromone $\tau_j^{(i)}$ on the variable $x_j^{(i)}$ (j=1,2,…,n, i=1,2,…,N) is updated as

$$\tau_j^{(i)} = \tau_j^{(i)} + \alpha \times (GN - k_i + 1) \times \tau_0 \tag{2}$$

where τ_0 is the initial pheromone value, $\alpha \in (0,1)$ is the reinforcement rate, and k_i is the rank of the i-th ant. In this way, the pheromones on the variables of the best GN ants are improved while the pheromones attached to the other ants are reduced.

2.2 *Selection*

After the initialization, a set of N values is obtained for each variable. If the random number p_1 is larger than the first selection probability q_1, the ant is directly copied into the next iteration. Otherwise, the ant is reconstructed by selecting values for each variable from the corresponding set. The selection follows the pseudo random rule of the ant colony system and includes exploitation and exploration [1]. The proportion between exploration and exploitation is determined by the second selection probability q_2. Suppose p_2 is a uniformly distributed random number generated for $x_j^{(i)}$ (j=1,2,…,n, i=1,2,…,N). If p_2 is smaller than q_2, exploitation is performed and the value with the largest pheromone is selected for $x_j^{(i)}$; otherwise, exploration is performed through a roulette wheel selection based on the probabilities calculated by

$$p(x_j^{(k)}) = \frac{\tau_j^{(k)}}{\sum_{k=1}^{N} \tau_j^{(k)}}, \; k=1,2,\ldots,N \tag{3}$$

As a summary, the random proportional rule can be written as

$$x_j^{(i)} = \begin{cases} \arg\max\limits_{1\le k\le N}(\tau_j^{(k)}), & p_2 \le q_2 \\ J, & \text{otherwise} \end{cases} \tag{4}$$

where J represents the variable value chosen by the roulette wheel selection.
Fig. 1 shows an example of the selection process reconstructing a new solution for ant 2. Thickness of color in the circles indicates the amount of pheromones. It can be seen that values with more pheromones have larger probability to be selected.

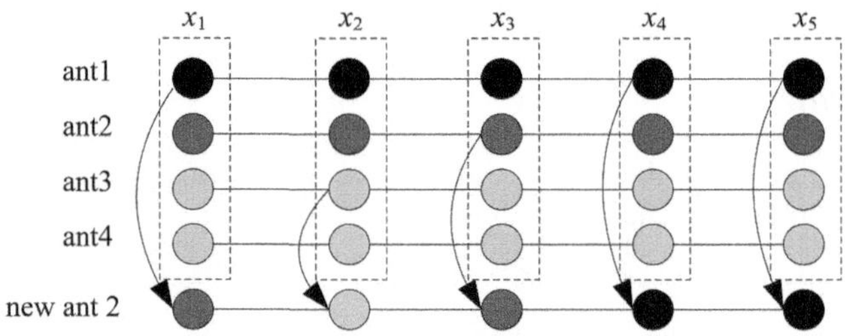

Fig. 1 An example of reconstructing solutions by the selection process

2.3 Dump Operation and Pheromones Update

The dump operation imports new materials for the algorithm. The worst *BN* ants are replaced by new ants with randomly generated solutions. After evaluating the new ants and sorting them from the best to the worst, the algorithm updates the pheromones of each ant as

$$\tau_j^{(i)} = \begin{cases} \rho\times\tau_j^{(i)} + \alpha\times(GN - k_i + 1)\times\tau_0, & i \le GN \\ \rho\times\tau_j^{(i)}, & \text{otherwise} \end{cases}, \ i=1,2,\ldots,N \tag{5}$$

where $\rho \in (0,1)$ is the evaporation rate. As $\tau_j^{(i)} \ge \tau_0$, only pheromones on the best *GN* ants are reinforced after evaporation while the pheromones of the other ants are reduced.

2.4 Random Search

The proposed algorithm assigns an n-dimensional radius $R(r_1,r_2,\ldots,r_n)$ to the best-so-far solution $X^{best}(x_1^{best}, x_2^{best}, \ldots, x_n^{best})$. During the random search, the variable value x_j in a new solution $X(x_1,x_2,\ldots,x_n)$ has equal probability to be any value of $[x_j^{best}-r_j, x_j^{best}), x_j^{best}, (x_j^{best}, x_j^{best}+r_j]$.
A total of θ new solutions are generated in the random search. The best new solution is compared with X^{best}. If the former is better, the radius will be increased to expand the range of random search. Otherwise, the radius will be reduced to shrink the range of random search. The increment and decrement of the radius are implemented as

$$r_i = \begin{cases} r_i \cdot \phi, & \text{shrinking} \\ r_i / \phi, & \text{expanding} \end{cases}, \quad i=1,2,\ldots,n \tag{6}$$

where $\phi \in (0,1)$ is the changing rate of radius. The adaptive searching radius helps improve the precision of the algorithm. Restricting the random search to the best-so-far solution can avoid wasting computational cost on useless exploitation around less desired points.

Fig. 2 summarizes the complete process of the above continuous ant algorithm. It can be observed that the algorithm iterates the steps of selection, dump operation and pheromone update, and random search until the termination criterion is satisfied. The best-so-far solution is returned as the result.

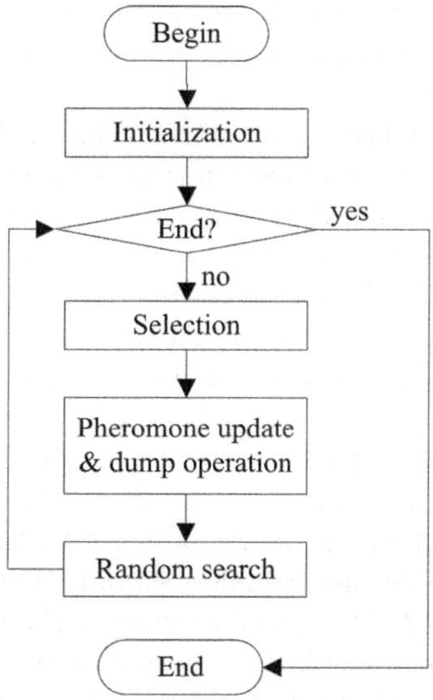

Fig. 2 The flowchart of the continuous ant algorithm employed in this paper

3 Implementation of the Agent-Based Parallel Ant Algorithm (APAA)

The proposed APAA divides the solution vector X of a problem into two sub-vectors X_1 and X_2, satisfying

$$X_1 \cup X_2 = X \tag{7}$$

and

$$X_1 \cap X_2 = \phi \tag{8}$$

In APAA, the two agents, namely A_1 and A_2, involve both X_1 and X_2 but only optimize one of them. For instance, A_1 only lays pheromones on the variables of X_1, and the operations in the ant algorithm are only performed on X_1, too. In other words, X_1 is the tunable vector of A_1 while X_2 is untunable. Similarly, A_2 optimizes X_2 with X_1 fixed. The two agents consider their untunable vectors only in the evaluation, which insures that the two agents evolve under the guidance of the same objective function in respect to X. During the migration, some good tunable vectors of one agent are chosen and transferred to the other agent as new untunable vectors. It is only by migration that the contents of an agent's untunable vectors change. Using shaded and blank boxes indicating untunable and tunable vectors respectively, Fig. 3 shows the framework of APAA. The bold arrows in Fig. 3 display the direction of migration.

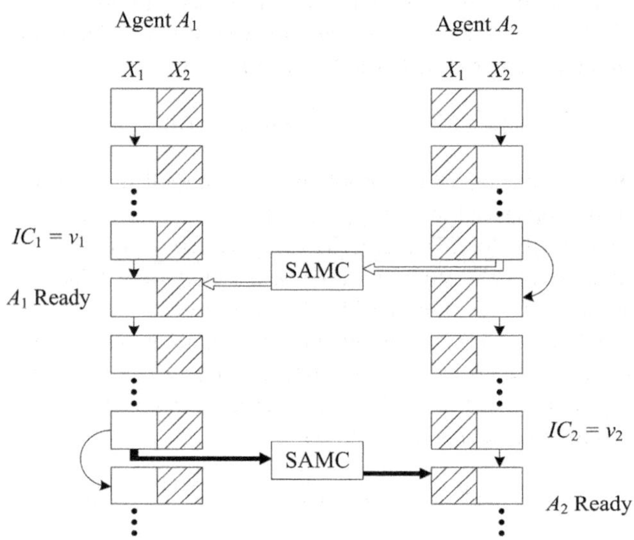

Fig. 3 The framework of APAA with the stagnation-based asynchronous migration controller (SAMC)

The key components of APAA include division of the solution vector and the migration controller, which will be illustrated in the following paragraphs.

3.1 Division of the Solution Vector

By dividing the solution vectors, the dimensions to be optimized by each agent decrease. Thus, the agents in APAA are able to focus on an optimization problem with a smaller scale, which benefits both solution quality and convergence speed. However, the division of the solution vector must be done in a careful manner; otherwise it will lead to devastating results [22].

Suppose that the optimal X_1 and X_2 depend on each other. A_1 optimizes X_1 according to its current untunable vector X_2. However, when the migration occurs, X_2 in A_1 will be replaced by the optimized X_2 from A_2. The optimization process has to be restarted and the previous work would be wasted. A similar situation can be found in A_2. Thus, the algorithm is not able to converge, or at least cannot achieve satisfying precision. In order to avoid the above situation, the solution vector should be divided in such a way that the two sub-vectors are the least correlated.

For the continuous problem considered in this paper, the solution vector is composed of real numbers. The sub-vectors can be obtained through careful observations on the objective function. However, situations in practical applications might be more complex. For those problems whose solution vectors cannot be well divided, the efficiency of the proposed APAA may be reduced.

3.2 Stagnation-Based Asynchronous Migration Controller (SAMC)

The migration controller determines the migration frequency and sets up the migration plans for both agents. In every agent, the untunable vectors establish the underlying environment for the optimization of the tunable vectors. This environment is renewed after migration, and the latest untunable vectors are expected to guide the optimization more efficiently. If the agent's optimization slows down, or even stagnates, the agent may advance the migration in the original plan. In addition, it may prefer a more diverse environment to activate the optimization process. Nevertheless, if the agent is enjoying a propitious optimization process, it may want to postpone the migration and fine-tune its tunable vector. The static migration controller with fixed migration frequency can not satisfy the above requests. In that case, the algorithm wastes computation in inefficient optimization and extra migration. The proposed SAMC is based on considering the stagnation degree in the optimization process to adaptively adjust the migration frequency. The migration can thus be called at more appropriate time.

Denote the migration interval of agent A_i as m_i (i = 1, 2). The migration can be advanced or postponed by decreasing or increasing m_i correspondingly. IC_i records the number of iterations in A_i since the last migration. As shown in Fig. 3, SAMC tags A_i as ready when IC_i exceed v_i. Then the migration occurs and A_i renews its untunable vectors with the selected tunable vectors of the other agent.

Since SAMC performs exactly the same on both agents, the subscripts of v_i and IC_i are ignored to facilitate the following descriptions. SAMC is composed of three rules, which adjust v under different situations. Suppose X^{pre} and X^{cur} represent the agent's best solutions until the previous and the current iterations. '$\succ$' indicates a dominant relationship between X^{cur} and X^{pre}. The rules of SAMC are summarized as follows.

Rule 1 – $X^{cur} \succ X^{pre} \Rightarrow v = v + \Delta v$.

Rule 1 is executed when no stagnation occurs during the agent's optimization process. v is increased by Δv and the agent has more time to fine-tune its tunable vector. Δv is defined as

$$\Delta v = \left\lceil \frac{\Delta v_{\max}}{IC} \right\rceil \tag{9}$$

which implies that Δv is limited to $[1, \Delta v_{\max}]$ and decreases as IC increases.

Rule 2 – $(X^{\text{cur}} = X^{\text{pre}}) \wedge (T \leq T_{\max}) \Rightarrow v = v$.

Rule 2 describes the situation that the stagnation occurs, but the stagnation time T remains under the upper limit $T_{\max}$. In that case, SAMC maintains the value of v until further situations are detected.

Rule 3 – $(X^{\text{cur}} = X^{\text{pre}}) \wedge (T > T_{\max}) \Rightarrow v = v - \Delta v$.

If the stagnation time T exceeds the upper limit $T_{\max}$, SAMC reduces v by Δv and advances the migration. Δv is defined as

$$\Delta v = \left\lceil \frac{\Delta v_{\max}}{1/(\gamma \cdot IC)} \right\rceil \tag{10}$$

where γ is a predefined constant in [0,1]. Eq. (10) tells that the latter the stagnation occurs, the more v decreases. This is because the optimization process is considered to have few chances to escape from stagnation when the time before migration is little.

Δv in (9) and (10) will accumulate when improvement or stagnation lasts for several iterations. Therefore, the migration plan changes more when the improvement or the stagnation continues for a longer period. However, no matter how v changes, it must stay within a predefined range $[v_{\min}, v_{\max}]$, which prevents the algorithm from over-frequent migration or acting as two isolated agents.

Besides adjusting the migration frequency, SAMC also adjusts the content of migration according to the stagnation degree. Suppose when the migration occurs in A_1, IC_1 is larger than the initial migration interval v_0. A_1 imports the best tunable vector of A_2 to renew its untunable vectors. If IC_1 is smaller than v_0, SAMC considers A_1 to have experienced an optimization process dominated by stagnation. Thus A_1 imports the potential tunable vectors – not just the best one - of A_2 to diverse its underlying environment of optimization, expecting to activate the optimization process.

With A_1 as an example, Fig. 4 shows the process of SAMC. At the very beginning of APAA, the relative parameters are initialized as $v_1 = v_2 = v_0$, $IC_1 = IC_2 = 1$. As can be observed from Fig. 4, the two agents optimize their tunable vectors alternately. v_1 and v_2 are adjusted in every iteration according to the three rules of SAMC. The optimization continues until one of the agents A_i satisfies $IC_i > v_i$. Then APAA pauses and brings the selected tunable vector(s) in the other agent to A_i as new untunable vector(s). After migration, v_i and IC_i of A_i are reset to their initial values v_0 and 1, respectively. Then the optimization goes on.

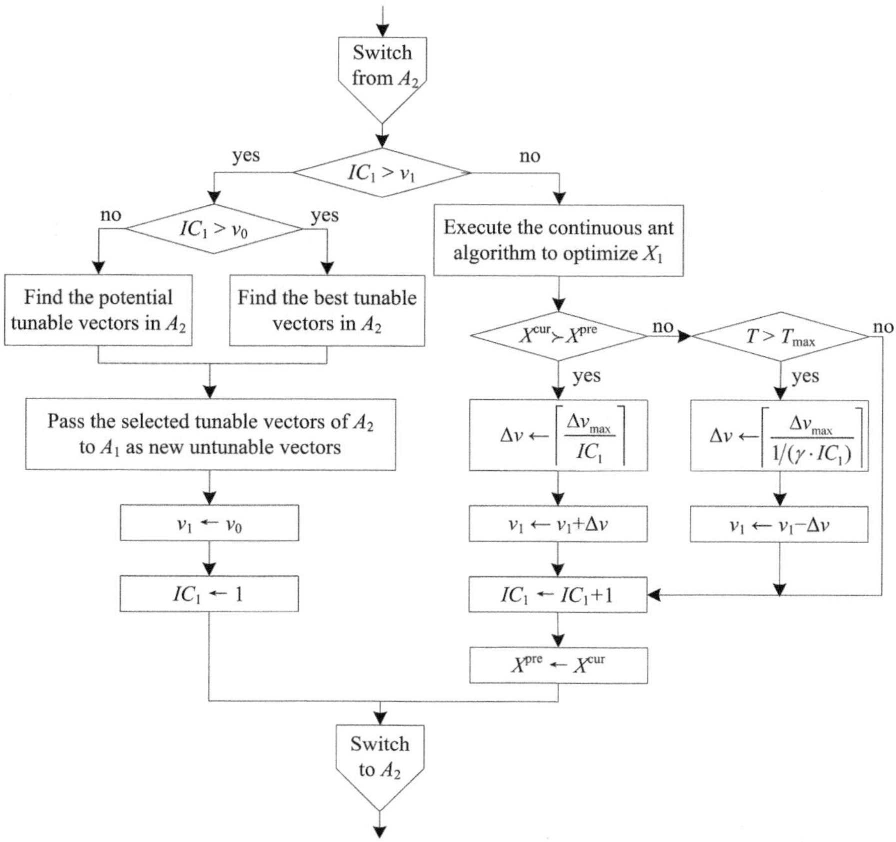

Fig. 4 Illustration of SAMC with A_1 as the example

4 Experiments and Discussions

This section compares APAA with SACO on a set of benchmark functions [23]. The performance of the algorithms is evaluated from both aspects of solution quality and convergence speed.

Table 1 tabulates the details of the test functions, including their equations, domains, and the optimal values (OPT). The symbol n in the tables represents the number of variables in the functions. For F_1, F_2, F_3, F_5, and F_6, n rises from 20 to 100 by a step of 20. For F_4, F_7, and F_8, n is specified as the values in the table. F_1 to F_3 are unimodal functions while the others are multimodal. For each function, an error bound ε is predefined. If the solution X found by a trial satisfies

$$|F_i(X) - OPT| < \varepsilon,\ i=1,2,\ldots,8 \tag{11}$$

the trial is considered successful. The column of MaxEval indicates the maximum number of the function evaluations spent in one trial.

Table 1 Test Functions

Equations	Domain	OPT	ε	MaxEval
$F_1 = \sum_{i=1}^{n} x_i^2$	$[-100,100]^n$	0	0.1	100,000
$F_2 = \sum_{i=1}^{n} \lvert x_i \rvert + \prod_{i=1}^{n} \lvert x_i \rvert$	$[-10,10]^n$	0	0.1	150,000
$F_3 = \sum_{i=1}^{n} \lfloor x_i + 0.5 \rfloor^2$	$[-100,100]^n$	0	0.1	20,000
$F_4 = \sum_{i=1}^{n} -x_i \sin(\sqrt{\lvert x_i \rvert})$ (n=20)	$[-500,500]^n$	−8379.67	0.025	100,000
$F_5 = \sum_{i=1}^{n} [x_i^2 - 10\cos(2\pi x_i) + 10]$	$[-5.12,5.12]^n$	0	0.1	20,000
$F_6 = -20\exp\left(-0.2\sqrt{1/n\sum_{i=1}^{n} x_i^2}\right) - \exp\left(1/n\sum_{i=1}^{n}\cos(2\pi x_i)\right) + 20 + e$	$[-32,32]^n$	0	0.1	20,000
$F_7 = -\sum_{i=1}^{n} \sin(x_i)\sin^{20}(i x_i^2/\pi)$ (n=100)	$[0,\pi]^n$	-99.2784	4.7	20,000
$F_8 = 1/n\sum_{i=1}^{n} (x_i^4 - 16x_i^2 + 5x_i)$ (n=100)	$[-5,5]^n$	-78.33236	0.015	20,000

4.1 Parameter Settings

The control parameters used in the sequential version of the continuous ant algorithm (SACO) are set as α=0.9, ρ=0.8, and q_1=q_2=0.5. In order to achieve favorable results, the total of ants (N), the number of new solutions generated by the random search (θ), and the changing rate of the radius in the random search (ϕ) are fine-tuned for each function. Table 2 shows the best values of N, θ, and ϕ for each function.

Table 2 The Best Values of N, θ, and ϕ for Each Function

Func	N	θ	ϕ
F_1	10	10	0.6
F_2	10	30	0.99
F_3	10	10	0.9
F_4	30	10	0.8
F_5	30	10	0.8
F_6	30	10	0.8
F_7	100	10	0.8
F_8	100	10	0.8

For the sake of fair comparison, the total of ants in APAA is set the same as that in SACO. That is, the number of the ants and the number of new solutions generated by the random search in each agent is half of the best N and θ, respectively. Besides, in APAA, the legal range of migration intervals is set as v_{min}=20 and v_{max}= 200. The initial value of migration intervals is set as v_0=30. The parameters γ, Δv_{max}, and T_{max} are set as 0.1, 10, and 5, respectively. All the parameter values are determined empirically and have been proven good by numerical results.
All experiments are carried out on a computer with a CPU of Pentium IV 2.8GHZ and a memory of 256 MB.

4.2 Comparison of Solution Quality

Table 3 compares the solution quality of the proposed APAA and SACO. The columns of 'Worst', 'Best', and 'AVG' represent the worst, best, and average results in 30 independent trials. 'ok%' is the rate of successful trials. The better values in the comparison are indicated in bold.

Table 3 The Comparison of Solution Quality

Func	n	SACO				APAA			
		Best	Worst	AVG	ok%	Best	Worst	AVG	ok%
F_1	20	1.04×10^{-279}	4.35×10^{-268}	1.72×10^{-269}	1	**0**	**0**	**0**	1
F_2	20	**2.27×10^{-16}**	0.0265511	8.85×10^{-4}	1	1.18×10^{-15}	**2.52×10^{-15}**	**1.70×10^{-15}**	1
F_3	20	0	0	0	1	0	0	0	1
F_4	20	-8379.658	-8379.658	-8379.658	1	-8379.658	-8379.658	-8379.658	1
F_5	20	0	0	0	1	0	0	0	1
F_6	20	4.14×10^{-15}	4.14×10^{-15}	4.14×10^{-15}	1	4.14×10^{-15}	4.14×10^{-15}	4.14×10^{-15}	1
F_1	40	1.50×10^{-149}	1.53×10^{-144}	8.99×10^{-146}	1	**8.46×10^{-191}**	**1.35×10^{-182}**	**4.67×10^{-184}**	1
F_2	40	4.62×10^{-11}	0.197	0.020	0.93	**2.26×10^{-14}**	**5.07×10^{-14}**	**3.56×10^{-14}**	**1**
F_3	40	0	0	0	1	0	0	0	1
F_5	40	0	0	0	1	0	0	0	1
F_6	40	5.13×10^{-7}	9.36×10^{-4}	4.16×10^{-5}	1	**3.23×10^{-8}**	**7.31×10^{-5}**	**7.67×10^{-6}**	1
F_1	60	5.86×10^{-151}	1.98×10^{-144}	1.53×10^{-145}	1	**3.24×10^{-192}**	**1.53×10^{-181}**	**5.14×10^{-183}**	1
F_2	60	1.74×10^{-5}	0.118	0.0216995	0.93	**2.32×10^{-14}**	**5.01×10^{-14}**	**3.39×10^{-14}**	**1**
F_3	60	0	0	0	1	0	0	0	1
F_5	60	0	0.995	0.0332	0.97	0	0.995	0.0332	0.97
F_6	60	9.08×10^{-7}	5.26×10^{-5}	1.56×10^{-5}	1	**5.13×10^{-8}**	**3.94×10^{-5}**	**6.01×10^{-6}**	1
F_1	80	2.57×10^{-149}	1.56×10^{-144}	5.83×10^{-146}	1	**2.49×10^{-189}**	**3.36×10^{-183}**	**1.52×10^{-184}**	1
F_2	80	3.54×10^{-12}	0.171	0.0235	0.97	**2.67×10^{-14}**	**6.46×10^{-10}**	**2.16×10^{-11}**	**1**

Table 3 (*continued*)

F_3	80	0	0	0	1	0	0	0	1
F_5	80	0	1.383	0.146	0.87	0	**0.995**	**0.100**	**0.9**
F_6	80	3.38×10^{-6}	1.67×10^{-4}	4.09×10^{-5}	1	**2.23×10^{-7}**	1.66×10^{-4}	**1.57×10^{-5}**	1
F_1	100	2.65×10^{-150}	1.26×10^{-142}	4.36×10^{-144}	1	**2.08×10^{-191}**	**1.66×10^{-181}**	**6.53×10^{-183}**	1
F_2	100	6.16×10^{-10}	0.0959	0.0199	1	**1.69×10^{-14}**	**3.64×10^{-11}**	**1.93×10^{-12}**	1
F_3	100	0	1	0.0667	0.93	0	**0**	**0**	**1**
F_5	100	0	**1.990**	0.417	0.6	0	3.978	**0.299**	**0.8**
F_6	100	8.56×10^{-6}	4.40×10^{-4}	8.07×10^{-5}	1	**3.69×10^{-7}**	**1.01×10^{-4}**	**1.16×10^{-5}**	1
F_7	100	-97.730	-96.912	-97.306	1	**-97.868**	**-97.256**	**-97.580**	1
F_8	100	-78.332	-78.332	-78.332	1	-78.332	-78.332	-78.332	1

It can be observed from Table 3 that the advantage of APAA extends as the number of variables increases. When n=20, APAA outperforms SACO on F_1 and F_2. When n rises to 40, 60 and 80, the advantage of APAA extends to F_{10}. When n=100, APAA obtains results better than, or at least as good as, SACO on all the functions. It can also be observed that the advantage of APAA on unimodal functions is more obvious than on multimodal functions. This is because when APAA is trapped in local optima, the two agents experienced serious stagnation, which causes over-frequent migration. Computation is wasted on these migrations until the agents find better tunable vectors.

F_1 and F_9, as the typical instances for unimodal and multimodal functions, are used to illustrate the above speculation. Fig. 5 shows the value of v_1 and the objective value with the number of migration in F_1 and F_9. As can be seen in Fig. 5 (a), the unimodal function F_1 does not have local optima and the objective value drops down without apparent stagnation. Correspondingly, the value of v_1 remains at the medium level of its legal range, which is above the initial value 30. The multimodal function F_9 has many local optima that trap APAA. Fig. 5 (b) shows that when the migration number is at about 120 to 170, the curve of the objective value is nearly flat, implying that the algorithm is experiencing serious stagnation. In this period, the value of v_1 remains at the lower bound of the legal range. Such a phenomenon confirms our speculation on the relationship between migration intervals and stagnation. Although the frequent migration caused by small migration intervals is expansive, APAA remains advantageous against SACO because of the decrement in the problem scale for each agent.

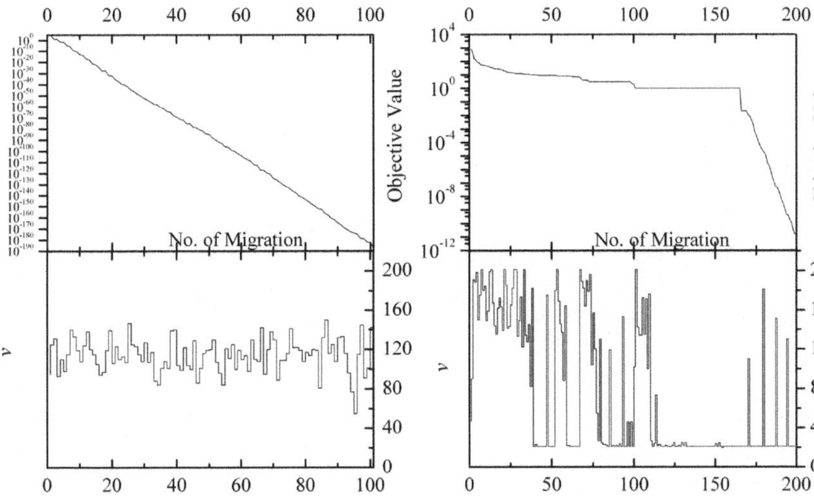

(a) F_1 of 100 dimensions (b) F_9 of 100 dimensions

Fig. 5 Illustration of the value of v and the objective value versus the number of migration

4.3 Comparison of Convergence Speed

The total of function evaluations spent by both agents to satisfy (11) measures the convergence speed, which is denoted by ErEval. Fig. 6 compares the values of ErEval between APAA and SACO on functions of different dimensions. In Fig. 6 (a)-(e), the horizontal axis is the number of test functions and the vertical axis indicates the value of ErEval. The shaded box stands for APAA while the blank box is for SACO. As can be seen, the shaded box is shorter than the blank box in most cases, indicating that APAA generally converges faster than SACO. The final graph in Fig. 6 depicts the values calculated by (12) on the z axis,

$$z=\text{ErEval}_{\text{SACO}}-\text{ErEval}_{\text{APAA}} \tag{12}$$

while the number and the dimension of test functions are indicated on x and y axes, respectively. This graph shows the trend of the difference in the convergence speed between the two algorithms when the dimension increases. In this graph, those bars higher than the surface of z=0 grows when the advantage of APAA becomes more obvious. In the opposite, the bars below the surface of $z = 0$ indicate that APAA has lower convergence speed than SACO. It can be observed that most of the bars stretch above the surface of z=0 as the dimension increases.

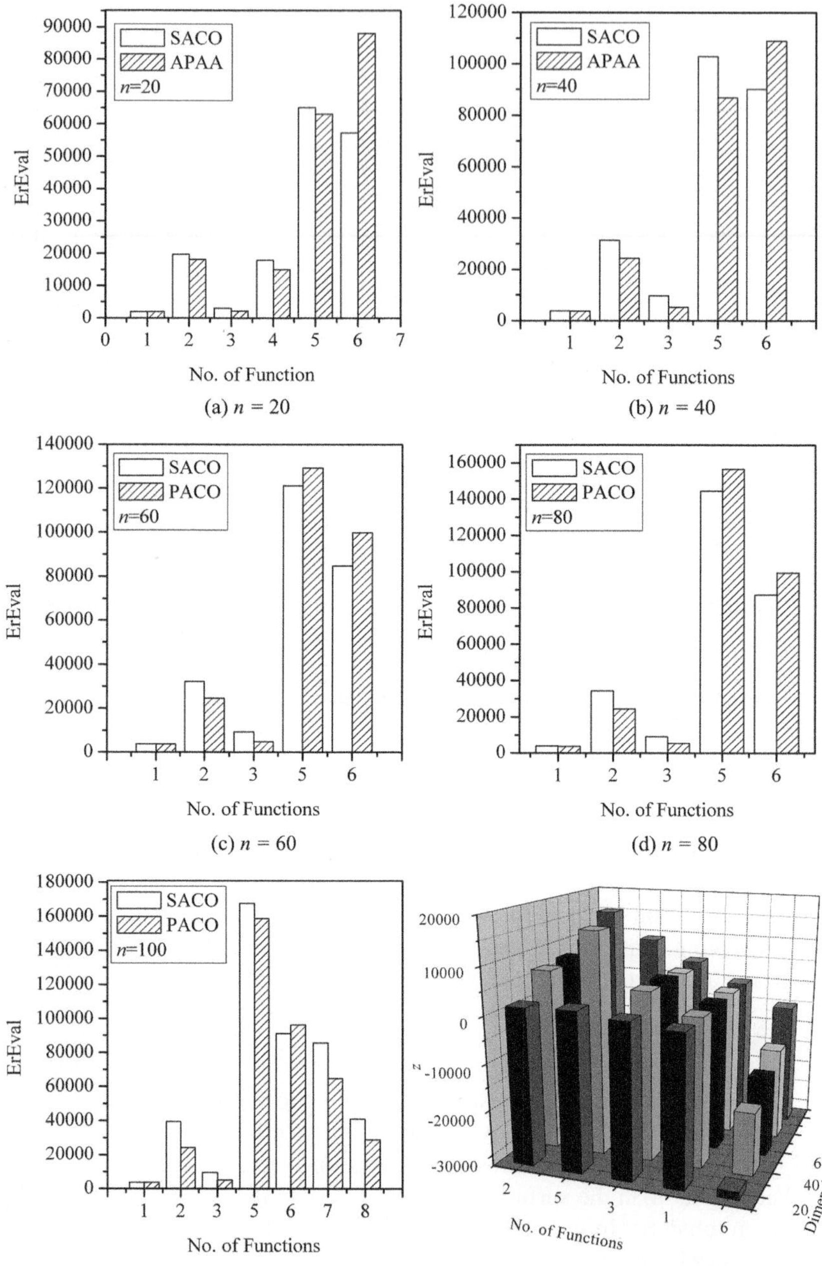

Fig. 6 The comparison of convergence speed on functions with dimensions (a) n=20, (b) n=40, (c) n=60, (d) n=80, (e) n=100, and (*f*) all

Fig. 7 shows the objective value as a function of the number of function evaluations on F_1, F_2, F_7, and F_8 of 100 dimensions. These functions are selected because the advantage of APAA is the most obvious on them. In these graphs, the dash line represents APAA while the solid line is for SACO. In Fig. 7 (a) and (b), the dash line always lies below the solid line, which means that APAA not only converges faster but also achieves better precision on the two unimodal functions F_1 and F_2. As for Fig. 7 (c) and (d), the dash line is below the solid line in the early phase. But in the end of the optimization process, the two lines overlap. Thus, APAA converges faster than SACO on these multimodal functions, yet the difference in solution quality is not significant.

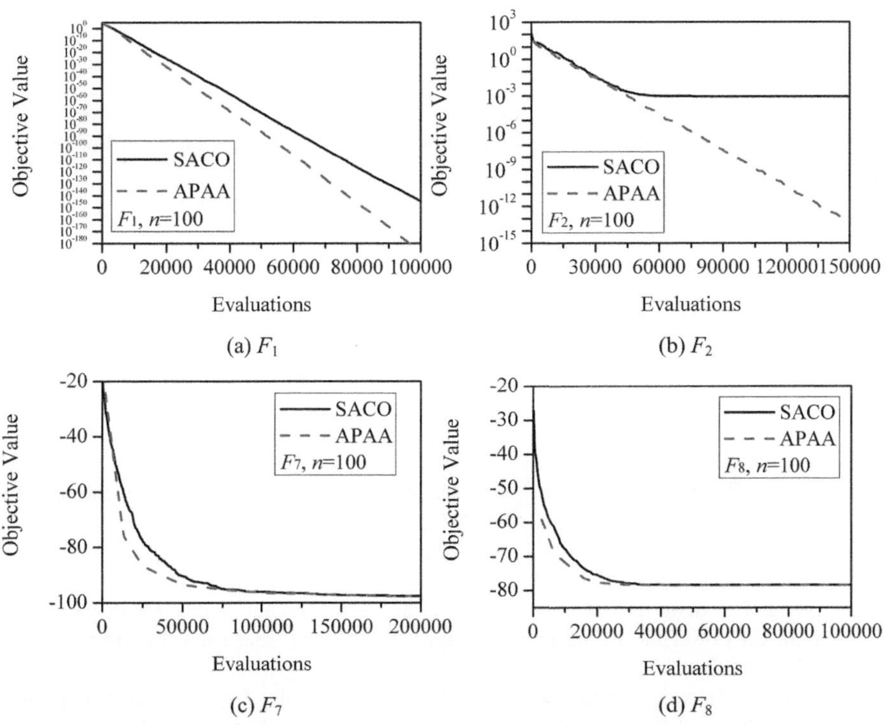

Fig. 7 The convergence lines of APAA and SACO on (a) F_1, (b) F_2, (c) F_7, and (d) F_8 of 100 dimensions

Concluded from the above, it can be said that APAA with SAMC generally outperforms SACO in solution quality and convergence speed. Moreover, the advantage of APAA becomes more obvious as the variables in the functions increase, which demonstrates the effectiveness and efficiency of the proposed parallel strategy in high dimensional space.

5 Conclusions

An agent-based parallel ant algorithm (APAA) with a stagnation-based asynchronous migration controller (SAMC) has been presented. The proposed APAA divides the problem and partitions the solution space. With different agents optimizing different solution sub-vectors synchronously, it realizes the goal of reducing the problem scale for each agent. Besides, the migration frequency in APAA is adaptively adjusted by SAMC according to the stagnation degree in the optimization process of each agent. Experiments on a set of different benchmark functions demonstrate the effectiveness and efficiency of the proposed APAA.

Acknowledgments

This work was supported by the National High Technology Research and Development Program ("863" Program) of China under Project 2009AA01Z208 and the National Science Foundation of China (NSFC) Joint Fund with Guangdong under Key Project U0835002.

References

1. Dorigo, M., Stützle, T.: Ant colony optimization. MIT Press, London (2004)
2. Chen, W.N., Zhang, J.: An ant colony optimization approach to a grid workflow scheduling problem with various QoS requirements. IEEE Trans. Syst., Man, and Cybern. Part C 39(1), 29–43 (2009)
3. Hu, X.M., Zhang, J., Xiao, J., et al.: Protein folding in hydrophobic-polar lattice model: a flexible ant-colony optimization approach. Protein & Pept. Lett. 15(5), 469–477 (2008)
4. Zhang, Y., Cai, H.C., Xiao, J., et al.: An ant colony system algorithm for the multicast routing problem. In: Proc of ICNC 2007, pp. 494–498 (2007)
5. Socha, K., Dorigo, M.: Ant colony optimization for continuous domain. Eur. J. of Operational Res. 185, 1155–1173 (2008)
6. Zhang, J., Chung, H.S.H., Lo, A.W.L., et al.: Extended ant colony optimization algorithm for power electronic circuit design. IEEE Trans. Electronic Engineering 24(1), 147–162 (2009)
7. Talbi, E., Roux, E., Fonlupt, C., et al.: Parallel ant colonies for combinatorial optimization problems. LNCS, vol. 1586, pp. 239–247. Springer, Heidelberg (1999)
8. Islam, M.T., Thulasiraman, P., Thulasiram, R.K.: A parallel ant colony optimization algorithm for all-pair routing in magnets. In: Proc. of IPDPS 2003, pp. 259–267 (2003)
9. Stützle, T.: Parallelization strategies for ant colony optimization. In: Eiben, A.E., Bäck, T., Schoenauer, M., Schwefel, H.-P. (eds.) PPSN 1998. LNCS, vol. 1498, pp. 722–731. Springer, Heidelberg (1998)
10. Randall, M., Lewis, A.: A parallel implementation of ant colony optimization. J. of Parallel and Distributed Computing 62, 1421–1432 (2002)
11. Chu, S.C., Roddick, J.F., Pan, J.S.: Ant colony system with communication strategies. Inf. Sci. 167, 63–76 (2004)

12. Michel, R., Middendorf, M.: An island model based ant system with lookahead for the shortest supersequence problem. In: Eiben, A.E., Bäck, T., Schoenauer, M., Schwefel, H.-P. (eds.) PPSN 1998. LNCS, vol. 1498, pp. 692–701. Springer, Heidelberg (1998)
13. Bullnheimer, B., Kotsis, G., Strauss, C.: Parallelization strategies for ant system. High Perform Algorithms and Softw in Nonlinear Optimization 24, 87–100 (1998)
14. Ellabib, I., Calamai, P., Basir, O.: Exchange strategies of multiple ant colony system. Inf. Sci. 177, 1248–1264 (2006)
15. Sawai, H., Adachi, S.: Effects of hierarchical migration in a parallel distributed parameter-free GA. In: Proc. of CEC 2000, vol. 2, pp. 1117–1124 (2000)
16. Lee, C.H., Park, S.H., Kim, J.H.: Topology and migration policy of fine-grained parallel evolutionary algorithms for numerical optimization. In: Proc of CEC 2000, vol. 1, pp. 70–76 (2000)
17. Zhang, J., Chung, H.S.H., Lo, W.L.: Pseduocoevolutionary genetic algorithms for power electronic circuits optimization. IEEE Trans. Syst., Man, and Cybern. Part C 36(4), 590–598 (2006)
18. Liu, J.: Autonomous agents and multi-agent system: exploration in learning self-organization, and adaptive computation. WorldScience, Singapore (2001)
19. Zhong, W.C., Liu, J., Xue, M.Z., et al.: A multiagent genetic algorithm for global numerical optimization. IEEE Trans. Syst., Man, and Cybern. Part B 34(2), 1128–1141 (2004)
20. Liu, J., Jing, H., Tang, Y.Y.: Multi-agent oriented constraint satisfaction. Artificial Intelligence 136(1), 101–144 (2002)
21. Zhang, J., Chen, W.N., Zhong, J.H., et al.: Continuous function optimization using hybrid ant colony approach with orthogonal design scheme. In: Wang, T.-D., Li, X., Chen, S.-H., Wang, X., Abbass, H.A., Iba, H., Chen, G.-L., Yao, X. (eds.) SEAL 2006. LNCS, vol. 4247, pp. 126–133. Springer, Heidelberg (2006)
22. Potter, M.A., Jong, K.A.D.: A cooperative coevolutionary approach to function optimization. In: Davidor, Y., Männer, R., Schwefel, H.-P. (eds.) PPSN 1994. LNCS, vol. 866, pp. 249–257. Springer, Heidelberg (1994)
23. Yao, X., Liu, Y., Lin, G.M.: Evolutionary programming made faster. IEEE Trans. Evolutionary Computation 3(2), 82–102 (1999)

An Attempt to Stochastic Modeling of Memetic Systems

Aleksander Byrski and Robert Schaefer

Abstract. The mathematical model of the biologically inspired, memetic, agent-based computation systems EMAS and iEMAS conformed to BDI (Belief Desire Intentions) standard is presented. The state of the systems and their dynamics are expressed as the stationary Markov chains. Such an approach allows for better understanding their complex behavior as well as their limitations. The contribution is concluded with selected experimental results obtained from the application of EMAS and iEMAS to the problem of global optimization for the popular benchmark functions and for computation-costly machine learning problems.

1 Motivation

Memetic algorithms grew up from the traditional genetic, population-based systems. Members of their populations contain and collect the additional knowledge about the solution during their life, not only the genotype, phenotype and fitness (e.g. as in SGA (Simple Genetic Algorithm) [25]). They are not the simple, passive individuals but rather the active, partially autonomous agents performing actions. Some of these action cover the classical genetic operators like mixing (genotype exchanging), or elimination of weak solutions (selection-like activities), some others tend directly to collect and improve their knowledge about solution. Basic ideas of memetic, agent-based approach may be found in [10, 19].

Memetic systems seem to have interesting features that may be perceived as an advantage over genetic algorithms, e.g. the life-time elongation of population member (agent) allows it to collect more information about the problem, than the individual can, sometimes without the multiple costly fitness computation. This knowledge may to speedup the search process and make it more effective. The synchronization

Aleksander Byrski · Robert Schaefer
Department of Computer Science, AGH University of Science and Technology,
Al. Mickiewicza 30, 30-059 Kraków, Poland
e-mail: olekb@agh.edu.pl, schaefer@agh.edu.pl

R.A. Sarker and T. Ray (Eds.): Agent-Based Evolutionary Search, ALO 5, pp. 179–202.
springerlink.com

among genetic operations performed on/by agents may be relaxed in comparison to the classic genetic strategy model. Sample results of memetic approach showing its advantage are found in [18, 15].

Another type of the hybrid approaches inspired by the nature are artificial immunological systems. They are relatively new techniques used for solving various problems e.g. intruder detection [8, 17, 21] or difficult search problems [11].

Although the memetics algorithmic have been intensively developed for about ten years, there is still lack of the rigorous mathematical formalism, that allows to establish some basic asymptotic features like guarantee of success for finding all global minimizers, proper type of convergence etc. Similar models exist for certain cases of genetic algorithms (see e.g. Vose model for SGA [25] or works of Rudolph concerning evolutionary algorithms [24, 23, 22]).

We try to formalize the simple class of memetic agent-based computational systems which hybridize evolutionary and immunological algorithms to achieve the possibility of a global optimizing problems with a complex fitness function. Immunological metaphor helps here to speed-up the selection process (in order to early remove unpromising individuals). Two particular class of systems will be considered: EMAS (Evolutionary Multi Agent Systems) and iEMAS (immunological EMAS). The idea of EMAS came from Cetnarowicz [9] and was further enhanced by Byrski and Kisiel-Dorohinicki (see e.g. [2, 3]).

The basic assumptions of EMAS ands iEMAS may be stressed as follows. Solutions are encoded by using the finite space of genotypes (e.g. binary strings of the prescribed lengths). Each agent being the population member contain single chromosome and the arbitrary chromosome may be contained only by the single agent. The agent is also associated with the unique phenotype and the fitness value. Active agents are gathered in the finite number of locations with the restricted capacities.

The knowledge about the problem solution gathered by the agent has the form of special resource called "life energy", being the real number from the range $(0,1]$. Life energy may change during the agent's life and has the crucial influence on its activities (actions). The fall of the life energy to zero eliminates the agent from the system. Each eliminated agent introduces the immunological agent (lymphocyte) that inherits its genotype. The matured lymphocytes help to eliminate the weakest agents from the system which creates the new kind of distributed selection. A new agent may be created by two parent agents. Its genotype is obtained from the parental ones by the mixing operation typical for the genetic algorithms (e.g. binary crossover and mutation). The child-agent inherits a part of life energy from its parents.

In genetic algorithms there is always a means for global timing, e.g. in simple genetic algorithm in every observable moment there is always one population of individuals (called generation). It changes completely (with help of variation operators) among these moments. This timing mechanism is immanent to these systems and can not be avoided.

In EMAS and iEMAS global timing is not necessary, state of the system is transformed by actions executed asynchronously by the agents. No mater how agents apply for executions of their actions asynchronously, the actions should be mutually excluded, because each action depends on and affects the whole state of the system.

The mathematical model developed for EMAS and iEMAS allows to:

- examining the knowledge gathered in the system by the analysis of the system state,
- examining the learning dynamics and asymptotic features (such as convergence, appropriately defined for our case).

As a consequence, the state of the system must express the knowledge about the problem collected by the system of agents. Technically it will be modeled as the incidence matrix, which assigns the life energy of each agent to the location currently occupied.

Actions are operations (random functions) that may change randomly the state of the system. The selection of the agent which will execute its action is expressed by the proper probability distribution, which may be changed according to the system's state.

However our work is in the beginning stage and reliable results of theoretical analysis (asymptotics) and experiments are still to be found, we have already precisely defined the space of system state which fulfills the above stated requirements (describes the knowledge and allows to perform asymptotic analysis). We have also proposed the precise way of describing the transition functions for actions performed on this system state space.

We would like to model the system as stochastic process with discrete time, so we introduced semi-timing mechanism (based on the existence of time checkpoints between two of them all agents operations are synchronized and approved). This model will contain many specific random distributions (being its parameters). They are not yet fully identified, however we assume that most of them should be very simple (uniform random distribution based). All possible activities of agents depend only on the system state that makes the stochastic process a markovian stationary system, what is very important for further analysis.

No matter how this modeling methodology was invented for the EMAS and iEMAS which are close in their architecture and algorithmic solutions to the classical memetic systems [20, 10], it may be easily extended to other systems of memetic type, for which the state of each agent can be unambiguously described by a single scalar parameter (as the life energy) whose value expresses its "maturity" in solving an optimization problem.

We present further development of the model described first in the paper [5] and extended in [6]. We update the system space in order to assure coherency (by removing unnecessary action of death and apoptosis and simplifying the other actions). We also introduce the constrained localizations capacities (limited number of agents).

Some important optimization tasks have already been solved by EMAS and iEMAS and yielded more effective results than certain classical approaches (e.g. optimization of neural-network architecture [4, 2, 3]). In contrast to the classical genetic approach in which the network is fully trained for each individual, the fully- and partially-trained as well as newly-introduced networks may exist at the same time in EMAS. The early removal of agents whose network would probably not yield appropriate results might significantly decrease the computation cost, what in fact is the most important advantage of the proposed iEMAS technique.

2 EMAS Definition

Agents in EMAS may be perceived as autonomous individuals. Every agent is capable of observing its environment by gathering information which it finds important, making decisions which affect its activity and performing actions which lead to changes in the overall state of the system (see e.g. [12, 13]).

We will focus on system that solves the global optimization problems which consist of finding all global minimizers $\arg\min\{\Phi(x)\}, x \in \mathscr{D}$ of the objective: $\Phi : \mathscr{D} \to \mathbb{R}_+$ where $\mathscr{D} \subset \mathbb{R}^N$ stands for the admissible, sufficiently regular set of solutions.

Every EMAS agent contains an immutable genotype, which stands for the encoded solution of the problem. Genotypes belong to the binary or real-number based finite genotype universum U. Agents are assigned to locations (analogous of "islands", see e.g. [7]) and may migrate among them. Genetic operations performed on the agent's genotypes, such as crossover and mutation, are similar to those used in classical evolutionary algorithms. Each agent is transformed asynchronously in the EMAS system. Selection mechanisms correspond to their prototype and are based on the existence of a non-renewable resource called *life energy*, which is gained and lost when agents perform actions [14].

Direct employment of different selection techniques (such as proportional or tournament-based) is impossible because of the asynchronous nature of the system and decomposition of the population. In order to enhance the possibilities of managing selection, especially to flexible change the selection pressure, the dedicated, agent-based techniques are introduced. One such technique of enhancing EMAS consists of the introduction of a new group of agents acting as lymphocytes [2]. They are responsible for recognizing (assuming some predefined affinity function) and penalizing (by decreasing the agent's energy or even making it inactive) agents with genotypes similar to the pattern possessed. Thus immunological EMAS (iEMAS) is created [3].

2.1 EMAS Structure

EMAS may be modeled as the following tuple

$$\langle U, Loc, Top, Ag, agsel, \omega, Act \rangle. \tag{1}$$

In the course of the section all symbols mentioned above will be defined.

EMAS contains a dynamic collection of agents that belong to the predefined finite set Ag. Every agent $ag_{gen} \in Ag$ contains exactly one potential solution of the given problem or its encoded representation (genotype), so there exists the bijection

$Ag \ni ag_{gen} \to gen \in U$. We restrict our considerations to the case of finite universa $\#U = r < +\infty$.

The state of a single agent is characterized by the tuple $Ag \ni ag_{gen} = \langle gen, e \rangle$, where $e \in [0,1]$ stands for the fraction of the total energy gathered by the agent.

Active EMAS agents are contained in locations described by a set of immutable integer labels $Loc = \{1,\dots,s\}$. The locations are linked by the channels along which agents may migrate from one location to another. The topology of channels is determined by the symmetric relation $Top \subset Loc^2$. We assume that the connection graph $\langle Loc, Top \rangle$ is coherent, and does not change during the system evolution.

2.2 EMAS State

Let us introduce the set of incedence and energy matrices Λ with s columns (number of all localizations) and r (number of all genotypes) rows. The columns $ince(i) \in \Lambda$ will contain energies of agents in i-th localization. In other words, if $ince(i,j) > 0$, it means that an agent denoted by gene $j \in U$ is active, its energy is $ince(i,j)$ and it is located in i-th localization.

We assume, that j-th column may contain at most $q_j \leq s$ values greater than zero, what denotes the maximum capacity of the j-th localization, and i-th row may contain at most one value greater than zero, what expresses that i-th agent may be present in only one localization at a time. Moreover $ince(i,j) \geq 0, \forall\, j = 1,\dots,r, i = 1,\dots,s$ and $\sum_{i=1}^{s}\sum_{j=1}^{r} ince(i,j) = 1$.

Gathering all these conditions, the set of incedence and energy matrices may be described in the following way:

$$\begin{aligned}\Lambda = \Bigg\{ & ince \in [0,1]^{s \cdot r} : \sum_{i=1}^{s}\sum_{j=1}^{r} ince(i,j) = 1 \\ & \text{and } \forall\, i = 1,\dots,s \sum_{j=1}^{r} [ince(i,j) > 0] \leq q_j \\ & \text{and } \forall j = 1,\dots,r \sum_{i=1}^{s} [ince(i,j) > 0] \leq 1 \Bigg\}\end{aligned} \tag{2}$$

where $[\cdot]$ denotes the value of the logical expression contained in the parentheses.

EMAS space of states consists of the set of incedence and energy matrices:

$$X = \Lambda \tag{3}$$

Depending on context, it will be convenient to describe the state of the system by both $x \in X$ or $ince \in \Lambda$. The advantage of this solution may be clearly seen in Section 3.

2.3 EMAS Behavior

Every agent starts its work in EMAS immediately after being activated. In every observable moment a certain agent gains the possibility of changing the state of the system by executing its action. There are two types of agents' actions of agents may be divided into two distinct types:

- global – they change the state of the system in two or more localizations, only one global action may be performed at a time,
- local – they change the state of the system inside one localization, only one local action for one localization may be performed at a time.

The following function is used do determine, which agent will be the next one to interact with the system

$$agsel : X \to \mathscr{M}(Ag) \tag{4}$$

where here and later $\mathscr{M}(\cdot)$ stands for the space of probabilistic measures.

After being chosen, the agent chooses one of the possible actions, then it checks whether the associated condition is true, if so, the agent performs the action. The agent suspends its work in the system after performing the action which results in its death. The cycle of an agent's life is presented in Figure 1.

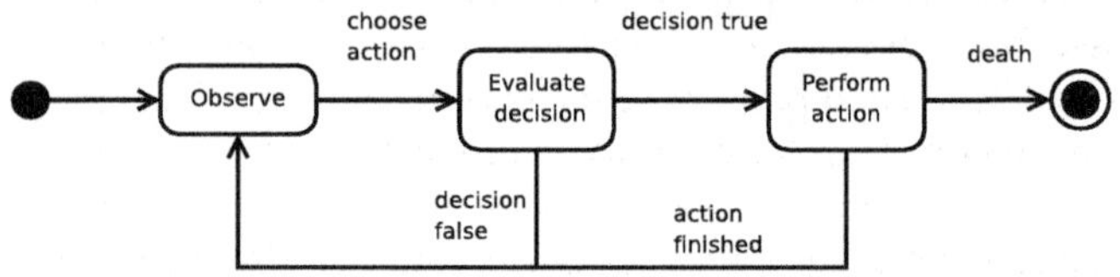

Fig. 1 Agent's state transition diagram

Every agent may perform actions contained in a predefined, finite set Act. The action, whose decision is to be evaluated by an agent, is chosen using the following function

$$\omega : U \times X \to \mathscr{M}(Act). \tag{5}$$

In the simplest case ω returns the uniform probability distribution over Act.

Every action $\alpha \in Act$ is the pair $(\delta_\alpha, \{\vartheta_\alpha^{gen}\}, gen \in U)$ where

$$\delta_\alpha : U \times X \to \mathscr{M}(\{0,1\}) \tag{6}$$

will denote the decision. The action α is performed with the probability $\delta_\alpha(gen,x)(\{1\})$ by the agent ag_{gen} in the state $x \in X$ i.e. when the decision δ_α is undertaken (δ_α is positively evaluated). Moreover

$$\vartheta_\alpha^{gen} : X \to \mathscr{M}(X) \tag{7}$$

defines the non-deterministic state transition caused by the execution of the action α by the agent ag_{gen}. The trivial state transition

$$\vartheta_{null} : X \to \mathcal{M}(X) \tag{8}$$

such that for all A being the measurable set in X and all $x \in X$

$$\vartheta_{null}(x)(A) = \begin{cases} 1 & \text{if } x \in A \\ 0 & \text{otherwise} \end{cases} \tag{9}$$

is performed with the probability $\delta_\alpha(gen,x)(\{0\})$ i.e. when the decision δ_α is not undertaken (δ_α is evaluated as zero).

2.4 EMAS Actions

Let us consider the following set of actions

$$Act = \{repr, get, migr\} \tag{10}$$

where *repr* activates an agent as the offspring agent in the system, *get* lets the better agent to take the part of the life energy from the worse agent and may make the agent with low energy inactive, *migr* denotes migration of agents between two locations.

Let us denote by l the location of the current active agent (indexed by the genotype *gen*) performing the action (i.e. $ince(l,gen) > 0$).

The decision of the energy transfer action *get* is based on the idea of agent rendezvous. Agents meet one of their neighbors (it chooses randomly one of its neighbors – agents from the same localization) and during the meeting, better agent (in the means of predefined fitness function that is problem-dependent) takes part of energy of the worse agent.

Observation 1. The value of the probability transition function for the action *get*

$$\rho_{get}^{gen} : X \to \mathcal{M}(X) \tag{11}$$

for the arbitrary state $x \in X$ and arbitrary measurable set $A \subset X$ is given by:

$$\rho_{get}^{gen}(x)(A) = \delta_{get}(gen,x)(\{0\}) \cdot \vartheta_{null}(x)(A) + \delta_{get}(gen,x)(\{1\}) \cdot \vartheta_{get}^{gen}(x)(A) \tag{12}$$

$$\delta_{get}(gen,x)(\{1\}) = \begin{cases} 1 & \text{if } NBAG_{gen} \neq \emptyset \\ 0 & \text{otherwise} \end{cases} \tag{13}$$

$$NBAG_{gen} = \{\{U \setminus \{gen\}\} \ni gen' : ince(l,gen') > 0\} \tag{14}$$

$$\vartheta_{get}^{gen}(x)(A) = \frac{1}{\#NBAG_{gen}} \left(\sum_{gen' \in BET_{gen}} \chi_A(x'(x, gen, gen')) + \sum_{gen' \in NBAG_{gen} \setminus BET_{gen}} \chi_A(x'(x, gen', gen)) \right) \tag{15}$$

$$BET_{gen} = \{NBAG_{gen} \ni gen' : FITN(gen') \geq FITN(gen)\} \tag{16}$$

$$x'(x,a,b) = ince' :$$

$$ince'(i,j) = \begin{cases} ince(i,j) - \Delta e \text{ if } j = a \text{ and } i = l \\ ince(i,j) + \Delta e \text{ if } j = b \text{ and } i = l \\ ince(i) \qquad \text{otherwise} \end{cases} \tag{17}$$

where $\chi(A)$ is the characteristic function of the set A. □

Proof. The decision of action *get*, δ_{get}^{gen} (defined by (13)) is based on condition, that there is at least one neighboring agent in the same localization, by checking the contents of the $NBAG_{gen}$ set (defined by (14)). $X \ni x = ince$ denotes the arbitrary state of the system when the decision is evaluated by the agent $ag_{gen}, gen \in U$.

This transition function uses the notion of the BET_{gen} set (defined in (16)) containing better agents than ag_{gen} being present in the same localization. The notions "better" and "worse" are based on the predefined fitness function $FITN : U \to \mathbb{R}$. The fitness function is related in some way to the objective Φ. In the simplest case $FITN(gen) = \Phi(code(gen))$, where $code : U \to \mathscr{D}$ is the encoding function.

In the case of positive evaluation of this decision, state transition described by (15) is performed. This formula comes from the Bayes theorem which allows to check all possible agents from $NBAG_{gen}$ decomposed into two non-intersecting subsets BET_{gen} and $NBAG_{gen} \setminus BET_{gen}$. For each agent contained in these sets different state transition is performed described by the function $x'(\cdot,\cdot,\cdot)$. The probability of choosing one agent from these sets is equal $\frac{1}{\#NBAG_{gen}}$.

The state transition function is constructed in the way described in (17). The incedence matrix $X \ni x' = ince'$ is obtained from $x = ince$ by changing two entries related to the pair of agents $ag_{gen}, ag_{gen'}$ that exchanged their energy. We assume, that $\Delta e = min\{ince(l,a), e_{tr}\}$ denotes the portion of energy that may be passed between agents during single *get* action (in both cases), $\Delta e > 0$ is the variable parameter that characterizes the action *get*.

The final formula (12) determining the probability transition resulted from the execution of action *get* is also obtained by using Bayes theorem.

It is to notify, that the explicite action of death *die* proposed in [6] was removed. It is not necessary anymore, because now the death occurs during performing of *get* action (when the agent has to return its all energy, it is removed from the system). □

The decision of the reproduction action *repr* is based on the idea of agent rendezvous (similar to *get*). Agent with sufficient energy meets one of their neighbors, however this time its energy must also exceed predefined threshold, and create an offspring agent based on their solutions (using predefined mixing operator). Part of parents' energy is passed to offspring agent.

Observation 2. The value of the probability transition function for the action *repr*

$$\rho^{gen}_{repr} : X \to \mathcal{M}(X) \tag{18}$$

for the arbitrary state $x \in X$ and arbitrary measurable set $A \subset X$ is given by:

$$\rho^{gen}_{repr}(x)(A) = \delta_{repr}(gen,x)(\{0\}) \cdot \vartheta_{null}(x)(A) + \delta_{repr}(gen,x)(\{1\}) \cdot \vartheta^{gen}_{repr}(x)(A) \tag{19}$$

$$\delta_{repr}(gen,x)(\{1\}) = \begin{cases} 1 & \text{if } ince(l,gen) > e_{repr} \text{ and } RPAG_{gen} \neq \emptyset \\ & \text{and } \sum_{i=1}^{r}[ince(l,i) > 0] < q_l \\ 0 & \text{otherwise} \end{cases} \tag{20}$$

$$RPAG_{gen} = \{NBAG_{gen} \ni gen' : ince(l,gen') > e_{repr}\} \tag{21}$$

$$\vartheta^{gen}_{repr}(x)(A) = \frac{1}{\#RPAG_{gen}}$$

$$\sum_{gen' \in RPAG_{gen}} \sum_{gen'' \in U} MIX(gen,gen')(\{gen''\})\chi_A(x'(x,gen,gen',gen'')) \tag{22}$$

if $\forall j : inc(c,j) = 0$

$$x'(x,a,b,c) = ince' : \tag{23}$$

$$ince'(i,j) = \begin{cases} ince(i,j) - \frac{e_0}{2} & \text{if } j \in \{a,b\} \text{ and } i = l \\ e_0 & \text{if } j = c \text{ and } i = l \\ ince(i,j) & \text{otherwise} \end{cases} \tag{24}$$

else (if $\exists k : inc(k,c) > 0$)

$$x'(x,a,b,c) = ince' : \tag{25}$$

$$ince'(i,j) = \begin{cases} ince(i,j) - \frac{e_0}{2} & \text{if } j \in \{a,b\} \text{ and } i = l \\ ince(i,j) + e_0 & \text{if } j = c \text{ and } i = k \\ ince(i,j) & \text{otherwise} \end{cases} \tag{26}$$

where $\chi(A)$ is the characteristic function of the set A. □

Proof. The decision of action *repr*, δ^{gen}_{repr} (defined by (20)) is based on condition, that there is at least one neighboring agent in the same localization, with sufficient energy to reproduce (higher than e_{repr}), what is verified by checking the contents of the set $RPAG_{gen}$ set (defined by (21)). $X \ni x = ince$ denotes the arbitrary state of the system when the decision is evaluated by the agent $ag_{gen}, gen \in U$.

In the case of positive evaluation (there exists $ag_{gen'}$ with sufficient energy) of this decision and there is enough space in the localization (the number of agents does not exceed q_l), state transition described by (22) is performed. This formula comes from the Bayes theorem which allows to check all possible agents from $RPAG_{gen}$. For each agent contained in this set different state transition is performed described by the function $x'(\cdot,\cdot,\cdot,\cdot)$. The probability of choosing one agent from these sets is equal to $\frac{1}{\#RPAG_{gen}}$.

Agent ag_{gen} that initiated the action with its neighbor $ag_{gen'}$ become parents and create an offspring agent ($ag_{gen''}$) using mixing operator $MIX : U \times U \to \mathscr{M}(U)$, which denotes the family of probability distributions associated with genetic mixing (crossover followed by mutation, see e.g. [25]). In particular $MIX(gen, gen')(\{gen''\})$ denotes the probability that gen'' is born from the parents gen and gen'.

The state transition function is constructed in the way described in (24) and (26). The incedence matrix $X \ni x = ince'$ is obtained from $x = ince$ by changing entries related to agents $ag_{gen}, ag_{gen'}, ag_{gen''}$. The part of parents' energy is passed to the offspring agent. When the offspring agent is active, its energy is raised by e_0, when it is inactive – it is activated, placed in the same localization as its parents (24) and its energy is set to e_0.

The final formula (19) determining the probability transition resulted from the execution of action $repr$ is also obtained by using Bayes theorem. □

The decision of the migration action $migr$ may be undertaken by an agent with enough energy to migrate and if there exists a localization that is able to accept it (its number of agents does not exceed maximum). When these conditions are met agent is moved from its localization to another one.

Observation 3. The value of the probability transition function for the action $migr$

$$\rho^{gen}_{migr} : X \to \mathscr{M}(X) \tag{27}$$

for the arbitrary state $x \in X$ and arbitrary measurable set $A \subset X$ is given by:

$$\rho^{gen}_{migr}(x)(A) = \delta_{migr}(gen,x)(\{0\}) \cdot \vartheta_{null}(x)(A) + \delta_{migr}(gen,x)(\{1\}) \cdot \vartheta^{gen}_{migr}(x)(A) \tag{28}$$

$$\delta_{migr}(gen,x)(\{1\}) = \begin{cases} 1 & \text{if } ince(l,gen) > e_{migr} \\ 0 & \text{otherwise} \end{cases} \tag{29}$$

$$ACCLOC_l = \{Loc \setminus \{l\} \ni l' : ((l,l') \in Top) \text{ and } (\sum_{i=1}^{r} [ince(l',i) > 0] < q_{l'})\} \tag{30}$$

$$\vartheta^{gen}_{migr}(x)(A) = \frac{1}{\#ACCLOC_l} \sum_{loc' \in ACCLOC_l} \chi_A(x'(x, gen, loc')) \tag{31}$$

$$x'(x,a,k) = ince' : ince'(i,j) = \begin{cases} 0 & \text{if } j = a \text{ and } i = l \\ ince(l,a) & \text{if } j = a \text{ and } i = k \\ ince(i,j) & \text{otherwise} \end{cases} \tag{32}$$

where $\chi(A)$ is the characteristic function of the set A. □

Proof. The decision of action *migr*, δ^{gen}_{migr} (defined by (29)) is based on condition, that the agent ag_{gen} has sufficient energy to migrate (higher than $e_{migr} \geq e_{repr}$) and there is at least one neighboring localization (l'), that is able to accept the agent which wants to migrate (number of agents in this localization does not exceed $q_{l'}$), what is stated by checking the contents of the set $ACCLOC_l$ set (defined by (30)). $X \ni x = ince$ denotes the arbitrary state of the system when the decision is evaluated by the agent $ag_{gen}, gen \in U$.

In the case of positive evaluation of this decision, state transition described by (31) is performed. The formula comes from the Bayes theorem which allows to check all possible localizations from $ACCLOC_l$. For each localization contained in this set, different state transition is performed described by the function $x'(\cdot,\cdot,\cdot)$. The probability of choosing one agent from these sets is equal to $\frac{1}{\#ACCLOC_l}$. Agent that initiated the action moves from its localization (l) to another one (l') that was uniformly chosen from the set $ACCLOC_l$.

The state transition function is constructed in the way described in (32). The incedence matrix $X \ni x' = ince'$ is obtained from $x = ince$ by changing two entries related to the position of agent ag_{gen} in the localization.

The final formula (28) determining the probability transition resulted from the execution of action *migr* is also obtained by using Bayes theorem. □

Observation 4. Probability transitions imposed by actions $get, repr, migr$ are markovian ones [1], because the probability transitions functions $\rho^{gen}_{get}, \rho^{gen}_{repr}, \rho^{gen}_{migr}$ given by the formulas (12)–(32) depend only on the current state $x \in X$ of the system. □

2.5 *EMAS Dynamics*

We intend to present the EMAS as the stationary Markov chain with the set of states X. First, we assume that all actions performed by agents are mutually exclusive in this way, that in every observable step, only one agents is chosen (using *agsel* function) and allowed to interact with the system (change its state). This approach to model agent systems may seem inadequate, because autonomy of agents has usually effect in their asynchronous behavior.

Observation 5. The function of the stochastic state transition of EMAS is given by:

$$\tau(x)(A) = \sum_{gen \in U} agsel(x)(\{gen\})\rho_{gen}(x)(A) \tag{33}$$

$$\rho_{gen}(x)(A) = \sum_{\alpha \in Act} \omega(gen,x)(\{\alpha\})\rho_{\alpha}^{gen}(x)(A) \tag{34}$$

□

Proof. Let $\{\rho_{gen}\}_{gen \in U}$ be the set of transition function which are related to the activity of the agent ag_{gen}, then the probability of passage from the state x to the state belonging the measurable subset $A \subset X$ caused by all actions invoked by the agent ag_{gen} is given by (34). This probability function depends on the function giving the probability of choosing the action α by the agent ag_{gen} and previously defined function ρ_{α}^{gen}.

Finally, the transition function for a single EMAS step is given by (33). This probability function depends on the function *agsel* that gives the probability of selecting ag_{gen} to perform the action and previously defined function ρ_{gen}.

Both formulas (34) and (33) are the simple consequence of applying the Bayes theorem. □

Observation 6. The stochastic state transition of EMAS given by formulas (33) and (34) is markovian one. □

Proof. All transition function of actions ρ_{α}^{gen} appearing in formula (34) are markovian (see Observation 4). Because the probability distribution ω depends only on the current state x for each *gen*, the transition function ρ_{gen} defined by (34) is also markovian. Moreover the probability distribution *agsel* depends on the current state only for each *gen*, that motivates the markovian features of the transition function τ given by (33). □

3 iEMAS Extension

Below we introduce the components of iEMAS formal model being extensions of the corresponding elements of EMAS model presented in sections 2.1—2.5.

3.1 iEMAS Structure

We will introduce the additional type of agents called "lymphocytes". Similarly to EMAS, iEMAS contains a dynamic collections of agents that belong to the predefined finite set Ag and dynamic collection of lymphocytes that belong to the finite set Tc. As in case of agents, lymphocytes are also unambiguously indexed by the genotypes from U, so that $\#Ag = \#Tc = \#U = \mathrm{r}$.

The lymphocytes have the same structure as the agents previously defined, however their actions differ (because their goals differ from agents' goals) and their total energy does not have to be constant.

The iEMAS may be modeled as the following tuple

$$\langle U, Loc, Top, Ag, Tc, typesel, agsel, tcsel, \omega, \varphi, Act, Tcact \rangle \tag{35}$$

where $U, Loc, Top, Ag, agsel, \omega, Act$ are as previously and remaining components will be defined in the course of the section.

The lymphocyte $Tc \ni tc_{gen} = \langle gen, p(gen) \rangle$ is composed of the genotype $gen \in U$ that plays the role of pattern that is used to detect similar genotypes by the means of predefined matching function $MATCH(\cdot)$ and its own energy $p(gen) \in [0, p_{max}]$.

3.2 iEMAS State

In addition to the EMAS state describing the localization and energy of the agents (see (3)) we need to consider the set of matrices containing similar information for lymphocytes. Yet there is no need to assure the constant total energy for lymphocytes. We describe this additional set of lymphocyte incedence and energy matrices in the following way:

$$\Gamma = \left\{ tcince \in [0, p_{max}]^{r \cdot s} : \forall\, j = 1, \ldots, s \sum_{i=1}^{r} [tcince(j,i) > 0] \leq lq_j \right.$$
$$\left. \text{and} \ \forall\, i = 1, \ldots, r \sum_{j=1}^{s} [tcince(j,i) > 0] \leq 1 \right\} \tag{36}$$

The space of iEMAS state is given by:

$$X = \Lambda \times \Gamma \tag{37}$$

3.3 iEMAS Behavior

We introduce two disjoint action sets, first one is designed for agents

$$Act = \{repr, get, migr\} \tag{38}$$

Outcome of actions *repr*, *migr* belonging to *Act* set is similar like in EMAS (see Section 2.4). In the case of iEMAS the domain of the decision and transition functions are formally extended (see (37)) because additional incidence and energy matrix for lymphocytes *tcince* is added. The decisions and the transition functions of these actions do not depend on and do not change the new state components in full extent. The effects of their operations are visible only on the part of the system state identical to EMAS. In particular, the changes of the state described before showing current state $x = ince$ and next state after the action $x' = ince'$, now have the form $x = (ince, tcince)$, $x' = (ince', tcince)$ respectively.

We also extend the trivial state transition ϑ_{null} in the same way.

The single action *get* inherited from EMAS is modified. When the agent performs this action, and its energy (or energy of evaluated agent) reaches zero, it activates the lymphocyte containing the genotype of inactivated agent.

The set of action designed for lymphocytes is

$$Tcact = \{give, kill\} \tag{39}$$

where *give* is the action of the lymphocyte used to decrease its own energy (performed during every activation), which in the end causes the lymphocyte to be deactivated (when its energy reaches zero) and *kill* is the action of removing the computational agent by a lymphocyte (performed when the genotype of the tested agent matches the pattern contained in the lymphocyte).

We introduce the function *typesel* choosing which type of agents will have possibility of performing the action:

$$typesel : X \to \mathcal{M}\{0,1\} \tag{40}$$

when 0 is chosen, one of agents will be activated, when 1 – lymphocyte.

The function choosing which agent will be activated *agsel* is similar like in EMAS (see (4)), but it now depends in some way on the extended state from X defined by (37). Now we introduce new function that will choose which lymphocyte will be activated:

$$tcsel : X \to \mathcal{M}(Tc) \tag{41}$$

Similarly as in case of *agsel* the probability $tcsel(x)(\{gen\})$ vanishes when the lymphocyte tc_{gen} is inactive in the state $x \in X$.

The function ω choosing the action for the active agent remains intact, though its domain changes (because of the new state definition, see (37).

The function choosing the action for the active lymphocyte has the following shape:

$$\varphi : X \to \mathcal{M}(Tcact) \tag{42}$$

3.4 iEMAS Actions

Let us describe by l (like in EMAS) the location of the active agent or lymphocyte (indexed by the genotype *gen*) performing the action (i.e. $ince(l,gen) > 0$ or $tcince(l,gen) > 0$).

Action *get* has to be redefined, because now it may be used to activate one of lymphocytes similar to the inactivated agent.

Observation 7. The value of the probability transition function for the action *get*

$$\eta_{get}^{gen} : X \to \mathcal{M}(X) \tag{43}$$

for the arbitrary state $x \in X$ and arbitrary measurable set $A \subset X$ is given by:

$$\eta_{get}^{gen}(x)(A) = \gamma_{get}^{gen}(x)(\{0\}) \cdot \vartheta_{null}(x)(A) + \gamma_{get}^{gen}(x)(\{1\}) \cdot \kappa_{get}^{gen}(x)(A) \tag{44}$$

$$\gamma_{get}(gen,x)(\{1\}) = \begin{cases} 1 & \text{if } NBAG_{gen} \neq \emptyset \text{ and } \sum_{i=1}^{r}[tcince(l,i) > 0] < lq_l \\ 0 & \text{otherwise} \end{cases} \quad (45)$$

$$\begin{aligned} \kappa_{get}^{gen}(x)(A) = \frac{1}{\#NBAG_{gen}} \Bigg(& \sum_{gen' \in BET_{gen}} ([ince(l,gen) \leq \Delta e]\chi_A(x'(x,gen,gen')) \\ & +[ince(l,gen) > \Delta e]\chi_A(x''(x,gen,gen'))) \\ + & \sum_{gen' \in NBAG_{gen} \setminus BET_{gen}} ([ince(l,gen') \leq \Delta e]\chi_A(x'(x,gen',gen)) \\ & +[ince(l,gen') > \Delta e]\chi_A(x''(x,gen',gen))) \Bigg) \end{aligned} \quad (46)$$

$$\begin{aligned} & x'(x,a,b) = (ince', tcince') : \\ & ince'(i,j) = \begin{cases} 0 & \text{if } j = a \text{ and } i = l \\ ince(l,j) + ince(l,b) & \text{if } j = b \text{ and } i = l \\ ince(i,j) & \text{otherwise} \end{cases} \\ & tcince'(i,j) = \begin{cases} p_{max} & \text{if } j = a \text{ and } i = l \\ tcince(i,j) & \text{otherwise} \end{cases} \end{aligned} \quad (47)$$

$$\begin{aligned} & x''(x,a,b) = (ince'', tcince) : \\ & ince'(i,j) = \begin{cases} ince(l,j) - \Delta e & \text{if } j = a \text{ and } i = l \\ ince(l,j) + \Delta e & \text{if } j = b \text{ and } i = l \\ ince(i,j) & \text{otherwise} \end{cases} \end{aligned} \quad (48)$$

where $\chi(A)$ is the characteristic function of the set A. □

Proof. The decision of action *get*, γ_{get}^{gen} (defined by (45)) is based on similar condition like corresponding *get* action in EMAS (see 2.4), i.e. there must be at least one agent in the localization, however this condition is extended by checking whether there exists a possibility of creating lymphocyte in the localization (number of lymphocytes must not exceed lq_l). $X \ni x = (ince, tcince)$ denotes the arbitrary state of the system when the decision is evaluated by the agent $ag_{gen}, gen \in U$.

In the case of positive evaluation of this decision, state transition described by (46) is performed. During the rendezvous, when the energy of "worse" agent (in the means of fitness function $FITN(\cdot)$) is lower or equal to Δe, it is inactivated and new lymphocyte is created. Otherwise, when the energy of "worse" agent is greater than Δe, its energy is simply decreased and no lymphocyte is created.

The state transition functions are constructed in the way described in (47) and (48). The incedence matrices $X \ni x' = (ince', tcince')$ are obtained from $x = (ince, tcince)$ by changing entries related to the energy of evaluating agents (ag_{gen},

$ag_{gen'}$) in such way that the energy of better one is raised, while the energy of worse one is set to zero. At the same moment the lymphocyte containing the genotype of the worse agent is created (its energy is set to maximum value p_{max}) and placed in the same localization.

In the second case, the incedence matrices $X \ni x'' = (ince'', tcince)$ are obtained from $x = (ince, tcince)$ by changing only entries related to the energies of the evaluating agents. They exchange the energy while the incedence matrix for lymphocytes remains intact.

The final formula (44) determining the probability transition resulted from the execution of action *get* is also obtained by using Bayes theorem. □

The action of decreasing the energy by the lymphocyte *give* is performed always when is chosen by the lymphocyte (the evaluation of the decision is always true). Finally, when the energy of the lymphocyte reaches zero, it is deactivated.

Observation 8. The value of the probability transition function for the action *give*

$$\eta^{gen}_{give} : X \to \mathcal{M}(X) \tag{49}$$

for the arbitrary state $x \in X$ and arbitrary measurable set $A \subset X$ is given by:

$$\eta^{gen}_{give}(x)(A) = \gamma^{gen}_{give}(x)(\{0\}) \cdot \vartheta_{null}(x)(A) + \gamma^{gen}_{give}(x)(\{1\}) \cdot \kappa^{gen}_{give}(x)(A) \tag{50}$$

$$\forall\ gen \in U \quad \gamma_{give}(gen, x)(\{1\}) = 1 \tag{51}$$

$$\kappa^{gen}_{give}(x)(A) = \begin{cases} 1 & \text{if } x'(x, gen) \in A \\ 0 & \text{otherwise} \end{cases} \tag{52}$$

$$x'(x,a) = (ince, tcince') :$$
$$tcince'(i,j) = \begin{cases} tcince(i,j) - \Delta p & \text{if } j = a \text{ and } i = l \\ tcince(i,j) & \text{otherwise} \end{cases} \tag{53}$$

□

Proof. The decision of action *give*, γ^{gen}_{give} defined by (51) returns always true, so the lymphocyte performs the action *give* always after choosing. $X \ni x = (ince, tcince)$ denotes the arbitrary state of the system when the decision is evaluated by the agent $ag_{gen}, gen \in U$.

The state transition described by (52) is performed. For each agent different state transition is performed described by the function $x'(\cdot,\cdot)$. The state transition function is constructed in the way described by (53). The incedence matrices $X \ni x' = (ince, tcince')$ are obtained from $x = (ince, tcince)$ by changing entry related to the energy of the lymphocyte tc_{gen}, it is simply lowered by the value of $\Delta p = min\{p(gen), p_{tr}\}$, where $p_{tr} > 0$ is the universal parameter that characterizes the action *give*. When the value of the lymphocyte's energy reaches zero, it remains inactive.

The final formula (50) determining the probability transition resulted from the execution of action *give* is also obtained by using Bayes theorem.

It is to notify, that the explicite action of apoptosis *apo* proposed in [6] was removed. It is not necessary anymore, because now the death occurs during performing of *give* action (when the agent has to return its all energy, it is removed from the system). □

The action *kill* is performed by the lymphocyte in order to remove probably non-promising agent from the system, when high affinity occurs between their genotypes. However, the notion of maturity of lymphocyte plays important role here. When the lymphocyte is immature, and it recognizes an agent that has relatively high energy, lymphocyte is removed from the system. When the lymphocyte is mature, its decision is authoritative, the agent is removed from the system and its energy is passed to one of its neighbors.

Observation 9. The value of the probability transition function for the action *kill*

$$\eta_{kill}^{gen} : X \rightarrow \mathcal{M}(X) \tag{54}$$

for the arbitrary state $x \in X$ and arbitrary measurable set $A \subset X$ is given by:

$$\eta_{kill}^{gen}(x)(A) = \gamma_{kill}^{gen}(x)(\{0\}) \cdot \vartheta_{null}(x)(A) + \gamma_{kill}^{gen}(x)(\{1\}) \cdot \kappa_{kill}^{gen}(x)(A) \tag{55}$$

$$\gamma_{kill}(gen, x)(\{1\}) = \begin{cases} 1 & \text{if } TNBAG_{gen} \neq \emptyset \\ 0 & \text{otherwise} \end{cases} \tag{56}$$

$$TNBAG_{gen} = \{U \ni gen' : ince(gen', l) > 0\} \tag{57}$$

$$\begin{aligned} \kappa_{kill}^{gen}(x)(A) = \frac{1}{\#SIM_{gen}} \sum_{gen' \in SIM_{gen}} \Bigg(& \frac{[tcince(l, gen) \leq p_{neg}]}{\#TNBAG_{gen}} \\ & \sum_{gen'' \in TNBAG_{gem}} \chi_A(x'(x, gen, gen', gen'')) \\ & + [(tcince(l, gen) > p_{neg}) \text{ and } (ince(l, gen') > e_{good})] \\ & \chi_A(x''(x, gen, gen')) \\ & + [(tcince(l, gen) > p_{neg}) \text{ and } (ince(l, gen') \leq e_{good})] \, \chi_A(x) \Bigg) \end{aligned} \tag{58}$$

$$SIM_{gen} = \{TNBAG_{gen} \ni gen' : MATCH(gen, gen') = 1\} \tag{59}$$

$$x'(x,a,b,c) = (ince', tcince):$$
$$ince'(i,j) = \begin{cases} ince(i,j) + ince(i,a) & \text{if } i = l \text{ and } j = c \\ 0 & \text{if } i = b \\ ince(i,j) & \text{otherwise} \end{cases} \tag{60}$$

$$x''(x,a) = (ince, tcince''):$$
$$tcince''(i,j) = \begin{cases} 0 & \text{if } i = l \text{ and } j = a \\ tcince(i,j) & \text{otherwise} \end{cases} \tag{61}$$

where $\chi(A)$ is the characteristic function of the set A. □

Proof. The decision of action *kill*, γ_{kill}^{gen} defined by (56) is based on checking (using $TNBAG_{gen}$ set defined in (57)) if there are any neighboring agents in the localization of the lymphocyte. $X \ni x = (ince, tcince)$ denotes the arbitrary state of the system when the decision is evaluated by the agent $ag_{gen}, gen \in U$.

If the decision is positively evaluated, the state transition described by (58) is performed. This formula comes from the Bayes theorem which allows to check all possible agents from the set SIM_{gen}. For each agent contained in this set different state transition is performed described by the functions $x'(\cdot,\cdot)$ and $x''(\cdot,\cdot,\cdot,\cdot)$ described below.

The lymphocyte chooses uniformly the agent from the SIM_{gen} set defined in (59). Depending on the energies of both lymphocyte and agent, following types of changes in the system state may occur:

- If the lymphocyte is mature (has finished its negative selection process) then it removes the desired agent – in other words, its energy is lower than p_{neg} $(0 < p(gen) < p_{neg} < p_{max})$.
- If the lymphocyte is immature (has not finished yet its negative selection process) and the agent recognized as similar is apparently good (its energy is higher than e_{good}) then the lymphocyte is removed.
- When the lymphocyte is immature and it recognizes the agent which is not good (its energy is lower than e_{good} then no changes in state are performed.

In order to determine whether the genotypes of agent and lymphocyte are similar, following surjective function is used: $MATCH : U \times U \to \{0,1\}$. This mapping is partial, what allows to removing not only already removed agents, but also similar ones.

The state transition function is constructed in the way described by (60) and (61). In the first case the incedence matrices $X \ni x' = (ince', tcince)$ are obtained from $x = (ince, tcince)$ by changing the entry related to the energy of the agent recognized by the mature lymphocyte. The agent is inactivated and its energy is passed to one of its neighbors (uniformly chosen). In the second case, $X \ni x'' = (ince, tcince'')$ is obtained from $x = (ince, tcince)$ by changing to zero entry related to the energy of the immature lymphocyte which becomes inactive.

The final formula (55) determining the probability transition resulted from the execution of action *kill* is also obtained by using Bayes theorem. □

Observation 10. Probability transitions imposed by actions *give*, *kill* are markovian ones [1], because the probability transitions functions η^{gen}_{give}, η^{gen}_{kill} given by the formulas (50)–(61) depend only on the current state $x \in X$ of the system. □

3.5 iEMAS Dynamics

We extend the model of EMAS dynamics in order to model the behavior of iEMAS.

Observation 11. The function of the stochastic state transition of iEMAS is given by:

$$\tau(x)(A) = typesel(x)(\{0\}) \sum_{gen \in U} agsel(x)(\{gen\})\rho_{gen}(x)(A) + \\ typesel(x)(\{1\}) \sum_{gen \in U} tcsel(x)(\{gen\})\eta_{gen}(x)(A) \tag{62}$$

$$\eta_{gen}(x)(A) = \sum_{\alpha \in Tcact} \varphi(gen, x)(\{\alpha\})\eta^{gen}_{\alpha}(x)(A) \tag{63}$$

□

Proof. In order to describe the stochastic character of iEMAS dynamics, we propose the state transition function described by (62). In this formula ρ_{gen} is given by the formula identical to (34), except the function of action selection $\omega(gen, x)$ depending now on the whole state x of iEMAS. The probability of the system state transfer depends on the outcome of the function *typesel* which selects agent or lymphocyte to perform the action. Then agent is selected using *agsel* or lymphocyte – using *tcsel*.

When the agent is chosen, ρ_{gen} function is used to compute the state transfer probability (see 2.5). Otherwise, when the lymphocyte is chosen, the probability of state transfer is computed by the stochastic state transition function defined in (63). This function depends on the φ which is used to compute the probability of choosing certain action by the lymphocyte, and on η^{gen}_{α} described below. □

Observation 12. The stochastic state transition of iEMAS given by formulas (62) and (63) is markovian one. □

Proof. All transition function of actions η^{gen}_{α} appearing in formula (63) are markovian (see Observation 10). Because the probability distribution φ depends only on the current state x for each *gen*, the transition function η_{gen} defined by (63) is also markovian. Next, the expression $\sum_{gen \in U} tcsel(x)(\{gen\})\eta_{gen}(x)(A)$ being the part of (62) defines also the markovian transition function, because the probability distribution *tcsel* depends only on the current state only for each *gen*. The expression

$\sum_{gen \in U} agsel(x)(\{gen\})\rho_{gen}(x)(A)$ is also markovian (see Observation 6). Finally, the transition function τ is markovian, because the probability distribution *typesel* depends only on the current state of the system. □

4 Experimental Results

The system was implemented using the AgE platform[1].

Although the main goal of this contribution is the presentation of EMAS/iEMAS formal model, we would like to include several interesting experimental results regarding examination of diversity and fitness of the algorithms EMAS, iEMAS and PEA (Parallel Evolutionary Algorithm being a classical simple genetic algorithm with introduced allopatric speciation [16, 7]) applied to the problem of evolution of classifying neural network architecture. The fitness presented in this section is simply the percent of accurate classifications for the examined data-sets, and the diversity is computed as average standard deviation counted for all genes (for all genotypes) in the population.

We have already performed the efficiency of EMAS, iEMAS and a certain PEA, and the results were presented e.g. in [5, 6]. These results allow to state, that our agent-based approaches are more efficient in solving examined problems, than PEA examined.

Their advantages are especially visible when the fitness evaluation is a costly, long term iteration process. In this section we present selected results concerning such a problem: the optimization of the architecture of classifying neural networks (multi-layered perceptrons).

The genotype of the agents consist of the parameters of the neural network such as the number of neurons in layers, the values of learning coefficients and momentum for the layer. During the creation of agent its genotype is decoded and neural network is created, then in the course of its life, agent performs the training of the network. The fitness function for the agent is based on the classification percent of its network.

The probability distributions $agsel, \omega, typesel, tcsel, \varphi$ used for selection of the agents and agent's actions were uniform.

The classified data-sets were taken from the well-known UCI Machine Learning Repository (Iris and Wine sets were used).

In Fig. 2 the fitness and diversity obtained for EMAS and PEA are shown. Both systems consisted of 3 demes including 15 individuals. Discrete crossover and uniform mutation variation operators were used. Tournament selection was used in PEA, being the most similar classical selection to the resource-based selection used in EMAS.

Comparing the results presented in Fig. 2, one may notify, that both EMAS and PEA produce similar fitness (see Fig. 2(a), 2(c)), though the standard deviation of

[1] AgE `http://age.iisg.agh.edu.pl` is agent-based, component-oriented computation platform, developed as open-source project at Intelligent Information Systems Group of AGH University of Science and Technology.

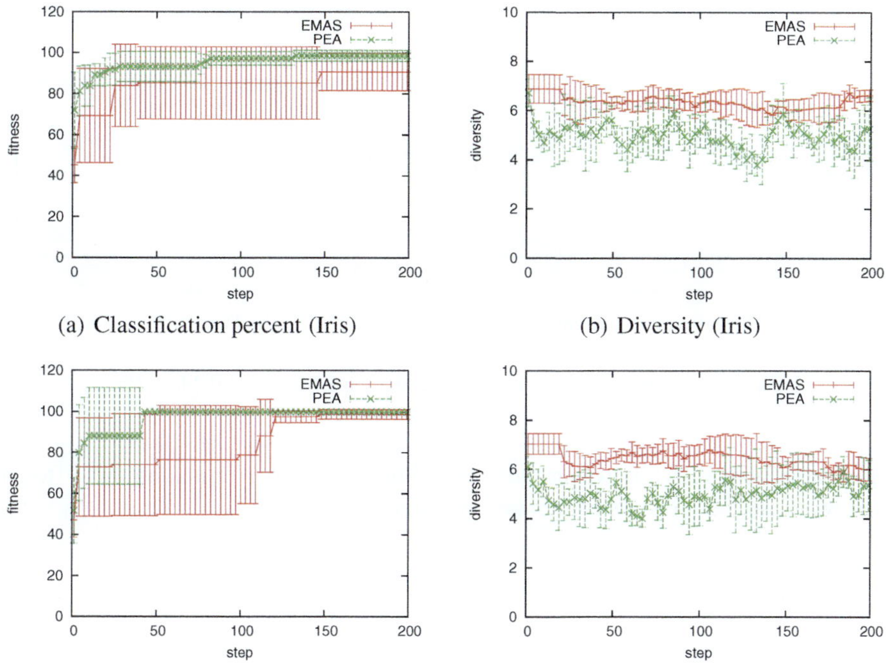

(a) Classification percent (Iris)

(b) Diversity (Iris)

(c) Classification percent (Wine)

(d) Diversity (Wine)

Fig. 2 Classification percent and diversity for EMAS and PEA (Iris and Wine problems)

the results for EMAS is greater than for PEA. This is caused by the higher diversity computed for EMAS in comparison to PEA (see Fig. 2(b), 2(d)), which is a higly desired result.

It is to notify that the average runtime of the system was about 10 minutes for EMAS, and about 3 hours for PEA. This is an effect of "intelligent" evolution which is performed in agent-based computational systems. Evolution in classical evolutionary systems is more random and these algorithms usually do not take full advantage of the information that may be discovered besides the genotypes and fitness values.

In Fig. 3 the fitness and diversity computed for EMAS and iEMAS were presented. Both systems consisted of 8 demes including 20 individuals. Discrete crossover and uniform mutation variation operators were used. Tournament selection was used in PEA, being the most similar classical selection to the resource-based selection used in EMAS.

As in the results described above, the fitness seems similar for EMAS and iEMAS, moreover, the standard deviation and the diversity computed for iEMAS is higher. The average runtime of EMAS was about 4 minutes, while for iEMAS was about 2 minutes. Regarding the configuration of the computation environment this time should be longer than the time of computation presented above (for EMAS

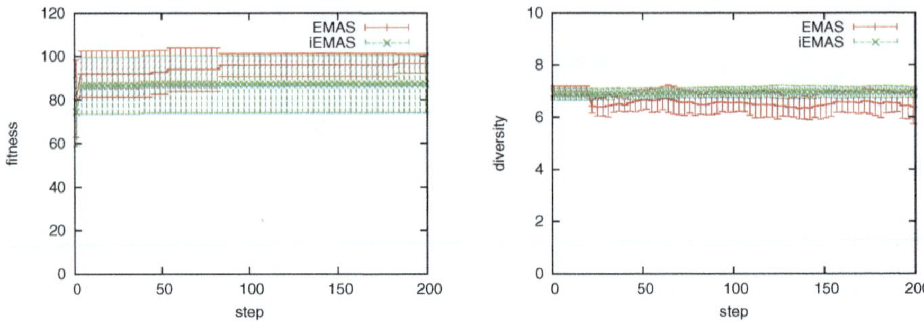

(a) Classification percent (b) Diversity

Fig. 3 Classification percent and diversity for EMAS and iEMAS (Iris)

compared with PEA), however these experiments were performed on the different hardware, so these results cannot be compared.

5 Conclusions

Memetic systems seem to have interesting features that may be perceived as an advantage over genetic algorithms, e.g. the life-time elongation of population member (agent) allows it to collect more information about the problem, than the individual can, sometimes without the multiple costly fitness computation. This knowledge may to speedup the search process and make it more effective. The synchronization among genetic operations performed on/by agents may be relaxed in comparison to the classic genetic strategy model.

EMAS and iEMAS are examples of systems very closely related to memetic ones. The main advantages of these systems over traditional genetic systems include higher computation effectivity shown in particular for the problems with costly fitness (see [3, 5]), which evaluation is performed as a long-lasting iterative process increasing the evaluation accuracy.

iEMAS metaphor may be used to further increase the efficiency (in the comparison with EMAS) by using special agents acting as lymphocytes that can early recognize and make the unpromising agents inactive on the basis of its rough fitness value, sparing a significant part of CPU time.

In the course of contribution the architecture of EMAS and iEMAS systems were described according to the BDI standard (see e.g. [13]). Next the proper space of states and the probability transition functions which allow to model EMAS and iEMAS stochastic dynamics were introduced. The resulted stochastic processes being the stationary Marcov chains allow to better understand the behavior of the complex systems under consideration as well as to study their constraints.

No matter how the general form of the Marcovian state transition functions were identified, we intend to study the asymptotic behavior with respect to the probability distributions of mixing operator MIX, agent selection functions $agsel$, ω, $typesel$, *tcsel*, φ and energy thresholds of agents and lymphocytes. One of the most challenging task will be to investigate, if there exists certain configurations of agent's energy parameters, for which every possible energetic state will be reachable after finite sequence of steps. This would prove the global search capabilities of the proposed systems as a consequence of ergodicity of markovian chain that is used for modeling.

No mater how agents apply for executions of their actions asynchronously, the actions should be mutually excluded, because all actions work on the whole state of the system, what is the drawback of the presented formal model. However we plan to extend our model in order to allow explicit parallel execution of certain subsets of actions in the near future.

References

1. Billingsley, P.: Probability and Measure. John Wiley and Sons, Chichester (1987)
2. Byrski, A., Kisiel-Dorohinicki, M.: Immunological selection mechanism in agent-based evolutionary computation. In: Klopotek, M.A., Wierzchon, S.T., Trojanowski, K. (eds.) Intelligent Information Processing and Web Mining: proceedings of the international IIS: IIPWM 2005 conference: Gdansk, Poland, Advances in Soft Computing, pp. 411–415. Springer, Heidelberg (2005)
3. Byrski, A., Kisiel-Dorohinicki, M.: Agent-based evolutionary and immunological optimization. In: Shi, Y., van Albada, G.D., Dongarra, J., Sloot, P.M.A. (eds.) ICCS 2007. LNCS, vol. 4488, pp. 928–935. Springer, Heidelberg (2007)
4. Byrski, A., Kisiel-Dorohinicki, M., Nawarecki, E.: Agent-based evolution of neural network architecture. In: Hamza, M. (ed.) Proc. of the IASTED Int. Symp. on Applied Informatics. IASTED/ACTA Press (2002)
5. Byrski, A., Schaefer, R.: Immunological mechanism for asynchronous evolutionary computation boosting. In: ICMAM 2008: European workshop on Intelligent Computational Methods and Applied Mathematics: an international forum for researches, teachers and students, Cracow, Poland (2008)
6. Byrski, A., Schaefer, R.: Formal model for agent-based asynchronous evolutionary computation. In: 2009 IEEE World Congress on Computational Intelligence. IEEE Computational Intelligence Society, IEEE Press, Trondheim (2009)
7. Cantú-Paz, E.: A summary of research on parallel genetic algorithms. IlliGAL Report No. 95007. University of Illinois (1995)
8. Centarowicz, K., Cieciwa, R., Nawarecki, E., Rojek, G.: Unfavorable behavior detection in real world systems using the multiagent system. In: Intelligent Information Processing and Web Mining (2005)
9. Cetnarowicz, K., Kisiel-Dorohinicki, M., Nawarecki, E.: The application of evolution process in multi-agent world (MAW) to the prediction system. In: Tokoro, M. (ed.) Proc. of the 2nd Int. Conf. on Multi-Agent Systems (ICMAS 1996). AAAI Press, Menlo Park (1996)
10. Cotta, C., Fernandez, A.J.: Memetic algorithms in planning, scheduling and timetabling. In: Dahal, P.C.K., Tan, K.-C. (eds.) Evolutionary Scheduling, pp. 1–30. Springer, Heidelberg (2007)
11. Dasgupta, D.: Artificial Immune Systems and Their Applications. Springer, Heidelberg (1998)

12. Jennings, N.R., Sycara, K., Wooldridge, M.: A roadmap of agent research and development. Journal of Autonomous Agents and Multi-Agent Systems 1(1), 7–38 (1998)
13. Jennings, N.R., Wooldridge, M.J.: Software agents. IEE Review, 17–20 (1996)
14. Kisiel-Dorohinicki, M.: Agent-oriented model of simulated evolution. In: Grosky, W.I., Plasil, F. (eds.) SOFSEM 2002. LNCS, vol. 2540, pp. 253–261. Springer, Heidelberg (2002)
15. Lozano, M., Martinez, G.: An evolutionary ils-perturbation technique. In: Blesa, M.J., et al. (eds.) HM 2008. LNCS, vol. 5296, pp. 1–15. Springer, Heidelberg (2008)
16. Michalewicz, Z.: Genetic Algorithms Plus Data Structures Equals Evolution Programs. Springer, Secaucus (1994)
17. Mitsumoto, N., Fukuda, T., Arai, F.: The immune mechanism, adaptation, learning for the multi agent system. Emerging Technologies and Factory Automation 6-10 (1994)
18. Molina, D., Lozano, M., Martinez, C.G., Herrera, F.: Memetic algorithm for intense local search methods using local search chain. In: Blesa, M.J., et al. (eds.) HM 2008. LNCS, vol. 5296, pp. 58–71. Springer, Heidelberg (2008)
19. Montes de Oca, M.A., Van den Enden, K., Stützle, T.: Incremental particle swarm-guided local search for continuous optimization. In: Blesa, M.J., et al. (eds.) HM 2008. LNCS, vol. 5296, pp. 72–86. Springer, Heidelberg (2008)
20. Moscato, P.: Memetic algorithms: a short introduction. In: New ideas in optimization, pp. 219–234. McGraw-Hill Ltd., UK, Maidenhead (1999)
21. Nishiyama, H., Mizoguchi, F.: Design of security system based on immune system. In: Tenth IEEE International Workshops on Enabling Technologies: Infrastructure for Collaborative Enterprises (2001)
22. Rudolph, G.: Evolution strategies. In: Bäck, T., Fogel, D., Michalewicz, Z. (eds.) Handbook of Evolutionary Computations, Oxford University Press, Oxford (1997)
23. Rudolph, G.: Models of stochastic convergence. In: Bäck, T., Fogel, D., Michalewicz, Z. (eds.) Handbook of Evolutionary Computations. Oxford University Press, Oxford (1997)
24. Rudolph, G.: Stochastic processes. In: Bäck, T., Fogel, D., Michalewicz, Z. (eds.) Handbook of Evolutionary Computations, Oxford University Press, Oxford (1997)
25. Vose, M.: The Simple Genetic Algorithm: Foundations and Theory. MIT Press, Cambridge (1998)

Searching for the Effective Bidding Strategy Using Parameter Tuning in Genetic Algorithm

Kim Soon Gan, Patricia Anthony, Jason Teo, and Kim On Chin

Abstract. Online auctions play an important role in today's e-commerce for procuring goods. The proliferation of online auctions has caused the increasing need to monitor and track multiple bids in multiple auctions. An autonomous agent that exploits a heuristic decision making framework was developed to tackle the problem of bidding across multiple auctions with varying start and end time and with varying protocols (including English, Dutch and Vickrey). This flexible and configurable framework enables the agent to adopt varying tactics and strategies that attempt to ensure that the desired item is delivered in a manner consistent with the user's preferences. However, the proposed bidding strategy is based on four bidding constraints which is polynomial in nature such that there are infinite solutions that can be deployed at any point in time. Given this large space of possibilities, a genetic algorithm is employed to search offline for effective strategies in particular class of environment. The strategies that emerge from this evolution are then codified into the agent's reasoning behaviour so that it can select the most appropriate strategy to employ in its prevailing circumstances. In this paper, the parameters of the crossover and mutation are tuned in order to come up with an optimal rate for this particular environment. The proposed framework is implemented in a simulated marketplace environment and its effectiveness is empirically demonstrated. The relative performance of the evolved bidding strategies is discussed in this paper.

Keywords: Genetic Operators, Genetic Algorithms (GAs), Crossover, Mutation.

1 Introduction

The advancement of the internet technology has brought a new method of trading online namely e-commerce. E-commerce (or electronic commerce) is any business transaction (buying and selling process) whose price or essential terms are negotiated over an online system such as the Internet, Extranet or Electronic Data Interchange network. In today's e-commerce market, online auction has acted as an

Kim Soon Gan · Patricia Anthony · Jason Teo · Kim On Chin
Universiti Malaysia Sabah Locked Bag No. 2073, 88999, Kota Kinabalu, Sabah, Malaysia
e-mail: g_k_s967@yahoo.com, panthony@ums.edu.my, jtwteo@ums.edu.my, kimonchin@gmail.com

R.A. Sarker and T. Ray (Eds.): Agent-Based Evolutionary Search, ALO 5, pp. 203–228.
springerlink.com

important tool in the services for procuring goods and items either for commercialize use or for personal use. Some of the well known auctions for procuring goods in today's online market include eBay.com, uBid.com and Yahoo! Auction. Problems like monitoring the process of auction, tracking bid and bidding in multiple auctions arise when the number of auctions increases. The user needs to monitor many auctions sites, pick the right auction to participate, and make the right bid in making sure that the desired item satisfies the user's preference. All these tasks are somewhat complex and time consuming. The task gets even more complicated when they are different start and end times and when the auctions employ different protocols. For this reasons, a variety of software support tools either by the online auction hosts or third parties are provided to assist consumers when bidding in online auctions. This software includes automated bidding and bid sniping. Automated bidding software or proxy bidders act on the bidder's behalf and place bids according to a strict set of rules and predefined parameters. Bid sniping software is to place bid a few minutes or seconds before an auction. However, these kinds of software have some shortcomings. Firstly, they are only for a particular auction with a particular protocol. Secondly, they only remain in the same auction site and do not move to other auction sites. Lastly, they still need the intervention of the user in that the user still needs to make decision on the starting bid (initially) and the bid increments. To address these shortcomings, an autonomous agent was developed that can participate in multiple heterogeneous auctions, that is empowered with trading capabilities and that can make purchases autonomously [1]. The agent utilizes a bidding strategy that is heavily influenced by two values namely k (is a constant that determines the value of the starting bid) and β (the rate of concession to p_r). These two values correspond to the polynomial function of the four bidding constraints namely the remaining time left, the remaining auction left, the user's desire for bargain and the user's level of desperateness. The range of k value is from 0 to 1 and the range of β value is from 0.005 to 1000. The possible combinations between these two values are endless and thus the search space for the solution strategies is very large. (This will be elaborated in detail in Section 4). Hence, genetic algorithms were used to find the nearly optimal bidding strategy for a given auction environment.

2 Genetic Algorithms

Genetic algorithms (or GAs) were first introduced by John Holland in 1975 [2]. Genetic algorithms are adaptive search method that model genetic evolution that is said to be the closest evolutionary algorithm (EA) that mimics the evolutionary process based on the Darwinian Theory. Genetic algorithms are a particular class of evolutionary computation that use techniques inspired by evolution biology such as inheritance, selection, crossover and mutation [3]. GA has been used as a search technique in computing to find true or approximate solutions to optimization and search problems.

GA creates a population of solution randomly known as individuals. Each individual is represented with a set of parameter which may be solutions to the problem domain. These parameters are known as gene and the combination of these genes are known as chromosome (individual). Classical representation scheme used binary number as the set of parameter and also called binary string. The number of parameters for an individual is normally fixed. However, studies have been conducted on variable length of chromosome. Besides binary number, chromosome can be represented in integer and real-valued parameter [4][5]. The individuals are then evaluated using fitness functions. The evaluation will assign a fitness score to individuals indicating how well the solution performed in the problem domain.

Crossover is a genetic operator that can assemble short low-order schemata of above average fitness to create a higher-order schemata of higher-fitness thus allowing the genetic algorithms to perform well based on the building block hypothesis [6]. The crossover process exchanges subparts of two chromosomes which mimics the biological recombination between two single chromosomes ("haploids"). This reproductive mechanism is a recombination in which its subcomponents of the parent's genomes are cloned and reassembled to create an offspring genome [7]. There are several common crossover operators that can be applied namely one-point crossover, two-point crossover, uniform crossover and multipoint crossover.

On the other hand, mutation is a genetic operator that is often known as the "background" operator in genetic algorithm [2]. The mutation operator ensures that the crossover operator has a full range of alleles so that the individuals are not trapped in local optima. Yet many experiments have pointed out that the mutation operator is a very important genetic operator in genetic algorithm [8][9]. Mutation is a mechanism that will alter one or more gene values in a chromosome from its initial state. This can result in entirely new gene values being added to the gene pool. Mutation serves as an enumerative function, producing alleles not previously tried. Without mutation, the evolution would be stagnated because no new possible variations are being created.

The selection operator is used twice. Firstly it is used before the crossover and mutation process to select the individuals to the mating pool. Secondly, it is applied after the crossover and mutation process to select the individuals to form the next generation population. The former selection is normally known as parental selection and the latter is known as survival selection. There are several selection approaches that can be applied in GA such as random selection, proportional selection, tournament selection, rank based selection and elitism. The genetic algorithms will repeat the above process except for the population initialization until the termination criteria are met. The termination criteria is when the maximum number of generation is reached, or an acceptable best fit individual has evolved, or the average and/or maximum fitness value do not change significantly over the past few generations. The evolving process terminates when the population converges. This is a condition where the population evolves over successive generations such that the fitness of the best and the average individual in each generation increase towards a global optimum [10]. The global optimum is defined as the highest peak in the search space [11].

In this paper, the bidding strategies are evolved using genetic algorithms and its performance is observed in terms of the average fitness, the success rate and the average payoff. The evolved strategies will be evaluated empirically in a simulated online auction environment. The main contribution of this paper is to compare and analyze the performance of the strategies evolved using genetic algorithms with the predetermined set of genetic operators' rate in an auction environment. By performing this process, a combination of genetic operators' probability can be discovered and this will be used to improve the bidding strategy for the online auctions.

The remainder of the paper is organized as follow. Section 3 describes the related work. The bidding strategy framework is discussed in Section 4. The experimental setting is discussed in Section 5. Section 6 discusses the experimental evaluation and finally Section 7 presents the conclusion and future work.

3 Related Work

A number of significant researches have been conducted to investigate the genetic operators in genetic algorithm. All these studies showed that both crossover and mutation operators play an important role in the algorithm. Crossover has shown to be provably essential in genetic algorithm [12]. A similar study investigates the effect of the crossover operator on the shaky-ladder hyperplane-defined function in a new type of environment. The study involved an investigation on the effect of the crossover behaviour in the dynamic environment which is quite similar to the environment described in this paper. The result showed that genetic algorithm was not able to perform nearly as well when crossover is used to a limited degree or turned off completely [13]. Besides that, the building block hypothesis proved that crossover is able to allow the genetic algorithm to perform well by assembling the short low-order schemata of above average fitness to create higher-order schemata of higher-fitness [6]. There are many researchers who are engaged in the investigation of the mutation operators in genetic algorithm. R. Hinterding [8] showed that mutation is a very important operator in GA and should not be treated as a trivial operator. Another investigation conducted proved that the universal heuristic such as 1/N and the error threshold can be improved with prior information of the landscape [9]. This study recommended a mutation rate that is less than 1/N and error threshold in the case where there are peaks that are quite close in fitness value. The study provided some insights on the behaviour of the genetic operator and showed that the mutation operator is a very important component in the genetic algorithm. An investigation on solving the motion planning problem using genetic algorithm was conducted by varying the mutation rate [14]. The results of the experiment showed that some mutation is definitely beneficial. However a high level of mutation also destroys the good properties of the new paths before they can propagate. The optimal combinations of operators' probabilities of applying these operators are problem dependent. Thus, simulations have to be conducted to find the optimal operator combinations [3]. Due to this many applications have implemented different genetic operators' rates [15] [16][17].

Many researches in auction have used genetic algorithm to design or enhance the auction simulation or bidding strategies to be more effective. Genetic algorithm has shown to be an effective method in making the bidding strategies more effective. ZIP, introduced by Cliff, is an artificial trading agent that uses simple machine learning to adapt to operate as buyers or sellers in online open-outcry auction market environments [18]. A standard genetic algorithm is then applied to optimize the values of the eight parameters governing the behavior of the ZIP traders which previously must be set manually. The result showed that GA-optimized traders performed better than those populated by ZIP traders with manually set parameter values [19]. This work is then extended to 60 parameters to be set correctly. The experiment has showed promising result when compared to the ZIP traders with eight parameters [20]. Genetic algorithm is then applied to optimize the auction market parameters setting. The result of this experiment also showed successful result. Many tests have been conducted on ZIP to improve the agent traders and the auction market mechanism using genetic algorithm [21][22]. Thus, ZIP was able to demonstrate that the genetic algorithm can perform well in evolving the parameters of bidding agents and the strategies.

In another investigation, a UDA (utility-based double auction) mechanism is presented [23]. In UDA, a flexible synchronous double auction where the auctioneer maximizes all traders' diverse and complex utility functions through optimization modeling based on genetic algorithm. It is a double auction mechanism based on dynamic utility function integrating the notion of, utility function and genetic algorithm. The GA-optimizer is used to maximize total utility function, composed of all participants' dynamic utility functions, and matches the buyers and sellers. The mechanism experimental result has empirical proof that it performs better than conventional double auction.

Genetic algorithm has shown to perform well in many of the research in auction environment. Thus, this chapter will also show how genetic algorithm is able to make the bidding strategies more effective.

4 Bidding Strategy Framework

The electronic marketplace simulation is a platform for the bidding agents to participate and perform bidding activities in multiple online auctions. To simulate the real auction environment, the auctions employs different protocols and run concurrently. This market simulation allows a flexible configuration of all the parameters in the auction which include the item sold, the number of bidders, the private valuation, the number of auction and other parameters. The market simulation is to evaluate the performance of the evolved bidding strategies. The marketplace simulator shown in Figure 1 consists of concurrent auctions that employ different protocols. These protocols include English, Dutch and Vickrey. All these auctions have a known start time and only English and Vickrey auctions have a known end time. The bidding agent is given a deadline (t_{max}) by when it must obtain the desired item and it is told the consumer's private valuation (p_r) for this item.

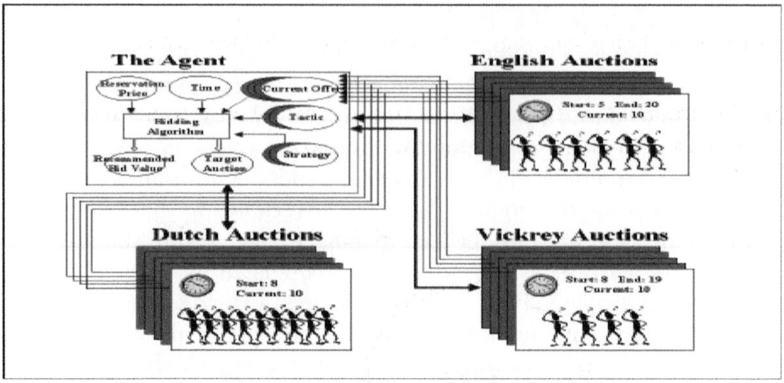

Fig. 1 The marketplace simulator

The marketplace is developed so that it is flexible and configurable to the number of auctions and with varying time duration. The marketplace is run for a fixed amount of time which varies between 21 and 50 time steps. The marketplace can run multiple English, Dutch and Vickrey auctions concurrently with variable start and end times. The number of English, Dutch and Vickrey auctions running in the marketplace is generated from the uniform distribution with a fixed range between 7 and 15, such that the total number of auctions running in the marketplace is between 20 and 45. The marketplace maintains information about each of the auctions that are running and this includes:

- The number of English, Dutch and Vickrey auctions running in the marketplace.
- The status of each auction (whether it is ongoing or closed).
- The highest bidder and the current highest bid for each English auction.
- The current offer for each Dutch auction.
- The bid histories for all the auctions.
- The winning bidders in all the auctions.

The marketplace announces the current bid values and the current highest bids for English auctions and the current offers for Dutch auctions at each time step. At the end of a given auction, it determines the winning bid and announces the winner. This set of information is used by the agent when deciding in which auction to participate, at what value to bid and at which time to bid.

while $(t \leq tmax)$ *and (Item_NA = true)*
Build active auction list;
Calculate current maximum bid using the agent's strategy;
Select potential auctions to bid in, from active auction list;
Select target auction as one that maximises agent's expected utility;
Bid in the target auction using current maximum bid as reservation price at this time;
Endwhile

Fig. 2 The bidding agent algorithm

The bidding algorithm for this framework is shown in Figure 2. Let Γ be a finite time interval and let t be the current universal time across all auctions, where $t \in \Gamma$. Let t_{max} be the time by when the agent must obtain the good and let A be the list of all the auctions that will be active before time t_{max}. At any time t, there is a set of active auctions $L(t)$ where $L(t) \subseteq A$. Let $E(t)$, $D(t)$, and $V(t)$ be the set of active English, Dutch and Vickrey auctions respectively, where $E(t) \cap D(t) = \emptyset$, $D(t) \cap V(t) = \emptyset$, $E(t) \cap V(t) = \emptyset$, and $E(t) \cup D(t) \cup V(t) = L(t)$. Each auction $i \in A$ will have its own start time, σ_i, and its own end time η_i where $i \in E(t) \cup V(t)$. Let ʅ be the agent's bid increment value, and p_r be its private valuation for the target item. Let *Item_NA* be a boolean flag to indicate whether the target item has already been purchased by the agent. Assume that the value of p_r is based on current reliable market prices observed from past auctions, and that the marketplace is offering the item that the agent is interested in. While the bidder agent have not obtain the desire item, the bidder agents need to build an active auctions list to keep track of current active auction. Active auction is defined as auction that ongoing or just started but not reach the end time yet. In order to build the active auction list, the bidder agent follows the algorithm below as depicted in Figure 3. $S_i(t)$ is a boolean flag representing the status of auction i at time t, such that $i \in A$ and $S_i(t) \in$ *(ongoing; completed)*. Each auction $i \in A$, has a start time σ_i, and its own end time η_i. The active auction list is built by taking all the auctions that are currently running at time t. In English and Vickrey auctions, any auction that has started but not reached its end time is considered active. $S_i(t)$ is used in Dutch auctions since the end time of this type of auction is not fixed.

```
for all i ∈ A
    if ((t ≥ σi ) and (t ≤ ηi ) or (Si (t) = ongoing)
        add i to L(t)
    endif
endfor
```

Fig. 3 Algorithm to build active auction list

After the bidder agent build the active auctions list, the bidder agent will start calculating the current maximum bid based on the agent strategy. The current maximum bid is defined as the bidding place the agent willing to bid at the current time that is less than or equal to the agent private valuation. Four bidding constraint are used to determined the current maximum bid namely the remaining time left, the remaining auctions left, the desire for bargain and the level of desperateness. The agent's decision-making model works as follows (see [24] for a complete description). The bidder agent builds an active auction list (auctions that have started but not reached their end times, denoted as $L(t)$) and gathers relevant information (e.g. start and end times, current bid values) about them. It then calculates the *current maximum bid* it is willing to make at the *current time*. This current maximum bid, by definition, will always be less than or equal to the private valuation. The current maximum bid is determined by considering four *bidding constraints* including the remaining time left, the remaining auctions left, the user's desire for a bargain and the user's level of desperateness. For each such bidding constraint, there is a corresponding function that suggests the value to bid based on that constraint at that time. These (polynomial) functions (based on [25]) are parameterized by two key values: k (range [0-1]) is a constant that determines the value of the starting bid and β (range [0.005 - 1000]) defines the shape of the curve (and so the rate of concession to p_r).

(1) and (2) are the formula corresponding to the remaining time tactic which considers the amount of bidding time the bidder agent has to obtain the desire item.

$$f_{rt} = \alpha_{rt}(t)P_r \tag{1}$$

where $\alpha_{rt}(t)$ is a polynomial function of the form:

$$\alpha_{rt}(t) = k_{rt} + (1 - k_{rt})\left(\frac{t}{t_{\max}}\right)^{\frac{1}{\beta}} \tag{2}$$

The remaining auction left tactic considers the number of remaining auctions that the bidder agent is able to participate in order to obtain the item utilizes the formula in (3) and (4).

$$f_{ra} = \alpha_{ra}(t)P_r \tag{3}$$

where $\alpha_{ra}(t)$ is a polynomial function of form:

$$\alpha_{ra} = k_{ra} + (1 - k_{ra})\left(\frac{c(t)}{|A|}\right)^{\frac{1}{\beta}} \tag{4}$$

The desire for a bargain tactic is the bidder agent that is interested in getting a bargain for obtaining the desired item. The current bid value is calculated based on the following expression:

$$f_{ba} = \omega(t) + \alpha_{ba}(t)(P_r - \omega(t)) \tag{5}$$

In the expression above, the variable $\omega(t)$ takes into account all the ongoing auctions along with the current bid value. The expression for $\omega(t)$ is calculated based on the formula as below:

$$\omega(t) = \frac{1}{|L(t)|}\left(\sum_{1 \le i \le L(t)} \frac{t - \sigma_i}{\eta_i - \sigma_i} v_i(t)\right) \tag{6}$$

where v_i is the current highest bid value in an auction I at time t, and $I \in L(t)$; σ_i, and η_i is the start and end time of auction i

The expression for $\alpha_{ba}(t)$ is defined as:

$$\alpha_{ba}(t) = k_{ba} + (1 - k_{ba})\left(\frac{t}{t_{\max}}\right)^{\frac{1}{\beta}} \tag{7}$$

The level of desperateness tactic is defined as the bidder agent's desperateness to obtain the targeted item within a given period Thus, the bidder agent possessing this behavior tends to bid aggressively. This tactic utilizes the same minimum bid value and the polynomial function as the desire for bargain tactic but with a minor variation to the β and k_{de} value.

There is a weight associated to each of these tactics and this weight is to emphasize which combination of the tactics that will be used to bid in the online auction. The final current maximum bid is based on the combination of the four strategies by making use of the weight. Thus, at any given point there are countless possibilities of these combinations of tactics.

All behaviours in between are also possible by setting the parameters appropriately. Figure 4 shows various combinations of the bidding constraints. In this case, (0.25, 0.25, 0.25, 0.25) indicates that equal weighting has been applied to all the bidding constraints, (0.50, 0.00, 0.50, 0.00) indicates that the agent only considers bidding constraints of the remaining time left and the desire for a bargain and places equal weight of 0.50 for both bidding constraints. It can also be seen that by varying the weights of the bidding constraints, different bidding patterns can be generated as shown in Figure 4.

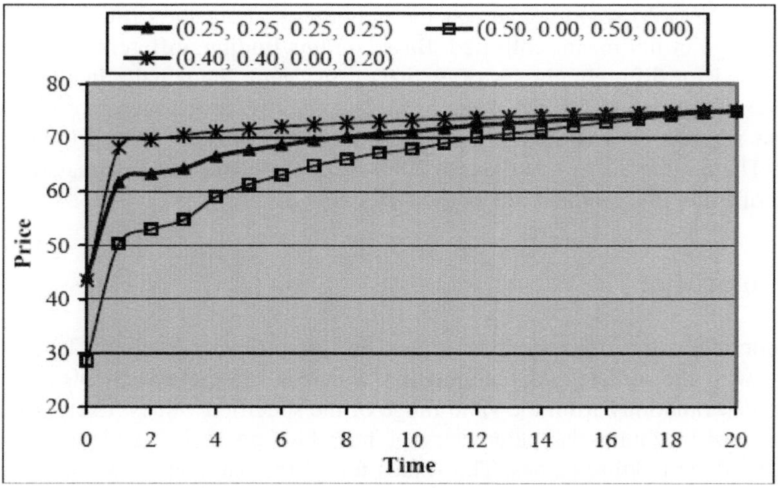

Fig. 4 Various combinations of the bidding constraints

The agent selects the potential auctions in which it can bid and calculates what it should bid at this time in each such auction based on the value of the current maximum bid. The auction with corresponding bid with the highest expected utility is then selected from the potential auctions as the target auction. Finally, the agent bids in the target auction.

A series of experiments were conducted in a controlled dynamic environment to test the efficiency (in terms of success rate and average payoff) of the agent's strategy. The number of strategies that can be employed is infinite because the bidding agent is heavily influenced by the strategy employed which in turn relates to the values of k and β in the given tactics and the weights for each tactic when these are to be combined. Thus, GA was used to evolve the best strategy to be used in the prevailing environment.

The performance of the individuals in the population is evaluated using the fitness function. The fitness function used to evaluate the performance of the individual in the population takes into account the utility of winning an auction i which is computed as

$$U_i(v) = \left(\frac{p_r - v}{p_r} \right) + c \tag{8}$$

where v is winning bid and c is arbitrary constant ranging from 0.001 to 0.005 to ensure the agent is awarded with some value in case the winning bid is equivalent to its private valuation. The individual is penalized if it fails to get the item. In this case the penalty incurred ranges from 0.01 to 0.05. An experiment was conducted to test whether penalising an agent with a higher value generates a better strategy than penalising an agent with a lower value. The experiment consists of evolving environment with varying penalty values of 0.01, 0.02, 0.03, 0.04, 0.05 and 0.07 and observing the average utility obtained for each penalty value. Analysis of variance (ANOVA) was used to test the hypothesis about the differences between the average utility means collected. Based on this finding, different penalty values were used for different classes of environment. Here, the penalty incurred is 0.01 to 0.05. This also ensures that the agent does not receive a negative utility. The fitness score is then computed by taking the average utility from a total of 2000 runs. The number of runs is fixed to 2000 times to decrease the estimated standard error of mean to a minimal level based on statistical analysis.

5 Algorithm

Genetic algorithm framework is applied in this problem domain. The algorithm starts with the initial bidder population which is represented by 50 individuals generated randomly from the valid range of the specified values. Each of the individuals will contain the parameters of four bidding tactics and the associated weight of the bidding tactics. The values for all the parameters are represented in floating points form. Figure 5 shows the encoding for a bidding strategy.

k_{rt}	β_r	k_{ra}	β_{ra}	k_{ba}	β_{ba}	k_{de}	β_{de}	w_{rt}	w_{ra}	w_{ba}	w_{de}

Fig. 5 Encoding of a bidding strategy

Each individual will be evaluated based on its fitness function. The selection process ensures the survival of the fitter individuals to next generation. Tournament selection is used to select 90% of the population to the next generation. In tournament selection, two individuals from the population is randomly chosen and compared. "Elitism" is applied in the algorithm to retain the best 10% individual of the current generation to the next generation to assure that the fitter individuals survive to the next generation. After the selection process, each individual will be subjected to crossover and mutation. The exchanging of genetic material process is performed using an extension combination operator [10][26][27], which works by taking the difference between two values of the crossover point, adding this difference to the higher (giving the maximum range) and subtracting it from the lower (giving a minimum range). The new values are then generated between the minimum and maximum range. Mutation follows the crossover process where this allows an increase in the variation of the population to be explored. The gene from the chosen individual is picked randomly and a small value (0.05) is added or subtracted, depending on the range limitation for that particular gene. The mutation process is only applied to the values of k and β for each tactic. The weights are not considered here because adding a small value to the weight requires a renormalization and will have very little effect on the agent's online behavior. The stopping criterion is after the population has converged. In this case, the population always converges before 50 iterations (typically, the value lies between 24 and 40). Thus, the stopping criterion is set to 50. The crossover and mutation rate applied are dependent to the problem domain to be solved.

The crossover (p_c = 0.6) and mutation rate (p_m = 0.02) applied in the previous work [24] is based on the recommended value from the literatures and may not necessarily evolve the best bidding strategies. Many of the researchers have conducted experiment to fine tune the control parameter in genetic algorithm to solve a particular problem. In this experiment different combinations of crossover and mutation are tested to discover the best combination of genetic operators' probabilities that will perform best in the algorithm. The main purpose of the experiment is to discover a combination of static genetic operators' probabilities which can improve the algorithm in evolving the most effective strategies.

6 Experimental Setting

The purpose of the experiments is to compare the performances of the predetermined set of crossover rates and mutation rates in genetic algorithm in searching for the nearly optimal bidding strategies. This experiment is to discover a combination of static genetic operators' probabilities which can improve the algorithm in evolving the most effective strategies. The experiment is divided into three parts. First, a set of experiments was conducted to compare the performance of the predetermined set of crossover rate in the genetic algorithm in the online auction framework. The mutation rate is fixed and so that the best crossover rate can be discovered. In the second part of the experiment, the crossover rate is fixed to discover the best mutation rate. The crossover rates used are shown in Table 1.

Table 1 Crossover values with a fixed mutation rate of 0.02

Crossover rate	Mutation rate
0.2	0.02
0.4	0.02
0.6	0.02
0.8	0.02

This experiment aims to discover the best static crossover rate for the genetic algorithm in exploring the bidding strategies. In order to do this, the crossover rate is varied with several values while the mutation rate is fixed to one value. The experiment evolves the strategies by using the crossover rate of 0.2, 0.4, 0.6 and 0.8 and fixing the mutation rate to $p_m = 0.02$. The crossover rate of 0.2 is chosen to be used as it is frequently seen in practice to be a good choice [7]. The crossover rate of 0 is excluded from the test as it was shown in the previous research that the nonzero crossover can perform better than zero crossovers and zero crossovers are essentially mutation-based algorithm [13]. GA with crossover has shown to perform better that the GA without crossover [28]. This testing set did not include the crossover rate of 1 as it is shown in the literature review that this crossover rate is too disruptive to be used [29]. The value of 0.6 is selected as the mid point between the 0-1 and based on the previous work and it has shown a promising result. Besides, $p_c = 0.6$ is also the crossover value as proposed by De Jong [15]. In order to balance the quantum of the investigation, it was decided to select 0.4 and 0.8 as the testing value which have also been used in several investigation [30][31][32][33][34][35]. These test values are also chosen based on some of the other literatures [15][16][17][36][37][38]. All the tested values mentioned are commonly used in genetic algorithm investigation for problem solving. In this section, the mutation rate is fixed to 0.02 since the purpose of the experiment is to study the effect of varying the crossover rate. The same mutation rate has been applied in the previous work based on the literature and also in other research [18][29][39].

In the second part of the experiment, the crossover rate is fixed while the mutation rate is varied. This experiment is performed to discover the best static mutation rate for the genetic algorithm in exploring the bidding strategies. The GA employed a fixed crossover probability of $p_c = 0.6$ which was applied in the previous work. The mutation rates selected for the testing is 0.002, 0.02 and 0.2. The testing values are chosen based on the literatures [27][36][37]. The value of 0.02

is selected since it was used in the previous work and has shown a promising result. Moreover, it is close to the Grefenstette setting along with the other works [16]. Also, other researches have conducted experiment by using this rate [40][41]. The mutation rates of 0.2, 0.02 and 0.002 are selected to balance the quantum of investigation supported by the literatures. Mutation rate of 0.002 is selected because it is closest to the De Jong recommended mutation rate $p_m = 0.02$ [15] and Schaffer recommended a mutation rate range in [0.005-0.01] [17]. Lastly, the mutation rate of $p_m = 0.2$ is selected primarily to investigate the effect of utilizing a higher mutation rate in the evolving bidding strategy since some of the work in genetic algorithm has employed the mutation rate of 0.2 [38][42]. The mutation rates used are shown in Table 2.

Finally, the crossover rate and the mutation rate discovered from the first and the second experiment will be combined to investigate whether it will evolve better strategies.

Table 2 Mutation values with a fixed crossover rate of 0.6

Crossover rate	Mutation rate
0.6	0.2
0.6	0.02
0.6	0.002

As mentioned, a simulated electronic marketplace has been developed in the previous work to test the performance of the strategies. This simulated electronic marketplace consists of a number of auctions that run concurrently. These auctions embody three different protocols namely English, Dutch and Vickrey. All these information are readily accessible to the bidding agent. This simulated environment will be used to test the performance of the bidding strategies. The genetic algorithm is also applied in the same environment. The parameters for the empirical evaluations are shown in Table 3. The parameters include the agent's reservation price; the agent's bidding time and the number of active auctions. The agent reservation price is the maximum amount that the agent is willing to pay for the item. The bidding time is the time allocated for the agent to obtain the user's required item. The active auctions are the list of auctions that is ongoing before time t_{max}.

Table 3 Configurable parameters for the simulated environment

Agent's reservation price	$73 \leq p_r \leq 79$
Bidding time for each auction	$21 \leq t_{max} \leq 50$
Number of active auctions	$20 \leq L(t) \leq 45$

7 Experimental Evaluation

The performance of the evolved strategies is evaluated based on three measurements:

1. The Average Fitness
This is the average fitness obtained by the population at each generation over 50 generations. The average fitness shows how well the strategy converges over time to find the best solution.

2. The Success Rate
This is the percentage of time that an agent succeeds in acquiring the item by the given time at any price less than or equal to its private valuation. This measure is important in determining how successful the agent is, since one of the agent's key tasks is to deliver the item to the user (when possible). Thus, this measure will determine the efficiency of the agent in terms of guaranteeing the delivery of the requested item.

3. The Average Payoff
Let p_r be the agent's private valuation of the target item, and let n be the number of times the marketplace was run. Here, the average payoff (p), is defined as

$$\frac{\sum_{1 \le n \le 100} \left(\frac{p_r - v_i}{p_r} \right)}{n} \tag{9}$$

where v_i is the winning bid for auction i. The average payoff is calculated by deducting the agent's bid value (the value at which it acquires the item) from the private valuation. This value is then divided by the agent's private valuation, summed and average over the number of runs, n. The agent's payoff is 0 if it is not successful in obtaining the item.

8 Results and Discussion

The average fitness for the varying crossover rates is shown in Figure 6. It can be seen that the point of convergence for the different crossover rate is similar. However, the population with crossover rates of 0.4 achieved a higher average fitness when compared to other population with crossover rates of 0.2, 0.6 and 0.8. This figure also shows that the population with a crossover rate of 0.4 performs best among other populations. When the crossover rate is too high such as 0.6 and 0.8, the high performance strings are eliminated faster than selection can produce improvements. On the other hand, a low crossover rate such as 0.2 may cause stagnation due to the lower exploration rate. The population with a crossover rate of 0.4 is able to maintain sufficient diversity of the population till the end of the run and the exploitation is best in this genetic algorithm environment compared to the others. This is an indication that the population evolved from the 0.4 crossover rate will have higher fitness and would most likely perform better than the individuals from the rest of the crossover rate. The next two experiments will be able to confirm this hypothesis.

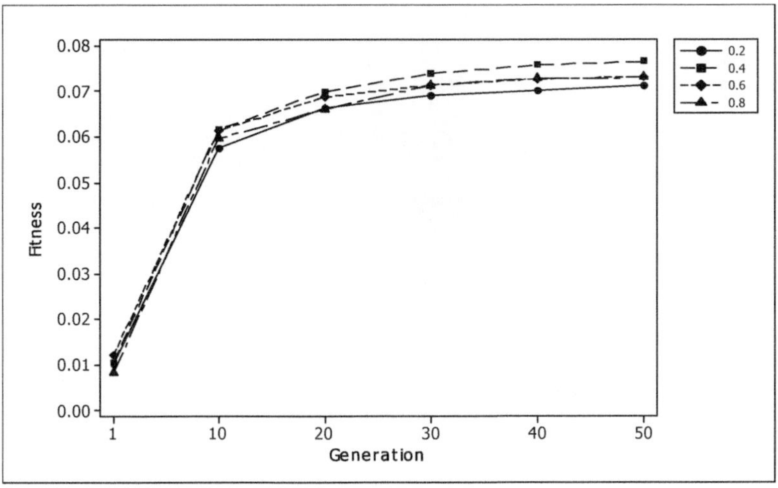

Fig. 6 Average fitness of population with varying crossover rates and a fixed mutation of 0.02

The success rate and the average payoff are important measurements for the evolving strategies. The higher value in these measurements indicates that the evolved strategies are performing well. Figure 7 shows the comparison of the success rate of the bidding strategies that were evolved using the different values of crossover rate. The result shows that the strategy evolved using a crossover rate of 0.4 performs the best overall. The bidder agents evolved from the crossover rate of 0.4 outperformed the agents from the other crossover rates by delivering a more promising success rate. The agent evolved from the crossover rate of 0.4 produces a 10% increase in success rate compared to the individuals evolved from the other crossover rates. Besides that, the strategy evolved using a crossover rate of 0.4 performs better than the strategy evolved using a crossover rate of 0.6. Here, it can be concluded that the strategy evolved from the 0.4 crossover rate can evolve better strategies and delivers higher success rate when bidding in online auctions thus improving the GA in searching for better bidding strategies.

Figure 8 shows the comparison of the average payoff between the individuals of the test set. As expected, all the bidder agents evolved from the 0.4 crossover rate outperformed the rest with a higher average payoff of 2%. This shows that the strategy evolved using the 0.4 crossover rate does not only generate a better average fitness but also evolves better effective strategies compared to the strategy evolved for the crossover rate of 0.6, 0.2 and 0.8. They gained higher profit when procuring the item. The result shown here confirms the hypothesis we have made earlier, where individuals in population with a better average fitness would performed best.

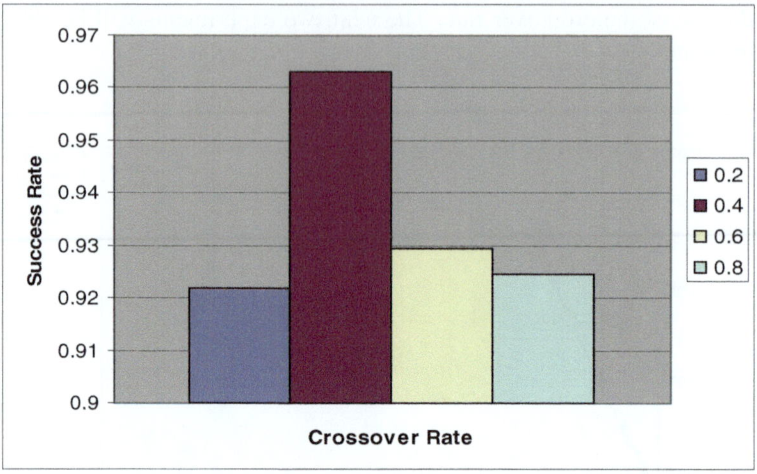

Fig. 7 Success rate for strategies evolved with varying crossover rates and fixed mutation rate of 0.02

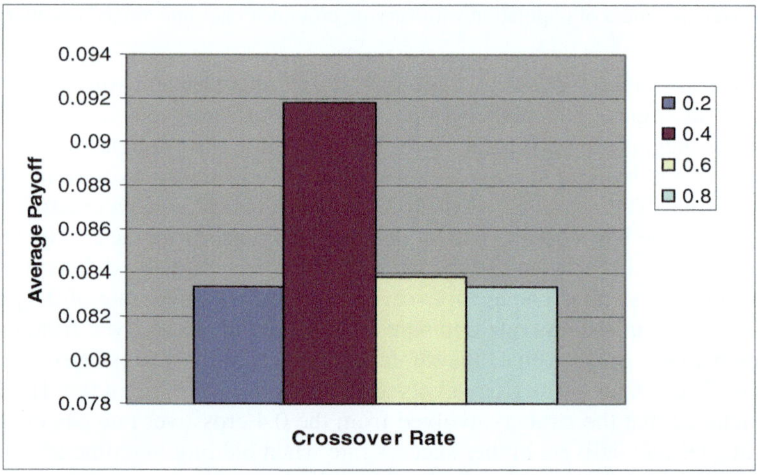

Fig. 8 Average payoff for strategies evolved with varying crossover rates and a fixed mutation rate of 0.02

Based on the results of the experiments that have been carried out, it can be concluded that the strategies evolved from the 0.4 crossover rate performed better than the strategies evolved from the other testing sets in terms of success rate and average payoff in an online auction environment setting. It also achieved a higher average fitness function during the evolution process. It performs better because it is able to maintain a sufficient diversity of the population and the exploitation in this rate is best in this genetic algorithm environment. Besides that, the strategies evolved with 0.4 crossover rate is able to achieve a balance between exploration and exploitation compared to the other testing sets.

Figure 9 shows the average fitness for the evolving bidding strategy with different mutation rates and a fixed crossover probability of 0.6 in genetic algorithm. The population with mutation rate of 0.02 achieved a higher average fitness when compared to the other populations with mutation rates of 0.2 and 0.002. Figure 9 also shows that the population with a mutation rate of 0.02 performs best among other populations. When the mutation rate is too high such as 0.2, it is too disruptive because the randomness will eliminate the good strings. On the other hand, a low mutation rate such as 0.002 may cause stagnation due to the low diversity of the rates. This result seems to indicate that the one with the highest fitness would evolve better bidding strategies as before.

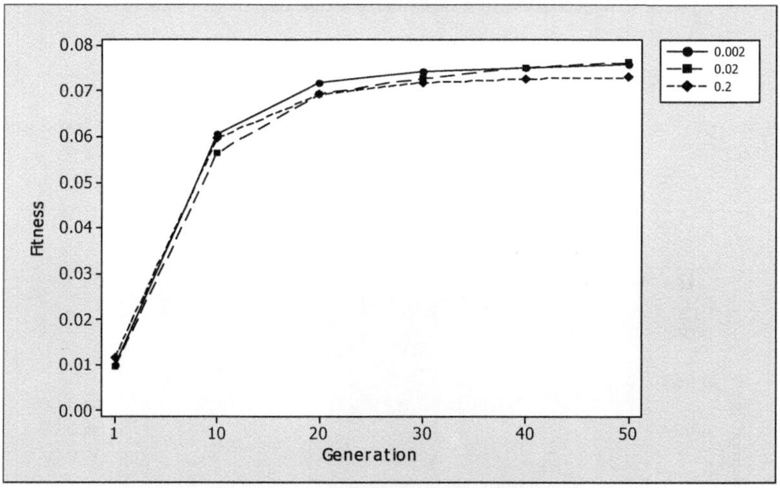

Fig. 9 Average fitness of population with varying mutation rates and a fixed crossover rate of 0.6

Figure 10 shows the comparison of the success rate between the bidder agents of the test set of 0.2, 0.02 and 0.002. The result shows that the bidder agent evolved using a mutation rate of 0.02 performs the best overall. The bidder agents evolved from the mutation rate of 0.02 outperformed the agents from the other mutation rate by delivering a more promising success rate. The individuals evolved from the mutation rate of 0.02 produces a 1.2% to 2.5% increase in success rate compared to the individuals evolved from the other mutation rates. Here, it can be concluded that the strategy evolved from the 0.02 mutation rate can evolve better strategies and delivers higher success rate when bidding in online auctions thus improving the GA in searching for better bidding strategies.

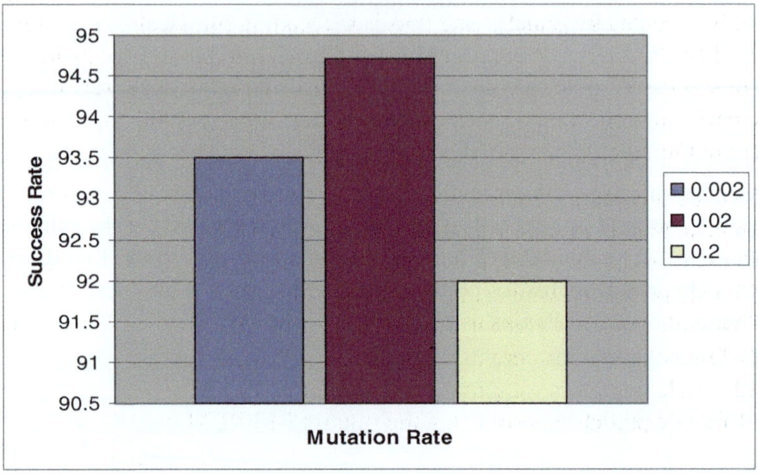

Fig. 10 Success rate for strategies evolved with varying mutation rates and a fixed mutation rate of 0.6

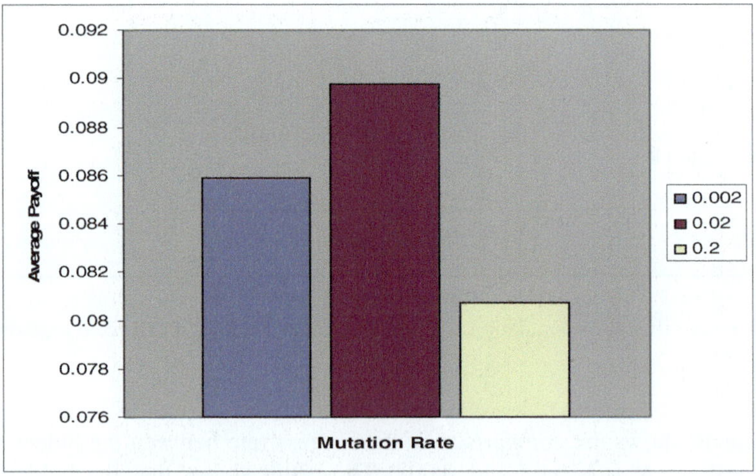

Fig. 11 Average payoff for strategies evolved with varying mutation rates and a fixed crossover rate of 0.6

Figure 11 shows the comparison of the average payoff between the bidder agents of the test set. As expected, all the bidder agents evolved from the 0.02

mutation rate outperformed the rest with a higher average payoff of between 2% and 4%. The strategy with highest fitness generates a better success rate compare to others. It can be observed that the mutation rate of 0.02 evolves better strategies when compared to other mutation rates (0.2 and 0.002). This rate is similar to the research outcome by Cervantes [9] in which a mutation rate below the 1/*N* and error threshold is recommendable. Based on the results of the experiments that had been carried out, it can be concluded that the strategies evolved with 0.02 mutation rate performed better than the strategies evolved with 0.2 and 0.002 mutation rates in terms of success rate and average payoff in an online auction environment setting. It also achieved a higher average fitness function during the evolution process.

Based on the experimental result obtained in, it was found that the best crossover rate is 0.4 and mutation rate is 0.02. Thus, by using the new combination of $p_c = 0.4$ and $p_m = 0.02$, the evolved strategies are expected to perform better since it will combine the advantages of both rates in exploring the solution. In order to confirm this statement, another set of experiments has been carried out by fixing 0.4 as the crossover rate and varying the mutation rate with the same set as in Table 2.

The result shown in Figures 12, 13 and 14, with the new crossover rate $p_c = 0.4$, $p_m = 0.02$ still performed the best in this particular combination. The result illustrates that with mutation rate of 0.02, the population can evolve to a fitter population compared to the other population with mutation rates of 0.2 and 0.002. As for the success rate and the average payoff, both experiments have demonstrated similar results where the mutation rate of 0.02 achieved the highest result.

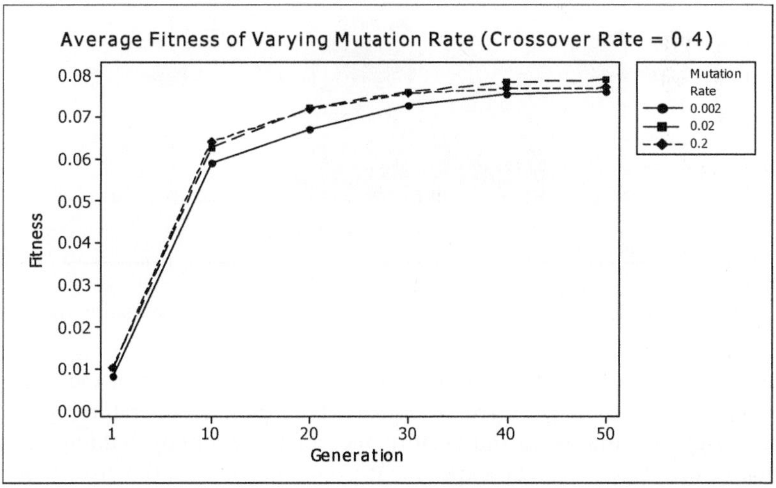

Fig. 12 Average fitness of population with varying mutation rates with 0.4 crossover rate

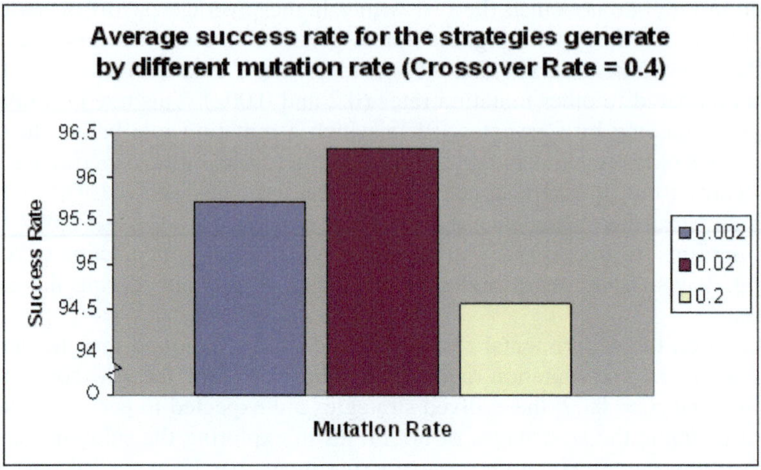

Fig. 13 Success rate for strategies evolved with varying mutation rates

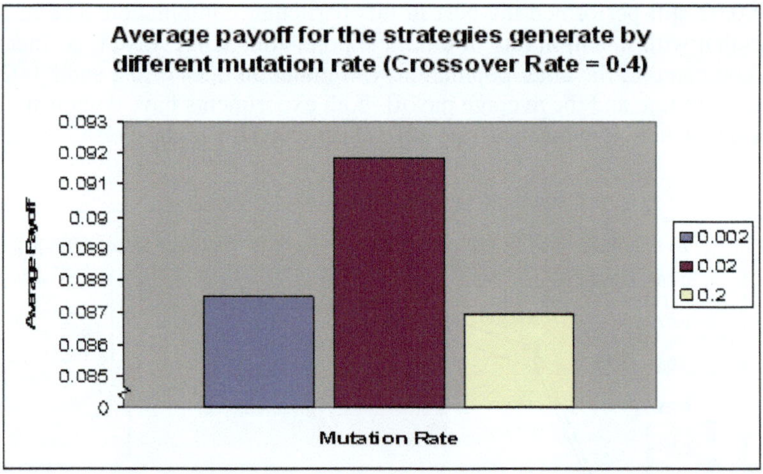

Fig. 14 Average payoff for strategies evolved with varying mutation rates

Hence, by combining the best crossover and mutation rates in the experiment would ultimately yield a better combination of genetic operators' probability which can greatly improved the performance of the evolving bidding strategies. To support this hypothesis, an experiment is conducted with the crossover rate p_c = 0.4 and mutation rate p_m = 0.02.

In the last stage of the experiment, the best value for the crossover rate and mutation rate discovered from the two previous experiments are combined to see whether it will evolve better individuals compared to the three best combinations which has been tested.

Table 4 Different genetic operator's probabilities compared with $p_c = 0.4$ and $p_m = 0.02$

	Crossover Rate	Mutation Rate.
Strategy 1	0.4	0.02
Strategy 2	0.6	0.02
Strategy 3	0.8	0.02
Strategy 4	0.6	0.002

Strategy 1 is the best crossover and mutation rate which were discovered by the varying the crossover and the mutation. Strategy 2, 3 and 4 are the best performers from the varying crossover and mutation experiments. Hence, in this experiment we compare the performance of the individuals generated using the crossover rate of $p_c = 0.4$ and mutation rate $p_m = 0.02$ with the other combinations of the genetic operators' probability used in the previous experiments. Table 4 shows the different combination of the crossover rates and mutation rates. As expected Figure 15 shows that the strategies evolved from the combination of the crossover rate of 0.4 and mutation rate of 0.02 achieves the highest average fitness compared to Strategy 2, 3 and 4. Strategy 2 is the original combination of the genetic operators' probability ($p_c = 0.6$ and $p_m = 0.02$) which was used in the previous work. Figures 16, 17 and 18 showed the comparison between the strategies evolved by using a combination of crossover rate 0.4 and a mutation rate of 0.02 and the combination of crossover rate 0.6 with a mutation rate of 0.02. The strategies evolved from the combination of the crossover rate of 0.4 and mutation rate of 0.02 produced better result in terms of the average fitness, the success rate and the average payoff.

The strategy which has higher average fitness will eventually generate higher success rate as shown in Figure 15. Figure 16 shows that the average payoff generated by Strategy 1 has the highest profit gain compared to Strategy 2, 3 and 4.

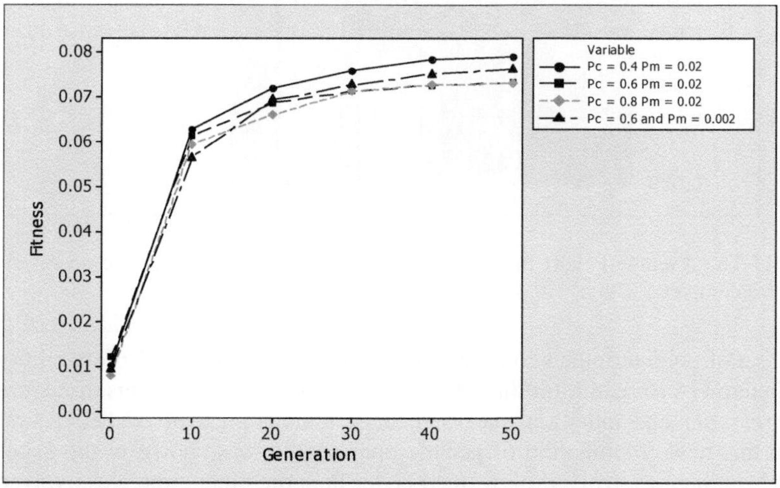

Fig. 15 Average fitness of population with different genetic operator's probabilities compared with $p_c = 0.4$ and $p_m = 0.02$

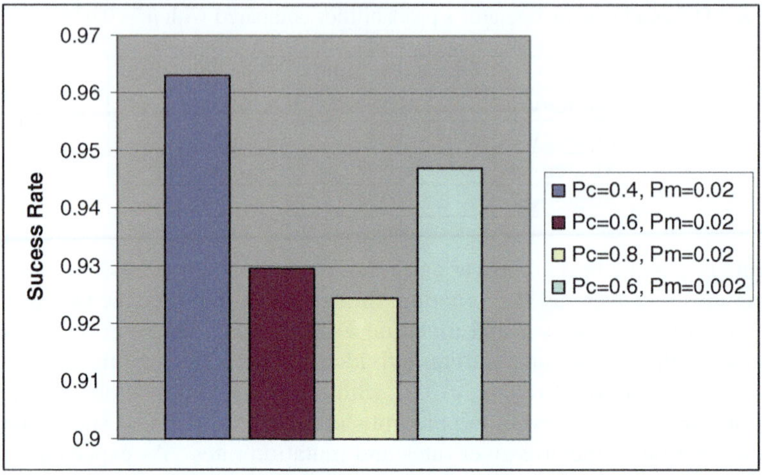

Fig. 16 Success rate for strategies evolved with different genetic operator's probabilities compared with $p_c = 0.4$ and $p_m = 0.02$

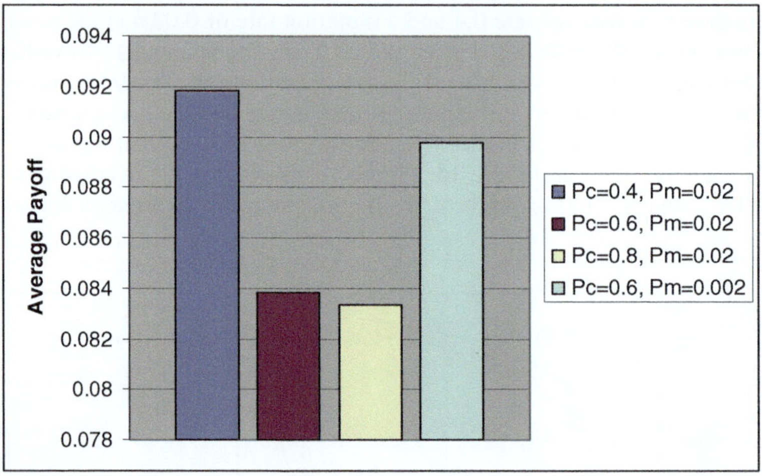

Fig. 17 The Average Payoff for strategies evolved with different genetic operator's probabilities compared with $p_c = 0.4$ and $p_m = 0.02$

A set of predetermined combination of the genetic operators' probabilities were applied to GA to search for the effective strategy. The experiments discovered better crossover and mutation rates and this combination evolved better strategies. With this new combination of genetic operators, it was shown in the experiment that the strategies evolved performed better than the other strategies evolved from the other combinations in terms of success rate and average payoff when bidding in the online auction marketplace.

9 Conclusion

This paper investigates the performance of various combinations of predetermined sets of genetic operators' rates in genetic algorithm on a flexible and configurable heuristic decision making framework that can tackle the problem of bidding across multiple auctions that apply different protocols (English, Vickrey and Dutch). As mentioned earlier, the optimal combinations of operators' probabilities of applying these operators are problem dependent. Thus, simulations have to be conducted to find the optimal operator combinations. In this paper, a set of predetermined combination of the genetic operators' probabilities were applied to GA to search for the effective strategy. The experiments discovered better crossover and mutation rates and the combination of these two values evolved better strategies as shown in the experiments. With this new combination of genetic operators, the evolved strategies performed better than the other strategies evolved from the other combinations in terms of success rate and average payoff when bidding in the online auction marketplace. By discovering a better combination of genetic operator's probabilities, this investigation also indirectly improves the performance of bidding strategies as shown in Figures 13, 14, and 15. This experiment has proven that the crossover rate ($p_c = 0.6$) and mutation rate ($p_m = 0.02$) which were applied in the previous work did not perform at its best in the framework. This investigation is simplified since the testing set applied to the genetic algorithm is still very limited and we have not covered all possible values. Trying all the different combination of crossover and mutation rates systematically is practically impossible. Besides that, the process of tuning the parameter is very time consuming. However, the experiments conducted are sufficient to demonstrate that tuning the parameters can evolve better strategies. However, these nearly optimal values may not be the best value to use in other domains in view of the fact that the tuning of the parameters is problem dependent. There are several areas that require further investigation. Firstly, the parameter tuning of the genetic operators can be further refined. Secondly, we would like to explore how well other variations of genetic algorithm perform in this framework such as adaptive genetic algorithms and self-adaptive genetic algorithms.

References

1. Anthony, P., Jennings, N.R.: Agent for Participating in Multiple Online Auctions. ACM Transaction on Internet Technology 2(3) (2001)
2. Holland, J.H.: Adaptation in Natural and Artificial Systems. The University of Michigan Press, Ann Arbor (1975)
3. Engelbrecht, A.P.: Computational Intelligence An Introduction. John Wiley & Sons, Chichester (2002)
4. Janikow, C.Z., Michalewiz, Z.: An experimental comparison of Binary and Floating Point Representations in Genetic Algorithms. In: Belew, R.K., Booker, L.B. (eds.) Proceedings of the 4th International Conferences in Genetic Algorithms, pp. 31–36. Morgan Kaufmann, San Francisco (1991)

5. Davis, L.: Hybridization and Numerical Representation. In: Davis, L. (ed.) The handbook of Genetic Algorithm, pp. 61–71. Van Nostrand Reinhold, New York (1991)
6. Goldberg, D.E.: Genetic Algorithms in Search, Optimization & machine learning. Addison Wesley, Reading (1989)
7. De Jong, K.A.: Evolutionary computation: A unified approach. MIT Press Book, Cambridge (2006)
8. Hinterding, R., Gielewski, H., Peachey, T.C.: The nature of mutation in genetic algorithms. In: Eshelman, L. (ed.) Proceedings of the Sixth ICGA, pp. 65–72. Morgan Kaufmann, San Francisco (1995)
9. Cervantes, J., Stephens, C.R.: Optimal mutation rates for genetic search. In: Proceedings of the 8th Annual Conference on GECCO, Seattle, pp. 1313–1320 (2006)
10. Beasley, D., Bull, D.R., Martin, R.R.: An Overview of Genetic Algorithms: Part 1, Fundamentals. University Computing 15(2), 58–69 (1993)
11. Coley, D.A.: An Introduction to Genetic Algorithms for Scientists and Engineers. World Scientific, New Jersey (1999)
12. Watson, R.A., Jansen, T.: A Building-Block Royal Road Where Crossover is Provably Essential. In: Genetic And Evolutionary Computation Conference Proceedings of the 9th annual conference on Genetic and evolutionary computation, London, England, pp. 1452–1459 (2007)
13. Rand, W., Riolo, R., Holland, J.H.: The effect of crossover on the behavior of the GA in dynamic environments: a case study using the shaky ladder hyperplane-defined functions. In: Proceedings of the 8th annual conference on, GECCO, Seattle (2006)
14. Eldershaw, Cameron, S.: Using genetic algorithms to Solve the motion planning problem. Journal of Universal Computer Science (April 2000)
15. De Jong, K.A.: The Analysis of the Behaviour of a Class of Genetic Adaptive Systems. Ph.D. dissertation. Department of Computer Science. University of Michigan, Ann Arbor
16. Grefenstette, J.J.: Optimization of control parameters for genetic algorithm. IEEE Transaction on Systems, Man and Cybernetics 16(1), 122–128 (1975/1986)
17. Schaffer, L., Caruana, R., Eshelman, L., Das, R.: A Study of Control Parameters Affecting Online Performance of Genetic Algorithm for Function Optimization. In: Schaffer, J.D. (ed.) Proceeding 3rd International Conference Genetic Algorithms, pp. 51–60. Morgan Kaufmann, San Francisco (1989)
18. Cliff, D.: Minimal Intelligence Agents for Bargaining Behaviours in Market Environment. Technical Report HPL-97-91. Hewlett Packard Laboratories (1997)
19. Cliff, D.: Genetic optimization of adaptive trading agents for double-auction markets. In: Proceedings Computing Intelligent Financial Engineering (CIFEr), pp. 252–258 (1998)
20. Cliff, D.: ZIP60: Further Explorations in the Evolutionary Design of Trader Agents and Online Auction-Market Mechanisms. IEEE Transactions on Evolutionary Computation (2006)
21. Cliff, D.: Evolution of market mechanism through a continuous space of auction types. In: Proceeding Congress Evolutionary Computation, pp. 2029–2034 (2002)
22. Cliff, D.: Visualizing search-spaces for evolved hybrid auction mechanisms. Presented at the 8th Int. Conf. Simulation and Synthesis of Living Systems (ALifeVIII) Conf. Beyond Fitness: Visualizing Evolution Workshop, Sydney (2002)
23. Choi, J.H., Ahn, H., Han, I.: Utility-based double auction mechanism using genetic algorithms. Expert System Applicationl, 150–158 (2008)

24. Anthony, P.: Bidding Agents for Multiple Heterogeneous Online Auctions. PhD thesis. University of Southampton, UK (2003)
25. Anthony, P., Hall, W., Dang, V.D., Jennings, N.R. : Autonomous agents for participating in multiple online auctions. In: Proc. of the IJCAI Workshop on EBusiness and the Intelligent Web, Seattle (2001)
26. Beasley, D., Bull, D.R., Martin, R.R.: An Overview of Genetic Algorithms: Part 2, Research Topics. University Computing 15(4), 170–181 (1993)
27. Anthony, P., Jennings, N.R.: Evolving bidding strategies for multiple auctions. In: Proceedings of the 15th ECAI, pp. 178–182 (2002)
28. de Silva, U.C., Suzuki, J.: On the Stationary Distribution of Gas with Fixed Crossover Probability. In: Genetic And Evolutionary Computation Conference Proceedings of the 2005 conference on Genetic and evolutionary computation, Washington DC, USA, pp. 1147–1151 (2005)
29. Alvarez, G.: Can we make genetic algorithms work in high-dimensionality problems? Stanford Exploration Project (SEP) report 112 (2002)
30. Taniguchi, N., Ando, N., Okamoto, M.: Dynamic Vehicle Routing and Scheduling With Real Time Travel Times on Road Network. Journal of the Eastern Asia Society for Transportation Studies 7, 1138–1153 (2007)
31. Ho, W., Ji, P., Dey, P.K.: Optimization of PCB component placements for the collect-and-place machines. The International Journal of Advanced Manufacturing Technology 37(7), 828–836 (2008)
32. Gen, M., Zhou, G.: A genetic algorithm for the mini-max spanning forest problem. In: Whitley, D., Goldberg, D.E., Cantu-Paz, E., Spector, L., Parmee, L., Beyer, H.-G. (eds.) Proceedings of the Genetic and Evolutionary Computation Conference 2000, p. 387. Morgan Kaufmann Publishers, San Francisco (2000)
33. Tsujimura, Y., Gen, M., Syarif, A.: Solving a Nonlinear Side Constrained Transportation Problem by Using Spanning Tree-based Genetic Algorithm with Fuzzy Logic Controller. In: Proceedings of the Congress on Evolutionary Computation 2002, vol. 1, pp. 546–551 (2002)
34. Kim, K., Murray, A.T., Xiao, N.: A multiobjective evolutionary algorithm for surveillance sensor placement. Environment and Planning B: Planning and Design, Pion Ltd., London 35(5), 935–948 (2008)
35. Kajisha, H.: Synthesis of Self-Replication Cellular Automata Using Genetic Algorithms. In: Proceedings of the IEEE-INNS-ENNS international Joint Conference on Neural Networks (IJCNN 2000), July 24 - 27, vol. 5, p. 5173. IEEE Computer Society, Washington (2000)
36. Scholand, A.J.: GA Parameter (1998), `http://eislab.gatech.edu/people/scholand/gapara.htm`
37. Back, T.: Optimal Mutation Rates in Genetic Search. In: Proceedings of the 5th International Conferences of Genetic Algorithms, pp. 2–8. Morgan Kaufmann, San Francisco (1993)
38. Bingul, Z., Sekmen, A.S., Palaniappan, S., Zein-Sabatto, S.: Genetic algorithms applied to real time multiobjective optimization problems. In: Proceedings of the 2000 {IEEE} SouteastCon Conference (SoutheastCON 2000), pp. 95–103 (2000)
39. Qin, L., Yang, S.X., Pollari, F., Dore, K., Fazil, A., Ahmed, R., Buxton, J., Grimsrud, K.: Genetic algorithm based neural classifier for factor subset extraction. Soft Computing - A Fusion of Foundations, Methodologies and Applications 12(7), 623–632 (2008)

40. Yamamoto, M., Zaier, R., Chen, P., Toyota, T.: Decision-making method of optimum inspection interval for plant maintenance by genetic algorithms. In: Proceedings of EcoDesign 2001: Second International Symposium on Environmentally Conscious Design and Inverse Manufacturing, pp. 466–469 (2001)
41. Shieh, H.M., May, M.D.: Solving the capacitated clustering problem with genetic algorithms. Journal of the Chinese Institute of Industrial Engineers 18(3), 1–12 (2001)
42. Dagli, C.H., Schierholt, K.: Evaluating the performance of the genetic neuro scheduler using constant as well as changing crossover and mutation rates. In: Proceedings of the 21st International Conference on Computers and Industrial Engineering, pp. 253–256 (1997)

PSO (Particle Swarm Optimization): One Method, Many Possible Applications

Federico Cecconi and Marco Campenní

Abstract. *Particle Swarm Optimization (PSO) is an optimization technique, deriving from the EO* [5]: the main features are the natural inspiration and the possibility to implement *PSO* onto different levels. This chapter is divided in three section: (1) the *PSO* definitions and relationship with *MAS* (Multi Agent Systems) framework; (2) three applications of *PSO* methods; (3) some general conclusions and perspectives. We try to show that *PSO* has a marked multidisciplinary character since systems with swarm characteristics can be observed in a variety of domains: the main argument in favor to *PSO* is proper the multidisciplinary character. Besides, *POS* can resolve multiobjective otpimization problems in efficient way, because *POS* naturally incorporates some concepts from *Pareto-Optimal* framework.

1 *PSO* and Evolutionary Search

Evolutionary optimization (*EO*) is a subject of intense interest in many fields of study. The generic goal is to determine values for model's parameters or state variables that provide the best possible solution to a predefined cost or objective function, or a set of optimal trade off values in the case of two or more conflicting objectives [12] [10] [6] [26] [13]. Numerous approaches have been proposed to efficiently find best solutions for one-objective function problems and Pareto-optimal solutions for complex multi objective optimization problems. In particular, evolutionary algorithms have emerged as a powerful approach for solving search and optimization problems involving multiple conflicting objectives. Beyond their ability to search intractably large spaces for multiple Pareto-optimal solutions, these

Federico Cecconi
LABSS-ISTC, Via Palestro 9 ROME Italy
e-mail: federico.cecconi@istc.cnr.it

Marco Campenní
LABSS-ISTC
e-mail: marco.campenni@istc.cnr.it

R.A. Sarker and T. Ray (Eds.): Agent-Based Evolutionary Search, ALO 5, pp. 229–254.
springerlink.com

algorithms are able to maintain a diverse set of solutions and exploit similarities of solutions by recombination. Although the multi-objective optimization problem has been studied quite extensively ([25] [15]), current available evolutionary algorithms typically implement a single algorithm for the population's evolution. Reliance on a single biological model of natural selection and adaptation presumes that a single method exists that efficiently evolves a population of potential solutions through the parameter space. However, existing theory and numerical experiments have demonstrated that it is impossible to develop a single algorithm for population's evolution that is always efficient for a diverse set of optimization problems. In recent years, memetic algorithms (also called hybrid genetic algorithms) have been proposed to increase the search efficiency of population based optimization algorithms. These methods are inspired by models of adaptation in natural systems, and use a genetic algorithm for global exploration of the search space, combined with a local search heuristic for exploitation. Memetic algorithms shown to significantly speed up the evolution toward the global optimal solution for a variety of real-world optimization problems. However, our conjecture is that a search procedure adaptively changing the way by means of which is generated the offspring and based on the shape and local peculiarities of the fitness landscape, will further improve the efficiency of evolutionary search. This approach is likely to be productive because the nature of the fitness landscape (objective functions mapped out in the parameter space, also called the response surface) often varies considerably between different optimization problems, and dynamically changes en route to the global optimal solutions.

Particle Swarm Optimization (PSO) is an optimization technique, deriving from the EO [5]: the main features are the natural inspiration (like evolutionary techniques of course) and the possibility to implement *PSO* onto different levels, from the agent's level (like a search problem in a physical environment), until solutions' space level.

We start the description of *PSO* from another bio-inspired group behavior model, the Ant colony optimization, a technique for combinatorial optimization, inspired from ant behavior [8]. Ant colony optimization was introduced as a technique for combinatorial optimization (for a recent review see [7]). The inspiring source of ant colony optimization model is the foraging behavior of real ant colonies. At the core of this behavior is the indirect communication between ants by means of chemical pheromone trails, which enables them to find short paths between their nest and food sources. This characteristic of real ant colonies is exploited in ant colony optimization algorithms in order to solve, for example, discrete and continuous optimization problems, and to manage tasks such as routing and load balancing in telecommunication networks and, more recently, in mobile ad hoc networks such as sensor networks. *The swarm optimization and the ant optimization have certain characteristics in common.*

Can we image a swarm of agents not only looking for resources but imitating behaviors each other too? The advantage using this approach is the flexibility and the scalability of algorithms: the chapter is largely dedicated to discuss this objective.

In *PSO* we define $\mathbf{x_k}$ a vector with the position of particle k, $\mathbf{v_k}$ the current velocity of particle k and the heading for particle k, $\mathbf{p_k}$ (in other words $\mathbf{p_k}$ is the best position found by particle k). We compute g, the index (*the ID*) of the particle that has found the best position. We define $Q(\mathbf{x})$ the quality of position $\mathbf{x}$. In other words, Q is either the value of objective functions or some valuation about objective functions in multi objective optimization.

$$g = argmax_k\, Q(\mathbf{p_k})$$

Now we write down the velocity dynamic in this form

$$\mathbf{v_k^{t+1}} = [\mathbf{wv_k} + \varphi_1(\mathbf{p_k} - \mathbf{x_k}) + \varphi_2(\mathbf{p_g} - \mathbf{x_k})]_{\mathbf{at\,time\,t}}$$

The equation is, essentially, the description of an acceleration dynamic: (a) in $w\mathbf{v}_k$ $w \in [0,1]$ is an inertial factor: particle tends to maintain direction. $\varphi_1(\mathbf{p}_k - \mathbf{x}_k)$ is the so-called *cognition* factor: particle should go toward private best; φ_1 is a random uniform variable distributed over $[0,2]$. $\varphi_2(\mathbf{p}_g - \mathbf{x}_k)$ is the social part of the dynamic: particle should go toward public best.

We update position with

$$\dot{\mathbf{x}}_k = \mathbf{v}_k$$

We could insert *PSO* in the family of hybrid algorithms[1]: in PSO (1) we apply *forces* to individuals in evolving populations to drive toward solutions (like a genetic algorithm) (2) we modify the direction of the evolutionary walk, using some kinds of intra-individual interactions, for example swarm imitation (go almost in the same direction of the nearest individual). So, in PSO the inspiration from the nature is twofold: on the one hand, PSO is a population-based search algorithm, inspired from the natural selection. On the other hand, the intra-individual interactions are inspired from a large collection of natural phenomena: for example swarm behavior (follow me !), social imitation, ant colony behavior, herd behavior. The essence of these systems is that they exhibit flexibility, robustness and self-organization. These features are central in multi objective optimization (see section 2.2 and 2.3).

This chapter is divided in three section: the section 2 containing the *PSO* definitions and relationship with *MAS* (Multi Agent Systems) framework. The 3, with three applications of *PSO* methods. Finally, the 4 section, with some general conclusions and future perspectives.

2 *PSO*: Algorithms Inspired by Nature *Twice*

There is a long history about the attempts to imitate nature developing artificial manufacts. Classical examples are the evolutionary computations [33] [21]. The assumption is that it's possible the imitation of the method the nature uses to generate new adaptations to new environments, the natural selection. If we substitute the natural environment with an artificial one, we can implements in stochastic algorithms the same methods that the natural selection uses during the evolution: (a) some kind

[1] This classification is not standard, see sub-section 2. We think that this kind of description leads to a correct identification of PSO's advantages and bounds.

of fitness (objective) function, (b) some description of the problem in terms of genotypes' populations; (c) some mechanism to reproduce most the more fit genotypes, (d) some kind of mutation's operator. So, *EO* imitates nature's robust way of evolving successful organisms. Those ones ill-suited to an environment die off, whereas fit ones reproduce. We could say that, in a some way, *PSO* is an *evolution of evolutionary algorithm paradigm.*

As we mention in the previous section, *PSO* uses a population of particles to cooperatively search for the best solution to a problem, represented by positions in a space, where each location has a "quality" ("goodness") determined by some objective function. Each particle *k* has a current position, a current velocity, and a *remembered* best location. In addition, all (some of) particles know the index of the particle that has found the best position so far. The velocity update rule is function of private information and social influence. Of course, the position is a function of velocity. Thus, *PSO* is a *double* natural algorithm: *PSO* imitates the evolutionary dynamic and the imitation dynamic between individuals. From this kind of double similiraty with the nature, derive two interesting features that we fully discuss in the rest of the chapter: flexibility in swarm-like problem, scalability from real agents to abstract layers of description.

From a more general point of view, we should insert *PSO* into the classification of optimization techniques. One dimension is the degree of objective function, another dimension is the linearity. PSO inherits from evolutionary optimization the substantial *indifference* regarding degree-lineraty properties of objective functions. From this point of view, *PSO* is not a classical gradient algorithm and the *form* of objective function surface matters only for the problems of local minima.

Now, we face up to the description of more common implementation details for *PSO*. The list will be incomplete: the study of PSO implementation is one of the main direction in PSO field [19]. (1) The topology. Is it always true that every particle knows the status of all the others? No. We should realize a topology of connections in such a manner that a particle sees only a subset of particles. For example, the topology may be a circle (the particle sees only two particles). In circle case the information is a flow. (2) The spatial extension. The particle may have a body. Using spatial extension it is possible to avoid some kind of premature convergence, in particular in problems with a lot of objective functions and a strong trade off between them. (3) Friction. If the velocity dynamic has a friction component we would obtain some advantages in terms of convergence stability. This is particularly true in problems with strong constraints, like micro economic framework (the prices have to be positive, and also the consumers' choises ......). (4) The integration with local search. The inspiration coming from learning in higher animals.

2.1 Definitions

As we seen before, *PSO* is a stochastic, population-based optimization algorithm. The original version was introduced by Kennedy and Eberhart [17] (a complete description of the method is in [18]). In *PSO*, a swarm of particles moves (*flies*) in an n-dimensional space attempting to discover the best solution. Now we try to implement the method in a quite simple problem. Suppose the following scenario: bugs are randomly searching food in an area. In the area, there are two pieces of food being searched but the bugs do not know where the food is. Instead, they know how far the food is. So what is the best strategy to find it? The effective one is to follow the bug which is nearest to the food. We use a netlogo example[2] to illustrate this central point in a classical *PSO* framework: the particles imitate the velocity properties each others.

See figure 1. The red circles are agents that will incorporate the solution of the problem, the x, y points that corresponds to max value for **f**. In figure 1 we show how the simulation starts. In figure 2 we show a state near to stability. The interesting feature is that some agents are trying to heading toward the smaller minimum: anyway, the distribution of particles indicates that the algorithm has correctly computed the relative dimension of the minima (in fact we find more particles around the larger minimum).

In this chapter we show some variations in a classical *PSO* framework to address some typical issues in a evolutionary search: (1) particles can emit some kind of signal to indicate a modification in own internal state. For example, particles can emit a signal if they modify their velocity; (2) the algorithms that control the velocity gradient could implement some rules linked with their position: we will see an example in this chapter in the microeconomy example. (3) We use a *PSO* to address a typical problem in artificial life field: the trade off between a tactical behavior (avoid a predator) and a strategical behavior (stay in a zone with a lot of resources). In this case we show that *PSO* can incorporate some constraint directly in the objective function, without degradation in the performance.

2.2 *PSO and Multiobjective Optimization Problems*

PSO is especially fit in multi-objective optimization problem (*MoOP*). So, we try to sketch the *MoOP*, and to describe how the *PSO* methods can tackle some main aspects of *MoOP*. In *MoOP*, we define an objective function

$$\mathbf{f} : \Re^{\mathbf{n}} \rightarrow \Re^{\mathbf{m}}$$

where n and m are respectively the dimension of the decision vector and the dimension of the objective vector. In minimization problems we would minimize

$$\mathbf{y} = (f_1(\mathbf{x}), \cdots, f_m(\mathbf{x}))$$

where **x** and **y** are decision and objective vectors. The essence of the problem is the conflict between different objective functions. For instance, the user may like to optimize a number of different concurrent objectives: in a micro economical context,

[2] To refer to this model please use: MacLennan, B.J. (2007). NetLogo Particle Swarm Optimization model. http://www.cs.utk.edu/ mclennan. Dept. of Electrical Engineering and Computer Science, Univ. of Tennessee, Knoxville.

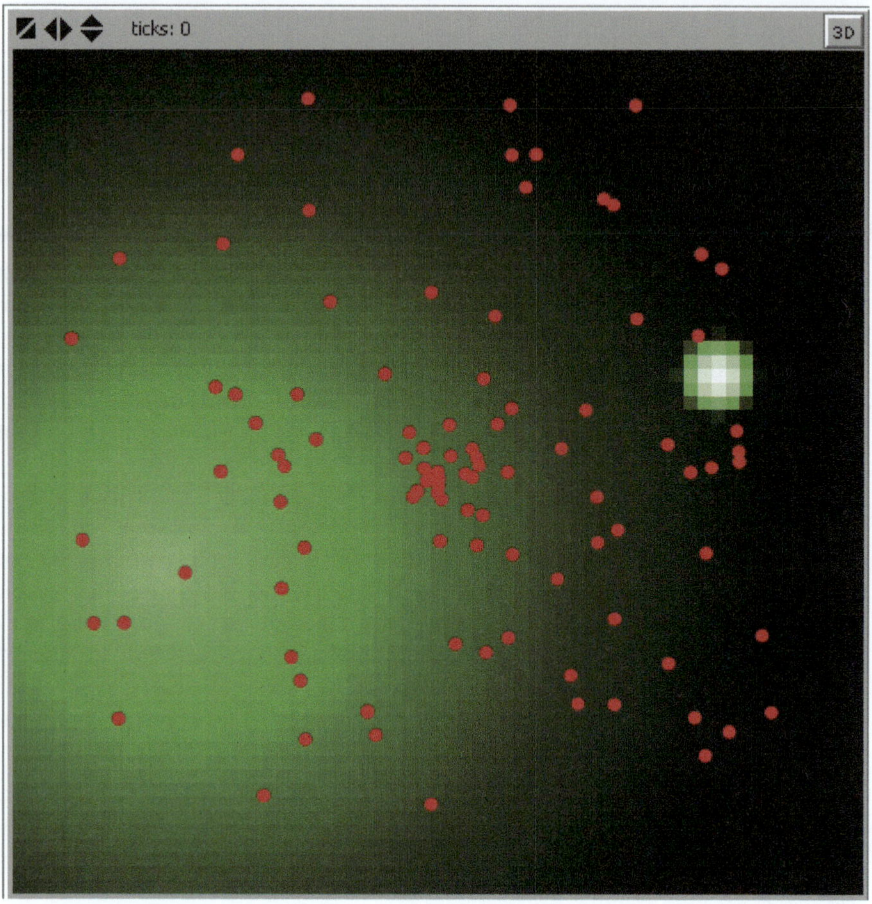

Fig. 1 The figure shows an example of two peaks objective function **f**. The gradient toward white indicates larger function values

it is a common situation trying to maximize the utility for the consumers and the profits for the firms. Otherwise, in evolutionary robotics, the artificial agents must explore an environment with efficiency and saving time, or we can study the conflict between tactical and strategical behavior in an artificial life framework, with a simple organism eating in different zones, with different quality of the food and different risk to meet a predator. In *MoOP* (and in general in optimization problems) the constraints play a role: we define functional constraints and side constraints. The functional constraints represent limitation on the system's variables that are limitations on the performance. For example the market clear constraint in a Walrasian market is the constraint that all the goods have to be used to satisfy the consumer consumption set (at a fixed price vector); the limitations that the good quantity have

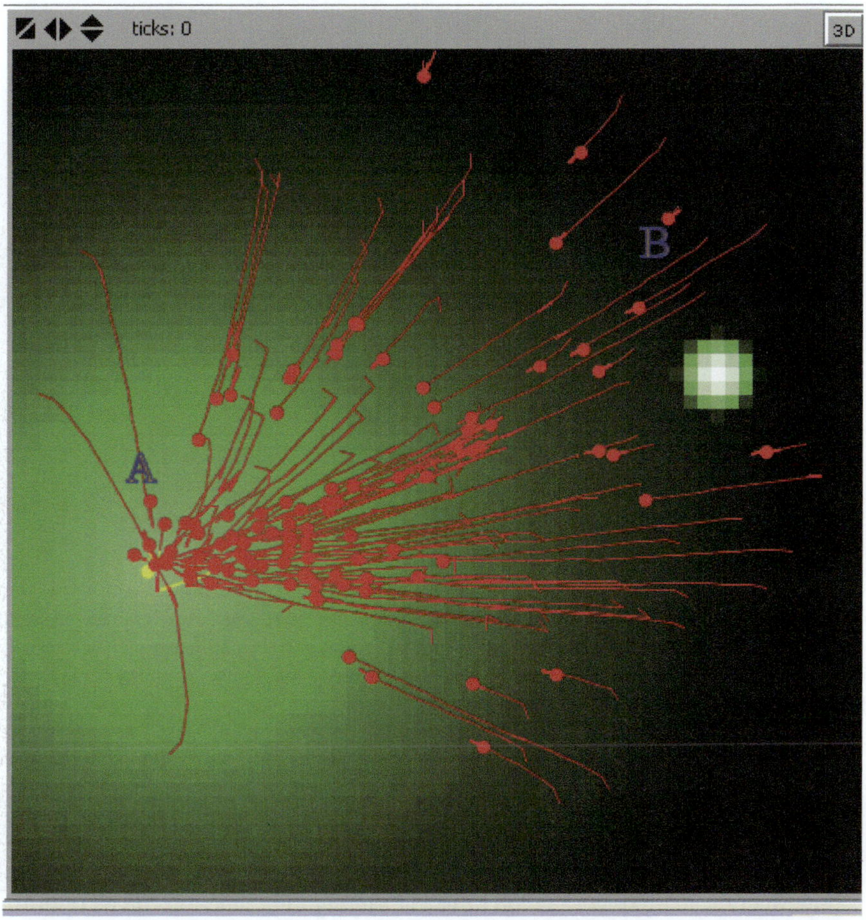

Fig. 2 We show a *PSO* dynamic: the agents (particles) in *A* are coming toward the big maximum of **f**. Some agents in *B* are trying to resist near the second, smaller, maximum

to be positive is a side constraint (i.e. we exclude the existence of negative good, like garbage).

Now, we return to the *MoOp*. Firstly, we recall the concept of Pareto frontier in an informal way. Given a decision vector **x**, a change of one element of **x** that can make improve the performance of at least one objective function without making worse any other ones, is called a Pareto improvement. An allocation is Pareto efficient or Pareto optimal when no further Pareto improvements can be made. The Pareto frontier in the space of solutions contains only Pareto optimal points.

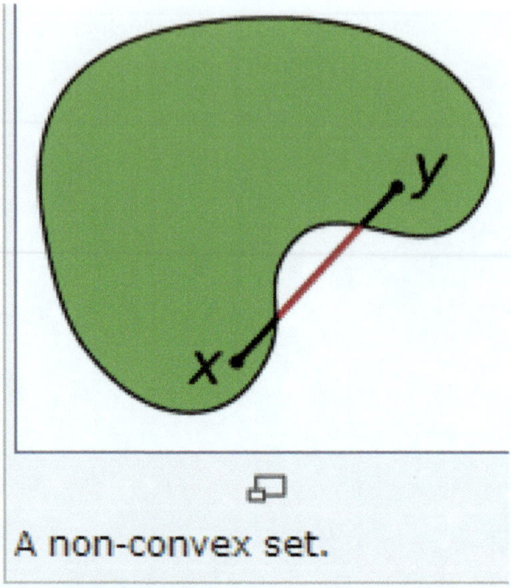

Fig. 3 An example of non-convex set. In spite of the huge confirmations about the problems of *aggregate strategy* in convex optimization problems, a great numbers of works in the field of evolutionary engineering and artificial life use methods based on a simple concept: they mix the multi objective functions into one, weighting the components one at the time. Modern handbooks about this issues are [23] and [29]

Finally, how can we define the minimization of a completive objective function, if the objective function conflict? We have three possibilities (see [28]):

1. *to aggregate*. We could mix the objective functions onto one: we define the resulting function an *aggregating function*. The aggregating function can be optimized using classical methods, gradient or evolutionary based. But we have a problem: if the aggregating function leads to a non-convex pareto frontier the optimization is often difficult (see figure 3).
2. *to use subpopulations*. Some authors use subpopulation to optimize each objective independently, using proportional selection. Crossover and mutation operators shuffle together the subpopulations. The only limit for this approach is the complexity.
3. *to incorporate pareto-optimality into the search*. The idea is elementary (... only after one person has spoken of it ...): we select the individual for the next generation using pareto dominance criteria, instead of the simple valuation of multi objective functions. This approach is very promising: the real difficult is the study of the correct ratio between number of non-dominated individuals and the parameters about offspring and parents.

In this chapter we suggest the idea that, in some cases, *PSO* can resolve *MoOP* in efficient way, because *PSO* incorporate some concepts from Pareto-Optimal framework, even though an evolutionary search algorithm, without the complexity of subpopulation schema. In section 4 we discuss which cases apper to be the *positive* for POS.

There is another interesting features of *PSO*: the possibility of different levels of descriptions. We start with the agent-in-the-environment level. In this case, the particles are agents, the velocity and the position are real dynamic variables and (using software simulation) we can introduce in the system other artificial variables, for example an emitted field, that the particles may perceive. Then, we go into a second level: the particles perform the role of some agents in the problem, for example the particles are economic agents in a microeconomic optimization scenario. Now, position and velocity behave like dynamic variables. We would introduce the concept of inertia, in the sense of the tendency to maintain some attitude respect to the model (for example attitude to risk). But, the particles are not a physical entities. Finally, we could use *PSO* at solution-of-the-problem level. In this case the position of the particles, and velocity too, are points in the solutions' space.

2.3 *PSO: A Population-Based Technique*

In a word, what is *PSO*? It is a simple, population based, stochastic optimization method. From the point of view of final users, the main questions is twofold: how can we make sure that we have convergent trajectories? do we really understand the behavior of PSO in terms of ... guaranteed convergence on global minima ...? A complete review about convergence of trajectories is in [31] and in PhD thesis of F. van der Bergh (see [2]).

We can resume the main result: it has been proved that particles converge to stable points. This result derives from the dynamic-style of PSO formulation. Trajectories of particles are trajectories in the classical dynamic sense. But, for the question of convergence on minima, the answers are much more messy. In this chapter we use an *empirical* approach to this problem; anyway we would try to skecth some general point about PSO minima serching properties.

What are the orgins of *PSO*? The initial objective was to simulate the graceful, unpredictable choreography of birds in a flock. Then, the flock's simulation is becomed the simulation of a *biased* flock, with birds that know where is the *best* position to catch. In this frame, we talk about i) cognitive components to indicate the capacity to quantify own performance and ii) social components to indicate the capacity to quantify the influence of the others on my performance. In this sense, we can use *exploration Vs exploitation* concept to define (a) the ability to explore (b) the ability to concentrate the search around a promising area. We could express this argument in terms of minima search: *PSO* has a great number of parameters which could directly influence the ratio ρ between exploration and exploitation. Thus, if a problem has a structure in such a manner that ρ is a critical parameter, *PSO* is a good candidate to resolve it. The more typical example is the search in peer-to-peer networks. In [20] authors show a performance of an ant colony optimization techinque in the framework of the searching of resources in a peer-to-peer network. The argument about the avoiding of exploitation-exploration dilemma using ant colony optimization can be replicate in *PSO*. We resume this point showing one of the analytic results in [31]. We may define the velocity update with

$$v_{ij}(t+1) = \chi\left[v_{ij}(t) + \varphi_1(y_{ij}(t) - x_{ij}(t)) + \varphi_2(\hat{y}_j(t) - x_{ij}(t))\right]$$

where $\hat{y}_j(t)$ is the global best position, $y_{ij}(t)$ is the particle best position and χ is

$$\chi = \frac{2\kappa}{\left|2 - \varphi - \sqrt{\varphi(\varphi - 4)}\right|}$$

and

$$\varphi = \varphi_1 + \varphi_2$$
$$\varphi_1 = c_1 r_1$$
$$\varphi_2 = c_2 r$$

With these assumpitons, Bergh find that

- with $\varphi \geq 4$ and $\kappa \in [0,1]$ the swarm is guaranted to converge
- κ controls the exploration-exploitation; with κ near zero we have fast convergence and local exploitation; with κ near 1 we observe a global search.

This direct link between searching and parameters is a chief asset of *PSO*.

Now, let's move on to another point about minima searching by *PSO* [22]: the five basic principles of swarm intelligence.

1. Proximity, i.e., the swarm must be able to perform simple space and time computations.
2. Quality, i.e., the swarm should be able to respond to quality factors in the environment.
3. Diverse response, i.e., the swarm should not commit its activities along excessively narrow channels.
4. Stability, i.e., the swarm should not change its behavior every time the environment alters.
5. Adaptability, i.e., the swarm must be able to change its behavior, when the computational cost is not prohibitive.

We focus principle four , stability point, …the swarm should not change behavior every time the environment changes …. This features for *PSO* is especially present if *PSO* is using some kind of environmental mark, for example fields, trails and so on. If the principle four is satisfied, *PSO* have the advantage over classical population-based algorithm when the problem incorporates some kind of environmental drift. A good example is the problem to find mimima in repetead games, in which the payoffs degrade during the time (you could imagine the dividing up of an ice). In these kind of problems (very studied in evolutionary game theory) *PSO* should be a good compromise.

Finally, we show the interaction beetween *PSO* and rough fitness landscape using an applet realized by Project Computing Pty Ltd (see [3]; using this applet, generations are represented by the following colours: green, yellow, magenta, blue, black and black is the older).

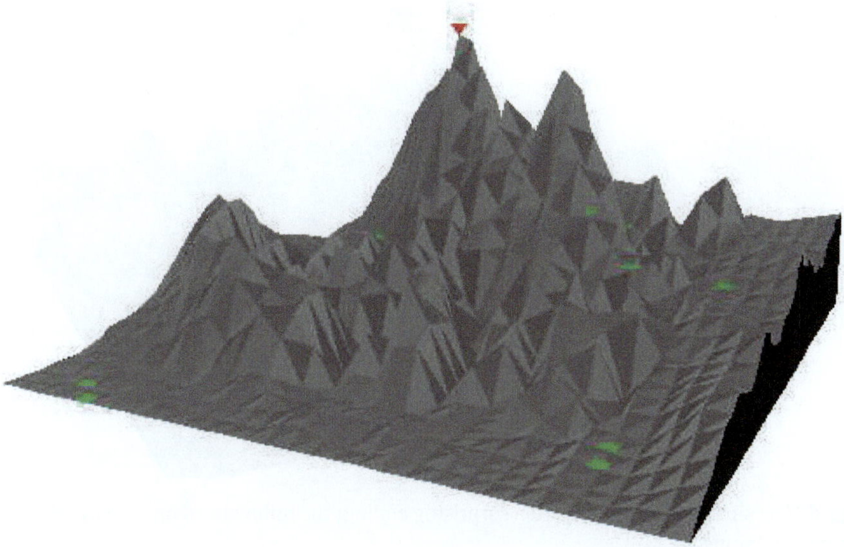

Fig. 4 The figure shows the behavior of *PSO* in a rough fitness landscape. Here is starting a simulation. Agents have to find the global maximum indicated by a red arrow

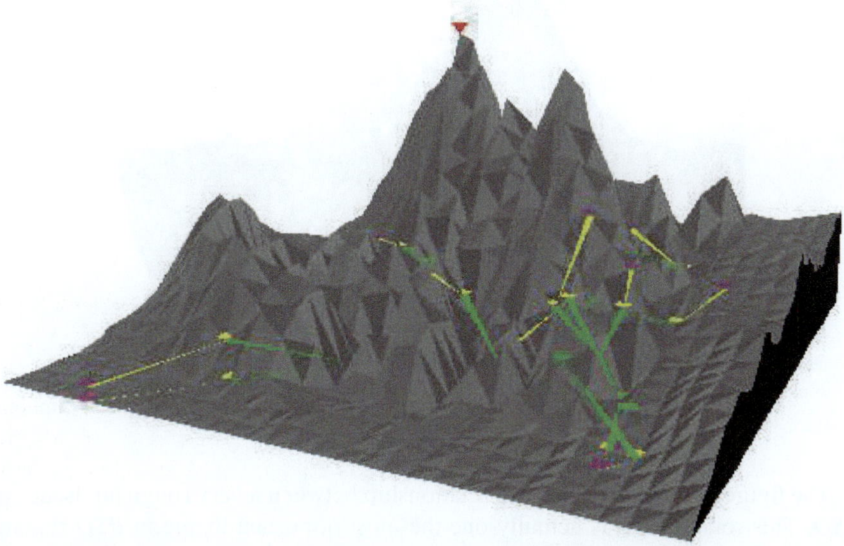

Fig. 5 In this figure we show the dynamic of PSO exploration. We observe two different trends: locally, agents look for maxima. The agents near the global maxima *drift* the population

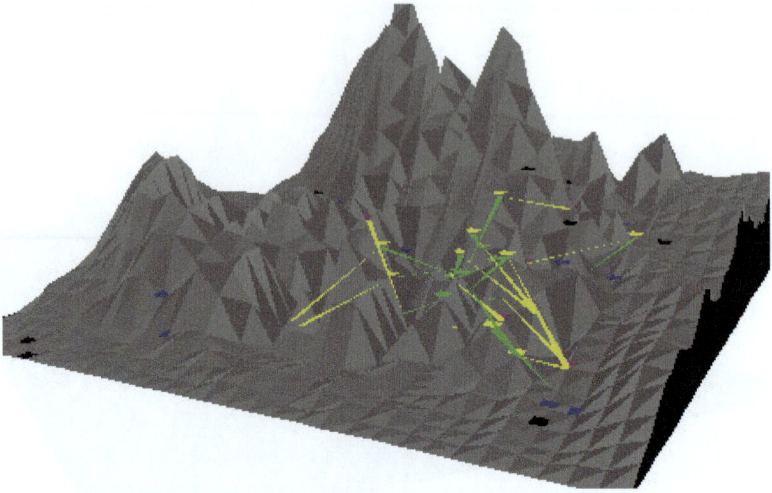

Fig. 6 The simulation goes toward a surprising ending: the influence of apparent global maximum is winning

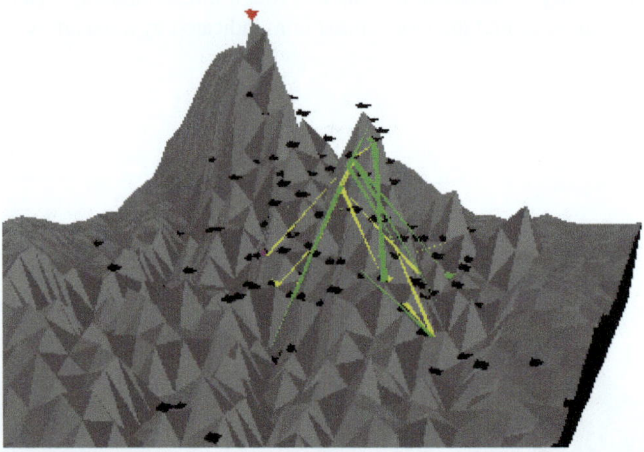

Fig. 7 The end. The population converge on a local maxima

The figure from 4 to 7 show the relationship between a very rough landscape and *PSO*. This relationship is actually one the most important theme in *PSO* research: what are the forms of fitness landscape make the initial conditions of the simutaion a critical factor for the convergence on global maxima?

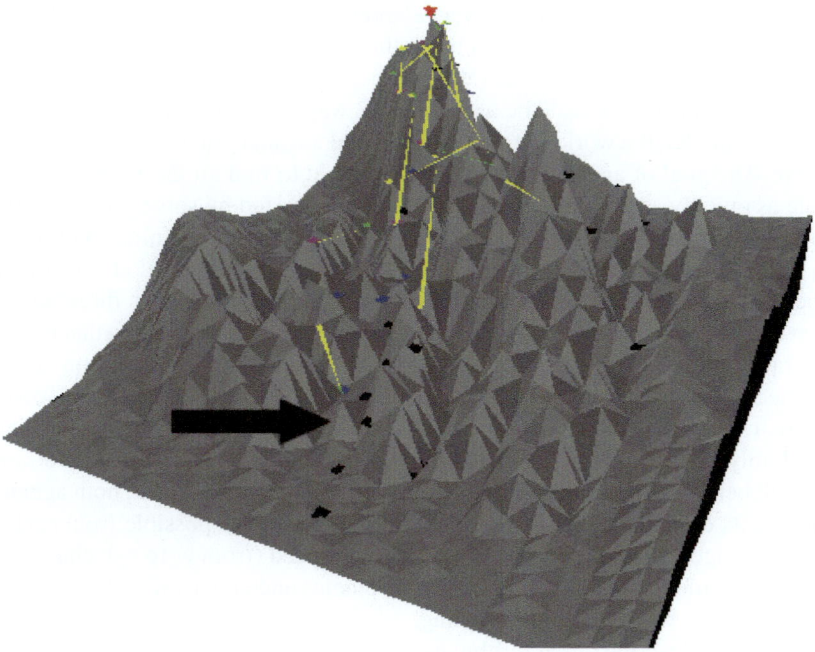

Fig. 8 We show a different simlation over the same landscape. The ending is completly different

3 Case Studies

3.1 Looking for Resources

Evolutionary robotics is an emerging area of research within the much larger field of autonomous robots. One of the primary goal of evolutionary robotics is to develop automatic methods and algorithms for creating intelligent autonomous robot controllers, and to do it in a way that does not require direct programming by humans (in an instructionist way). The primary advantage of robot design methods that do not require hand coding or in depth human knowledge is that they might one day be used to produce controllers or (even whole) robots that are capable of functioning in environments that humans do not understand well [24].

Our interest is focused on the implementation of very simple agents in a simulated environment. The main idea is to be able to realize robots that can move in a complex environment for exploring it and for finding useful resources; to create true artificial life concerns not only with intelligent control, but further with the capacity of the robot to find its own resources through foraging (looking for food, which includes both energy and spare parts). This is related to autonomous foraging, a concern within the sciences of behavioral ecology, social anthropology, and human behavioral ecology; as well as robotics, artificial intelligence, and artificial life [9]

[30] [11]. These robots should be very elementary as to the components; they should have only an engine to move own body, the possibility to emit a signal (e.g. a magnetic one) and a certain number (at least one) of sensors to detect the signal emitted by other robots (this should be a very simple way to allow them to communicate).

In our model, the world is a two-dimensional toroidal grid on which simple agents move. The goal of agents is to explore the world to find all the food sources available as quickly as possible. Each agent is characterized by some properties that are different for each individual: a speed v_1 (basic), a speed v_2 (minor and other than v_1, that the agent adopts when is in an area where there is food), a direction, a speed of rotation (relative to what the agent abruptly changes its direction), a flag that indicates the interaction of the agent with an environment's area that has special characteristics (i.e. the presence of food). The simulator updates position in function of the velocity as in a classic *PSO*. The resources of food are distributed worldwide in such a way as to create a few islands more rich in food and small amounts of food randomly distributed. In the model there are some simple rules that determine the overall behavior of the agents. If an agent A is close to an agent B, both agents must change its direction of movement so as to move away as possible from each other. When an agent finds the food: changes its own speed (from v_1 to v_2); changes its direction in order to make more circular movements and emits a signal that propagates as a magnetic wave in the surrounding.

Then, we describe our algorithm using a pseudo-code notation (see algorithms 1, 2, 3, 4).

Algorithm 1. The general algorithm. The number of agents is N. A is the population. h is the heading of the agents.

repeat
  Agents interact one each others.
  Agents interact with resources.
  Agents interact with field.
  for $i = 1$ to N **do**
    $A_i.h \leftarrow A_i.hr * \rho$ {This is the part of *PSO* algorithm managing the social interactions between agents. The heading of the agents is a function of r, the angular velocity. r is set by the interactions with others agent}
  end for
until

r is the angular velocity of the agents: it can be set to three values, $-1,1,0$. $r = 0$ indicates no angular rotation, the sign indicate the direction of rotation. ρ is a parameter that sets the strength of social interactions.

In this case *invert* means that the rotation of one the interacting agents is inverted: the inversion happens only if the initial rotation is different by zero (see algorithm 2).

When agents meet resource the velocity decrease and the heading is changed (see algorithm 2).

Algorithm 2. Agents interact one each others.

```
for i = 1 to N do
  if There is some agents in the interaction's range (choose one of them) then
    j ← one of the two agents interacting.
    invert A_j.r
  end if
end for
```

Algorithm 3. Agents interact with resources.

```
if Agent i stays on a patch with resources then
  to decrease velocity. to modify heading with an uniform casual variable [−30,30]
end if
```

Algorithm 4. Agents interact with field.

```
for i = 1 to N do
  to follow field's gradient. {In this part of the PSO algorithm each agent modifies its
  heading to follow the max gradient}
end for
```

When agents meet field, they try to follow the gradient of the field.

In our model, the signal is different from the chemical signal emitted by termites to report a source of food to theirs comrades; even if in our model the signal is subject to evaporation over the time, (a) the spread is in concentric relation to the origin and (b) the signal can added (if there is another) with the signal produced by another agent in another part of the world; (c) the signal is emetted if the agent modifies its velocity (the methaphor is the electromagnetic field). Once a signal is emitted, the other agents follow the gradient of the signal, thereby obtaining informations about the location of the source of food.

We compared the behavior exhibited by our model with that obtained using traditional alternative models: i) a model of random walk, in which agents move randomly in the world, and ii) a model of flock in which each agent moves trying to minimize its distance by another agent and not to maximize it (as happens in our model).

The results show that our model offers always a better performance than that obtained using a random walk model. In the case of the comparison with the model of flock, PSO_{field} shows significantly better performance, under specific conditions: if the ecosystem agents-food-world presents a small source of food (between 10 and 50 units of food) and agents exploring the world are few (10 agents), the agents of our model can successfully explore the world searching for food in a significant time less than the agents of the traditional model of flock, managing to concentrate its efforts in areas where there is some food and non-dispersing energy searching for food in the rest of the world (see figure 11 and 10).

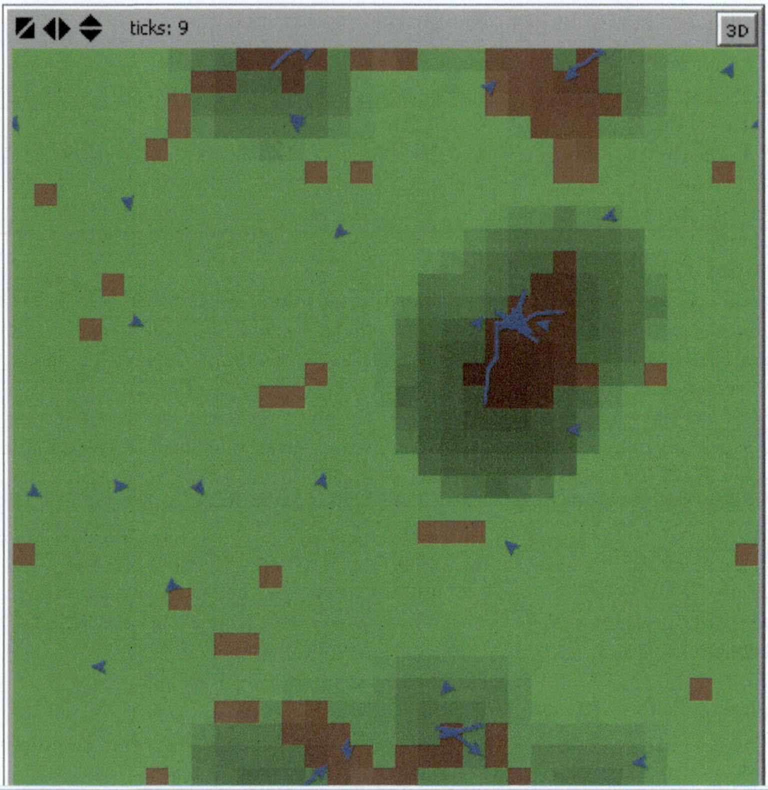

Fig. 9 A snapshot of PSO_{field}. It shows the distribution of the resources of food and the agents emitting field (the gray gradient) when modify their velocity

3.2 Microeconomy: General Equilibirum Theory

The tradition of General Equilibrium Theory (GE) started with [32]. It also was the main starting point for economics to become mathematical science. But it took a while for the theory to take off. Trying to solve problems like existence and stability of a general equilibrium, were mainly tackled in the 40s until the 70s of the 20th century (Arrow, Debreu, McKenzie and others, see [27]). Actualy, the theory suffers from some limitations, some are intrinsic to the theory, and secondly the omission of features, which are regarded as important characteristics of real-world economies. To mention some of them: transaction cost, informational problems, externalities (things the market does not account for, e.g. noise, pollution, etc.) and public goods. These kind of problems have been faced with some improvements in the semantyc of GE assumptions: for example we would be able to describe externalities like goods with negative consuptions bundle.

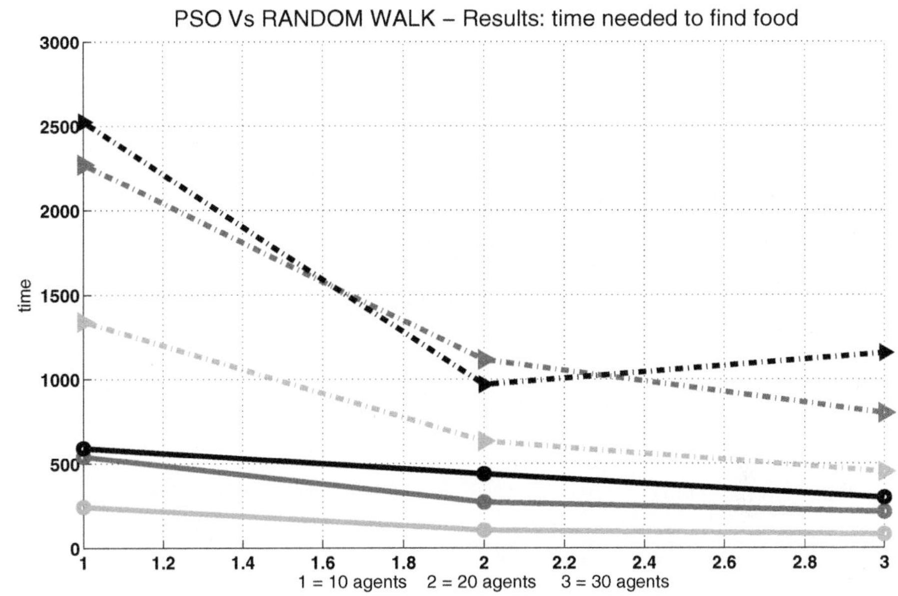

Fig. 10 The figure shows the results obtained by PSO_{field} compared with that obtained by a classic random walk. In this figure the circles indicate the performance of PSO_{field}; light gray lines correspond to 10 units of food; gray lines to 30; black ones to 50

The intrinsic problems of GE are a very important field for economic theory. So, we will discuss about one of the intrinsic problem of GE: the so-called lexicographic preference problem. We will show that the PSO approach permits to describe in an explicit way the real dimension of the lexicographic problem and to indicate some possible solutions.

We start with a generale introduction to GE. We study an economy composed by I consumers, J firms in which there are L commodities. The concepts of commodities (goods) is very large; for example we can suppose an economy with I consumers and two commodities; a private good, say labour, and a pubblic good. A private good can be consumed only once, so if consumer i consumes it, then it is unavailable for consumption by others. A pure pubblic good can be made available at no extra cost for every consumers. Each consumer i is associated with a consumption set $X_i \subset \Re^L$ and with a preference relation.

To fix this concepts, we imagine two consumers and two goods, apples and money.

$i = 1$ is real simple agent: one unit of money is equivalent to one apple. Clearly, it would increase money and increase apple eating. Before to a choice between two different combinations of goods, it sums the quantity of apples and the quantity of money and prefers the combination with higher value. For example it prefers

$$\langle 4\, apples, 2\, unit\, of\, money \rangle$$

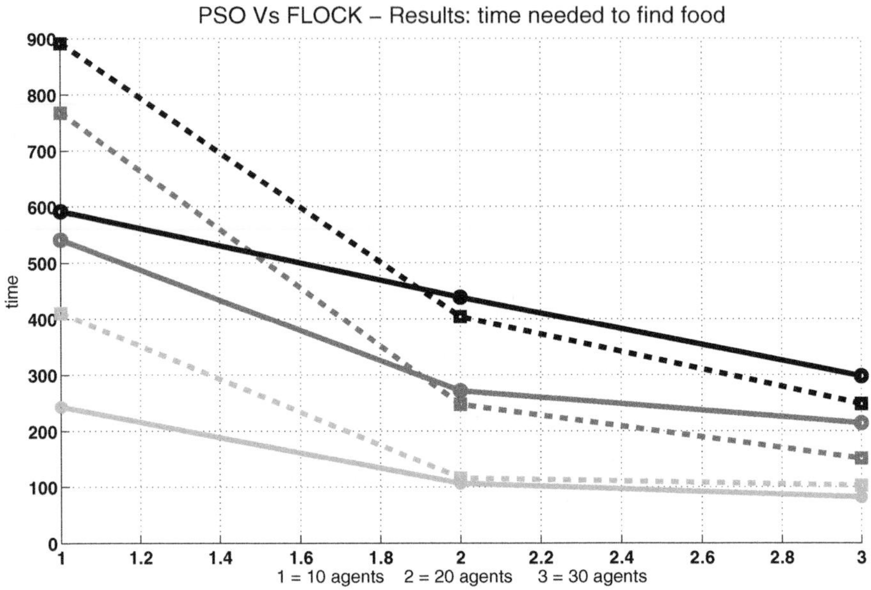

Fig. 11 The figure shows the results obtained by PSO_{field} compared with that obtained by a classic *flock* algorithm. The circles indicates the performance of PSO_{field}; light gray lines correspond to 10 units of food; gray lines to 30; black ones to 50. The performance is always better with a small number of agents for each amount of food (10, 30, 50 units)

and

$$\langle 5,3\rangle$$

to $\langle 3,2\rangle$.

The second one, $i = 2$, has more complex preferences: it values more apple than money. So, it obtains a larger payoff from $\langle 5\alpha,3\rangle$ respect to $\langle 3\alpha,5\rangle$ for $\alpha = 2$.

Our economy has initial resources of commodities, the economy's endowments, $\tilde{\omega} \in \Re^{\mathbf{L}}$. In our example we have $\tilde{\omega} = \langle \mathbf{0}, \mathbf{100}\rangle$, i.e. 0 apples and 100 unit of money. Finally we have only one firm (if you prefer a vendor), that has a production set $y = \langle y_1, -y_2\rangle$; the firm consumes money and offers apples. The economy is completly defined by a price vector $\mathbf{p}$ with the prices for apples and money. We use x to indicate a consuption set for consumers, y to indicate a production set for firm.

The GE define the concept of *Walrasian equilibrium*:

> we can say that the economy is a *Walrasian equlibiurm* if we can find an allocation (x^*, y^*, p^*) that is: (a) the utility of consumers is a maximum, (b) the profit of the firm is a maximum, (c) all offered apple are consumed and all of the money used (the market is clear).

To clear these issues, we use a formal description to unify the constraints (for example we must define the concept of budget, the consumers' choises have to stay in a fixed budget, i.e. the total cost for goods cannot exceed the total endowments). We have to compute these constraints and optimize these functions

1. For every firm (one in our case) we have to maximize the profit, that is $p \cdot y \leq p \cdot y^*$ for all y, i.e. the feasible production bundles of the firm.
2. For every consumer i, x^* is maximal for $\succ_i$ in the budget set, i.e. the product by p and x^* have to stay in the budget limitations. To compute budget limitations, we use an initial endowment for the consumer plus some amount of resources coming from firm's sharing, for example the wages. In our model there is no firm's sharing.
3. The market have to be clear: $\sum_i x_i^* = \omega + y^*$, where ω is the initial endowment, i.e. all consumers and the firm must be able to achieve their desired trades at the going market prices.

We can use a classical minmax algorithm to find one solution of our apple vs. money problem. Starting from $x_1 = 1,10$, $x_2 = 1,10$, $p = 1,1$, $y = 2,-80$, we (a) know that the consumer sets are in budget, (b) the utility for consumers is $u_s = 11,12$, (c) the profit for firm is 0, (d) and naturally the market is clear. At the end of optimization process: (a) consumers remain in budgets, (b) the utility is increased to $u_e \approx 16,20$, (c) the profit for firm is became ≈ 78. The *Walrasian equilibirum* is

$$x_1 \approx 4,13, x_2 \approx 4,13, p \approx 17,2, y \approx 8,-74$$

These are typical results in this conditions: in fact we find that the consumers' sets are identycal but the price for apple is greater that the price for money. This fact reflects the different preference algorithm between two consumers. Besides, the production rate (i.e. the ratio between offered apples and absoluted money input) is 0.025 at the beginning of the process, 0.1081 at the end of the optimization: the firm increase the production rate.

The GE approach permits to define the most important microeconomic theorem, the so-called First Fundamental Welfare Theorem

> Every GE is Pareto efficient (given that preferences are not (locally) satiated).

This result shows that using the GE framework to modeling economic interactions, we obtain the Pareto efficience, i.e. we have the outcome that, if economic agents play a GE, the final result will be an economy with no waste.

In other words, the GE framework is the mathematical side of the Adam Smith assumption about …*invisible hand* … that would be the responsible for the efficient resources' allocation of the market. The importance of this result is unquestionable. But, that is true if …*preferences are not (locally) satiated* …. What this means? Merely, the budget constraint only tells us what the consumer at most can buy. But does she/he have to buy/consume that much? No, preferences might be satiated. For example, you should write a lexicographic preference like this: …I prefer a to b if the first good in a is over the value θ and is greater than the first good in b, otherwise the relationsheep is indifference …. In this case, the preference is not satiated. If a consumer computes its utility function using $1 - exp(-a)$ as expression for a, we said that the preference rule is lexicographic for that good.[3]

As it is easy to imagine, there is an huge quantity of work about these points. In this chapter we show that PSO could tackle this critical point from an interesting point of view. Using PSO we obtain a convergence toward a Walrasian equilibiurm with also using a quasi-lexicographic preference algorithm.

Let we know describe the model.

We have three consumers and one firm. The market is composed of 4 goods. The difference between the consumers is the preference rule. First consumer is strongly lexicographic; in fact the utility function is

$$\beta(1 - exp(-\lambda x_{1,1}))x_{2,1}^2 x_{3,1} + x_{4,1}$$

the indexes for x indicate (first) the good index, (second) the consumer index. In this case we observe a consumer that (a) has a lexicographic valuation on first good, and a polinomial valuation about the others. The second consumer's preference rule is

$$\beta(1 - exp(-\lambda x_{1,2}))(1 - exp(-\lambda x_{2,2}))x_{3,2} + x_{4,2}$$

For the second consumer the lexicographic preference insists on two goods. The third consumer has a lexicographic preference on three goods.

Now, the problem is: why is *PSO* an useful approach to find one Walrasian equilibiurm for this model? For two different reasons. The first: *PSO* permites to explore solutions with a triggered amount of convergence and remaing into the feasible zone of the parameters. The feasible zone is, in this case, very complex: we have constraints about the parameters (prices have to be positive, quantities have to be positive and so on) and about the relationship between parameters (for example cumsumer have to stay in budget). The second reason is more important: if we use a *PSO* approach we can insert these constraints into the dynamic in a very natural way. We focus the attention on this second reason.

We have a population of 50 vectors, each vector indicates a position in the solutions' space with 16 dimensions: the dimensions from 1 to 12 define the consumption set for the agents ($3\,agents \times 4\,goods$). The last three dimensions define the vector price. The *PSO* algorithm is a classic local/social population based search, with a difference: we modify the velocity vector using an algorithm that incorporates the constraints. We try to explain the methods. In a classic *PSO* we modify velocity using social and local influence. We mantain the social influence, but to modify locally the velocity vector, we use these rules [4]:

> if a consumer is out of the budget, we reduce the velocity componenent of consumption bundle (the consumer tries to willing less amount of the goods) and reduce the velocity

[3] This is a simplification. The correct definition implies the use of quasi-lexicographic relationship.

[4] With *increase* and *decrease* we denote the algebric sum of random quantity on the vector of velocities.

component for the price (the algorithm tries to find a feasible price vector, respect to budget);

if a consumer is reducing its utility, we increase a lot the velocity component of consuption bundle for the lexicographic good;

if the firm is decreasing ist utility, we increase the velocity component for the prices.

Thus, the constraint (stay in budget) plays the same rule of local search for optimal solution (trying to increase the price if the profit for the firm is decreasing).

The results are very encouraging:

- The *PSO* method leads to the respect of the constraints: 3/4 of agents stay in budget; the velocity change to stay them in budget works.
- All of *inbudget* agents improved both their utility and firm's profit.
- 1/2 af agents are Pareto-Optimal: if they try to find better solutions starting from the points they reached in the solutions' space, they can not find them.
- The *PSO* method seems to lead spontaneously to Pareto-Optimal solutions.

3.3 Tactical vs. Strategic Behavior

This third application is the most abstract of the three. In this model, the *PSO* is not used to study the behavior of agents that interact with an environment with resources (as in the first model), or the behavior of agents operating in a virtual world (as in the microeconomy case, see the second application). In this case, the *PSO* is used to optimize the value to be assigned to the Markov matrixes which describe the behavior of an agent who has to solve a multi objective problem.

In this case we could define our approach an *artificial life* problem [4] [16].

Artificial life studies the logic of living systems in artificial environments. The goal is to study the phenomena of living systems in order to come to an understanding of the complex information processing that defines such (complex adaptive) systems. Complex adaptive systems (CAS) are special cases of complex systems. They are complex in that they are diverse and made up of multiple interconnected elements and adaptive in that they have the capacity to change and learn from experience. The term complex adaptive systems (CAS) was coined at the interdisciplinary Santa Fe Institute (SFI), by John H. Holland, Murray Gell-Mann and others [14]. CAS ideas and models are essentially evolutionary, grounded in modern biological views on adaptation and evolution. The theory of complex adaptive systems bridges developments of systems theory with the ideas of generalized Darwinism, which suggests that Darwinian principles of evolution can explain a range of complex material phenomena, from cosmic to social objects. A CAS is a complex, self-similar collection of interacting adaptive agents. The study of CAS focuses on complex, emergent and macroscopic properties of the system.

In CAS framework we have a classic multi-objective problem, that we can call the *strategic Vs tactical behavior* problem. A living organism has to choose if it is better to eat in a poor environment or in a fertile land. This is an elementary strategy problem. Then, in the fertile land there is a dangerous predator: to avoid the predator is a tactical problem. In our model the behavior of the agent must take into account both the objectives to be achieved: (1) the agent has to eat most food as possible, (2) the agent must not be attacked by the predator (which may appear with a certain probability). The agent then has to balance a strategic behavior (to search for food) with a tactical behavior (to avoid the predator).

The ideal situation that the model tries to capture is as follows. An agent is in an open space where it can escape to a possible predator and where there is a certain amount of food $q_1 = 1$. Near this portion of the world where the agent is, there is an enclosed space (say, a cave) in which there is a quantity of food $q_2 = 5$, with $q_2 > q_1$. In this cave, however there is a probability p_1 that a predator could attack the agent (according to another probability p_2), causing some damage $d = -10$. The goal of the agent is, first, to maximize the amount of found food and, second, to minimize the number of encounters with the predator. The heart of the model consists of a fitness function.

The fitness function has as argument a vector of six dimensions consisting of the values that the two matrixes describing the possible behavior must have. For example

$$A = \begin{matrix} .2 & .8 \\ .3 & .7 \end{matrix}$$

$$B = .1 \; .9$$

A is the matrix describing the behavior of the organism if the predator is faraway. The first row indicates if the organism remains in the poor land ($p = .2$) or goes to the fertile one ($p = .8$). The second row contains the behavior from the fertile land to the poor one (the first element, $p = .3$, indicates if the agent goes from the fertile land to the poor one, or the agent remains in the fertile land, $p = .7$). B is the behavior from fertile land to the poor one if the predator is present. The values .1 and .9 mean: with probability .1 organism goes to the safe, poor land; with probability .9 the agent remains in the fertile one. The implementation of the fitness function provides minimization of the damage; the damage has to be intended as a balance between found food and aggressions suffered by a predator.

This optimization problem has some special characteristics. The sum of the parameters that constitute the rows of the matrixes A and B must always do 1 (Markov matrixes). Moreover, the values that the elements of A and B can assume, must be normalized between 0 and 1 (as to the rows of the matrixes A and B, values are always divided by their sum and taken in absolute value, if negative).

Initially we used the *GATOOL* of Matlab to test the performance of our model with a traditional genetic algorithm. We used the fitness function that we implemented and the same constraints already described above. In a second stage we used a PSO tool (see [1]) to compare the performance of our *PSO* model with one obtained by a traditional genetic algorithm implementation.

We found some interesting results.

We varied the values of p_1 and p_2 (the probabilities that: a predator appears, p_1, and this predator could inflict a damage, p_2) to investigate different possible scenarios. So, we imagined five different scenarios:

1. both probabilities p_1 and p_2 have a low value: in this world there are not so many predators ($p_1 = .2$) and however they are not so dangerous ($p_2 = .2$);
2. p_1 has a low value and p_2 has a high value: in this world there are not so many predators ($p_1 = .2$), but they are very dangerous ($p_2 = .8$);
3. p_1 and p_2 have an intermediate value, the most stochastic one: in this world the probability to meet with a predator (p_1) and the probability that this predator could be dangerous (p_2) are the same: $p_1 = p_2 = .5$;
4. p_1 has a high value and p_2 has a low value: in this world there are many predators ($p_1 = .8$), but they are not so dangerous ($p_2 = .2$);
5. p_1 has a high value and p_2 has a high value too: in this world there are many predators ($p_1 = .8$) and they are very dangerous ($p_2 = .8$) (this is the worst possible world);

We run several simulations and we found some cases in which the *PSO* implementation of our model works better (namely for scenarios 1, 2, 5): this means that the value reached by fitness function is higher. The matrix *RESULTS* shows the values reached by *PSO* model's implementation (in the first column) and *GATOOL* model's implementation (in the second one); each row coincides with each different scenario.

$$RESULTS = \begin{matrix} 399.96 & 379.04 \\ 226.16 & 215.08 \\ 173.32 & 172.00 \\ 337.12 & 337.88 \\ 99.36 & 85.20 \end{matrix}$$

In the rest of the scenarios (3 and 4) the value reached by the *PSO* implementation of our model and by the *GATOOL* are almost the same. The most important result is that the behavior suggested by the two different implementations (*PSO* and *GATOOL*) could be very dissimilar. This last result indicates that different approaches can find different solutions (sometimes equivalent) for same problems. Future investigations in this direction could explain why.

4 Conclusions

In this chapter we have tryed to show that swarm optimization has a marked multidisciplinary character since systems with swarm characteristics can be observed in a variety of domains. The main argument in favor to *PSO* is proper the multidisciplinary character: in 2.2 section we said that in some cases, *POS* can resolve multiobjective otpimization problems in efficient way, because *POS* naturally incorporates some concepts from *Pareto-Optimal* framework. Now, we can specify this argument using the concept of multidisciplinary character.

It is customary to divide swarm intelligence research into two areas according to the nature of the systems under analysis: natural swarm intelligence research, where biological systems are studied; artificial swarm intelligence, where human artifacts are studied. An alternative and somehow more informative classification of swarm intelligence research can be given based on the goals that are pursued: we can identify a scientific and an engineering stream. The goal of the scientific stream is to model swarm intelligence systems and to single out and understand the mechanisms that allow a system as a whole to behave in a coordinated way as a result of local individual-individual and individual-environment interactions. On the other hand, the goal of the engineering stream is to exploit the understanding developed by the scientific stream in order to design systems that are able to solve problems of practical relevance. We argument that PSO is an useful techique to resolve complex optimization problems (a typical engineering stream) when the system has a strong characterization like a natural swarm system (the characterization like a natural swarm system is a typical scientific stream). *Thus, we can use a PSO technique if the system is naturally swarm-like, and we can value if a natural system has swarm charateristic using some kind of PSO.* The microeconomic case in this chapter is a perfect example. In that case, we have a system that searchs for equilibrium, and we know that kind of equilibiurm is pareto-optimal. We call that kind of systems naturally swarm-like, because while the algorithm search (a) to remain in constraints, (b) to optimize the utilities and the profit, solutions could interact, for example moving solutions out of constraints into more interesting zone of the solutions' space. On the other hand, *PSO* is useful to discover swarm charateristic in systems, and this features is clear in the first applicative case. Finally, we recall an other interesting aspect for *PSO* (with a direct link to multidisciplinary character): the tendency to permit different levels of analysis. The three applications in this cpather are a good proof.

Now, we try to sum up some conclusions about *PSO*: *PSO* is a population based stochastic optimization technique for the solution of continuous optimization problems. As we shown in the previous sections, in PSO a set of software agents called particles search for good solutions to a given continuous optimization problem. Each particle is a solution of the considered problem and uses its own experience and the experience of neighbor particles to choose how to move in the search space (from physical to the solutions' space). At each iteration of the algorithm, each particle moves with a velocity that is a weighted sum of three components: the old velocity, a velocity component that drives the particle towards the location in the search space where it previously found the best solution so far, and a velocity component that drives the particle towards the location in the search space where the neighbor particles found the best solution so far. The swarm is composed of many individuals; the individuals are relatively homogeneous (i.e., they are either all identical or they belong to a few typologies); the interactions among the individuals are based on simple behavioral rules that exploit only local information that the individuals exchange directly or via the environment (stigmergy); the overall behaviour of the system results from the interactions of individuals with each other and with their environment; the group's behavior self-organizes. The characterizing property of a swarm intelligence system is its ability to act in a coordinated way without the presence of a coordinator or of an external controller. In nature can be observed many examples of swarms that perform some collective behavior without any individual control over the group, or aware of the overall group's behavior. Notwithstanding the lack of individuals in charge of the group, the swarm as a whole can show an intelligent behavior. This is the result of the interaction of spatially neighboring individuals that act on the basis of simple rules.

These definitions are only the tip of the iceberg: we think that a lot of work is possible, in particularly about the topics that we have discussed here: the use of landmarks (like field), the level analysis and the relationship between *PSO* and fitness landscape.

References

1. `http://psotoolbox.sourceforge.net`
2. `http://www.cs.up.ac.za/cs/fvdbergh/publications.php`
3. `http://www.projectcomputing.com/index.html`
4. Adami, C., Ofria, C., Collier, T.C.: Evolution of biological complexity. Proc. Natl. Acad. Sci. U.S.A. 9, 4463–4468 (2000)
5. Alec Banks, A., Vincent, J., Anyakoha, C.: A review of particle swarm optimization. Natural Computating 7(1), 109–124 (2007)
6. Bäck, T.: Evolutionary Algorithms in Theory and Practice: Evolution Strategies, Evolutionary Programming, Genetic Algorithms. Oxford University, New York (1996)
7. Dorigo, M., Stuzle, T.: Ant colony optimization. MIT Press, Cambridge (2004)
8. Bonabeau, E., Dorigo, M., Theraulaz, G.: Inspiration for optimization from social insect behavior. Nature 406, 39–42 (2000)
9. Emlen, J.M.: The role of time and energy in food preference. American Naturalist 100, 603–609 (1996)
10. Fonseca, C., Fleming, P.: An overview of evolutionary algorithms in multiobjective optimization. Evolutionary Computation 3(1), 1–16 (1996)
11. Giraldeau, L.A., Caraco, T.: Social Foraging Theory. Princeton University Press, Princeton (2000)
12. Goldberg, D.: Genetic Algorithms in Search, Optimization and Machine Learning. Addison-Wesley, Reading (1989)
13. Taha, H.A.: Operations research. New Delhi (2005)
14. Holland, J.H.: Emergence: From Chaos to Order. Addison-Wesley, Reading (1998)
15. Hu, X., Eberhart, R.C., Shi, Y.: Particle swarm with extended memory for multiobjective optimization. In: Proceedings of the 2003 IEEE Swarm Intelligence Symposium, SIS 2003 (2003)
16. Johnston, J.: The Allure of Machinic Life: Cybernetics, Artificial Life, and the New AI. MIT Press, Cambridge (2008)
17. Kennedy, J., Eberhart, R.: Particle swarm optimization. In: Proceedings of the IEEE International Conference on Neural Networks (1995)
18. Kennedy, J., Eberhart, R., Shi, Y.: Swarm Intelligence. Morgan Kaufmann, San Francisco (2001)
19. Liang, J.J., Suganthan, P.N.: Dynamic multi-swarm particle swarm optimizer with local search. In: The 2005 IEEE Congress on Evolutionary Computation (2005)

20. Michlmayr, E.: Self-organization for search in peer-to-peer networks: The exploitation-exploration dilemma. In: Proceedings of the 1st International Conference on Bio inspired Models of Network, Information and Computing Systems, BIONETICS 2006 (2006)
21. Miller, J.H., Page, S.E.: Complex Adaptive Systems: An Introduction to Computational Models of Social Life. Princeton University Press, Princeton (2007)
22. Millonas, M.M.: Swarms, Phase Transitions, and Collective Intelligence. IEEE Press, Los Alamitos (1994)
23. Nocedal, J., Wright, S.: Numerical Optimization. Springer, Heidelberg (2006)
24. Nolfi, S., Floreano, D.: Evolutionary Robotics. MIT Press, Cambridge (2001)
25. Parsopoulos, K.E., Vrahatis, M.N.: Particle swarm optimization method in multiobjective problems. In: Proceedings of the 2002 ACM symposium on applied computing (2002)
26. Ravindran, A., Phillips, D.T., Solberg, J.J.: Operations Research - Principle and practice. John Wiley & Sons, New York (2001)
27. Reny, P.J., Jehle, G.A.: Advanced Microeconomic Theory. Addison-Wesley, Reading (2000)
28. Sarker, R., Liang, K.H., Newton, C.: A new multiobjective evolutionary algorithm. European Journal of Operational Research (2002)
29. Sarker, R.A., Newton, C.: Optimization Modelling: A Practical Approach. Taylor & Francis/CRC Press (2007)
30. Stephens, D.W., Krebs, J.R.: Foraging theory. Princeton University Press, Princeton (1986)
31. van der Bergh, F., Engelbrecht, A.P.: A study of particle swarm optimization particle trajectories. Information Sciences, 176(8) 937, 971 (2006)
32. Walras, L.: Elements of Pure Economics, or the theory of social wealth (1874)
33. Zomaya, A.Y.: Handbook of Nature-Inspired and Innovative Computing: Integrating Classical Models with Emerging Technologies. Springer, Heidelberg (2006)

VISPLORE: Exploring Particle Swarms by Visual Inspection

Namrata Khemka and Christian Jacob

Abstract. We describe VISPLORE, a visualization and experimentation environment for population-based search algorithms. Using particle swarm optimization (PSO) as an example, we demonstrate the advantages of an interactive visualization tool for multi-dimensional data. VISPLORE greatly supports the analysis of time dependent data sets, as they are produced by evolutionary optimization algorithms. We demonstrate various multi-dimensional visualization techniques, as built into VISPLORE, which help to understand the dynamics of stochastic search algorithms.

> A picture is worth a thousand words. An [interactive] interface is worth a thousand pictures.
>
> *Ben Shneiderman*, University of Maryland

An interactive interface can be a very useful tool for analyzing and visualizing large solution domains that are common for population-based search algorithms. Users need both an overview of the data as well as flexibility in analyzing specific portions of the data. Interactive toolkits also assist users, who have limited expertise with respect to the underlying search algorithm, to perform certain data manipulation and analysis. For instance, evaluating the massive data generated by population-based algorithms, such as Particle Swarm Optimization (PSO) [12], can

Namrata Khemka
University of Calgary, Dept. of Computer Science, 2500 University Drive NW, Calgary, Alberta T2N 1N4, Canada
e-mail: namrata.khemka@gmail.com

Christian Jacob
University of Calgary, Dept. of Computer Science, Dept. of Biochemistry and Molecular Biology, 2500 University Drive NW, Calgary, Alberta T2N 1N4, Canada
e-mail: cjacob@ucalgary.ca

R.A. Sarker and T. Ray (Eds.): Agent-Based Evolutionary Search, ALO 5, pp. 255–284.
springerlink.com

be significantly benefited by the use of an efficient interactive toolkit such as *VISPLORE*, as we will demonstrate. Particle swarms perform a parallel search in multi-dimensional problem domains. They are highly parallel, stochastic in nature, and they produce a large amount of data during their runs, which makes them challenging to analyze. Extracting useful information from hundreds and thousands of solutions to gain insights into an algorithm's performance is usually a non-trivial task [17].

In the past, we have worked on a real-world problem called the "Soccer Kick Simulation" [15] [16], where the goal was to find optimized settings (via Particle Swarm Optimization) for control parameters for a kinematic model of 17 leg muscles, such that a kicked ball travels as far and as fast as possible. The model included 56 parameters (dimensions). In the course of our investigations and experiments we realized the importance of visualization in order to efficiently and successfully tackle such a large-scale optimization problem. The soccer kick simulation (or in general any real-world system) produces a large number of potential solutions (over $120,000$ individuals) each having 56 dimensions. Monitoring and making sense of such large groups of dynamic real-time data poses various challenges. Presenting this information to the user as raw data (a series of numbers) would make it difficult if not impossible for the user to observe and understand the progression of the search algorithm. Normally, we evaluate the quality of the optimizers by the fitness of their solutions generated. A similar approach was taken for the soccer kick simulation, where we focused on the overall performance and gross population statistics, such as comparing the solutions found through different runs, the number of fitness function evaluations that were required to find a good solution, and the convergence rate of the algorithms. This methodology is a "black box" approach that solely focuses on the actual outcome, but mainly ignores the behavior of the algorithms.

Visualization methods can help us to analyze, in great depth, the potential solutions and results discovered. Visualization of particle swarm algorithms can therefore allow the user to make inferences that are not easy to accomplish otherwise [18]. Visualizing multi-dimensional individuals, analyzing the extent to which an algorithm has explored the search space, the effects of inherent parameters, analyzing gross population statistics, determining clusters and building blocks, and finding the successful combination of parameter values can influence our choice of experiments. We also gain further insights into particle swarm systems and the search space of an optimization task in general. The main purpose of visualization is to gain insights, discovering new patterns, offering knowledge about the explored regions within the multi-dimensional search spaces, determining the successful combinations of parameter values, finding the parameter values where the population has converged, or understanding how partial solutions are created. We have developed a toolkit to interactively visualize the solution spaces generated from particle swarm experiments. *VISPLORE* aims at the discovery of patterns, relationships, and clusters, rather than locating specific solutions.

Our chapter is organized as follows. Section 1 discusses some of the previously developed visualizations for population-based search systems. We explain our particular implementation of particle swarms in Section 2. Section 3 highlights different interactive visualization techniques from our *VISPLORE* toolkit, implemented in *Mathematica* [9] and particularly developed for exploration of particle swarms. We also discuss the search mechanism in Section 4, different views in the *VISPLORE* toolkit in Section 5, and how to customize plots generated in *VISPLORE* in Section 6. We explore the strengths of *VISPLORE* on the foxholes benchmark function in Section 7. We also examine and analyze the results of the visualization techniques on the soccer kick simulation in Section 8. Finally, we conclude our work in Section 9.

1 Related Work

A widely used and straight-forward form of visualization in population-based methods consists of two-dimensional population statistics graphs. These include the fitness of the best individual, the average fitness value of all individuals, and the worst fitness values at each timestep. These plots give a good indication of whether the algorithm is improving over time and provide insights about the overall behaviour (i.e., convergence or divergence) of the algorithms. An overview of these common forms of summary plots is provided by Pohlheim [20] that observes both the "course" and the "state" of an evolutionary algorithm. Visualizations introduced by Pohlheim include the fitness of individuals, distances between individuals, and certain statistics that can be tracked over a single generation or multiple experiments.

Attempts to visualize higher dimensions have led to the development of 'multi-dimensional scaling' techniques that transform search spaces of high dimensionality into lower dimensions. The most well-known technique in the realm of visualizing multi-dimensional population-based techniques is Sammon mapping [20]. This method places points on a two-dimensional canvas, which represent points in the higher dimensional space. The idea is to then iteratively move points closer together in the two-dimensional space if they are close together in the multi-dimensional space. Although this technique works, it has quadratic time complexity with respect to the number of points, and so can become demanding for large search spaces [8]. To overcome this problem, search space matrices [4] are used to map all possible individuals of a genetic algorithm onto a two-dimensional canvas such that the Hamming distance between neighboring points is minimized. This mapping is simple and the algorithm scales linearly with the number of points in the search space [8]. We can also display multi-dimensional data through the use of glyphs. Chernoff faces [11] are an example of glyphs where datasets are represented by a stylized drawing of a human face. Each attribute in the data maps to different items on the face (mouth, nose, separation between eyes, etc.). Although by using Chernoff faces we can easily distinguish features, it does not scale well for higher dimensions. Methods to visualize tree structures of genetic programming have been explored by Daida et al. [5]. Techniques for inspecting the structure of the evolved trees and methods for visual inspection of the tree structures at the level of the entire population are discussed by Daida. These methods allow for examination and understanding of the dynamics of the genetic programming trees. This technique has the capability of visualizing ten million nodes; but it is restricted to trees that have a binary structure.

The *GEATbx*, Genetic and Evolutionary Algorithm toolBoX, provides global optimization capabilities in *Matlab* [3] by using genetic algorithms and evolution strategies. *GEATbx* builds on the work done by Pohlheim [20] by providing the summary graphs for visualizing the convergence behaviour of the algorithms. *GEATbx* also supports multi-objective optimization, constraint handling, and multi-population evolution. *GEATbx* fails to provide data manipulation and interpretation methods, a user interface (e.g., changing inherent parameters, allowing users to visualize phenotypes and genotypes of individuals), and most importantly data visualization techniques. One of the early research in creating offline data visualization systems for evolutionary algorithms is the work of Wu et al. called *VIS* (A Genetic Algorithm Visualization Toolbox) [23]. *VIS* visualizes genotype information of an individual in details, provides parental information of the individuals, and allows a user to see which genetic operators resulted in the creation or removal of a selected individual. Another visualization tool for genetic algorithms, *GONZO*, was developed by Collins [4]. This system gives population summary graphs along with the genotype and parental information of an individual. *GONZO* displays the search space by using matrices. The key advantage of this system is that it allows for both online (while the genetic algorithm is in progress) and offline (after the genetic algorithm has completed) visualization, thereby allowing the user to interactively modify the inherent parameters. The *GAVEL* system [8] was developed relatively recently by Hart and Ross, and is very similar to *VIS*. *GAVEL* improves upon the VIS system because it adds the capability of viewing the ancestry of single genes, in addition to the ancestry of the individuals in the population. However, *GAVEL* discards any information not concerned with the creation of the final 'best' individual. The *Evolvica* [11] framework demonstrates the basic principles of evolutionary computation and optimization techniques. This *Mathematica*-based [9] framework allows the user to experiment with four different evolution-based algorithms (evolution strategies, evolutionary programming, genetic algorithms, and genetic programming). *EvolVision* [7], a Java-based visualization add-on to *Evolvica*, allows the user to view ancestry trees for individuals in the population. *EvolVision* supports different individual representations: schematic (an individual is displayed in an abstract form, e.g., a simple label), phenotype view, and genotype view. Recent work in [1] proposes an intelligent exploration technique for genetic algorithms. This approach uses self-organizing feature maps to analyze data of the evolution process and utilizes the obtained information to enhance the search strategy. It also provides visual insights into the state and the course of the evolution.

Each of the frameworks discussed above have their strengths. However, none of these systems provide an exploratory and inquiry platform for particle swarms. These systems also do not answer the visualization questions addressed earlier related to gaining further insights into the performance of particle swarms. The *VISPLORE* toolkit, fills this need; it provides streamlined access to the data generated by population-based methods. We discuss the basic PSO algorithm in Section 2 and the details of the *VISPLORE* toolkit in Section 3.

2 Particle Swarm Optimization

Particle Swarm Optimization (PSO) is a search and optimization technique, introduced by James Kennedy and Russell Eberhart in 1995 [12] [14] [22] [21]. The underlying motivation for PSO comes from the behaviours seen in social organisms that live in large groups (like flocks of birds or schools of fish). The search is performed by a population of particles[1] i with location vectors $\mathbf{p}_i$. The ith particle is represented as $\mathbf{p}_i = (p_{i1}, \ldots, p_{ij}, \ldots, p_{iD})$, where D is the search space dimension. Each particle $\mathbf{p}_i$ is a possible candidate solution in D-dimensional hyper-space. Each individual also keeps track of its velocity $\mathbf{v}_i = (v_{i1}, \ldots, v_{ij}, \ldots, v_{iD})$, that determines its step size for the next generation, i.e., how far a particle will travel and the direction of movement. Every particle 'flies' through the search space in discrete timesteps according to its own past experiences ($\mathbf{p_i}^{best}$) and the past experiences of the entire group ($\mathbf{p_g}^{best}$). The quality of a solution is determined by a pre-defined fitness function f. Further details about the PSO algorithm are provided in Table 1.

3 The *VISPLORE* Toolkit

Although the implementation of a PSO algorithm is straight-forward, analyzing the results of a particular particle swarm optimizer can be quite difficult. This is to some extent due to the fact that particle swarms produce a large amount of data, therefore they can not be analyzed easily by hand. A data visualization component for swarm computation can provide the user with valuable insights on how to improve the PSO algorithm and consequently lead to a better understanding of the high-dimensional fitness landscape.

The main goal of *VISPLORE* is to allow one to easily visualize data produced from a population-based algorithm. Those wishing to create interactive visualizations should not have to worry about the implementation details of the optimization nor the visualization technicalities of a population-based algorithm. Instead they should only have to supply the necessary fitness function components needed by the optimizers to evaluate the solutions. Note that *VISPLORE* is composed of a collection of *Mathematica* [9] notebooks (interactive documents containing *Mathematica* code). We use a simple unimodal sphere function (Table 2) for our experiments in this section. The parameter settings for the sphere function used here are: $\mu = 10$, $v_{ij} \in [-0.1, 0.1]$, $p_{ij} \in [-5.14, 5.14]$, $\varphi_1 = 0.1$, $\varphi_2 = 1.0$, and $d = 5$. In *VISPLORE*, visualization techniques are divided into categories according to the criteria discussed in the following sections (Figure 1).

3.1 Visualization of a Particle

The most important element of particle swarms are the particles, i.e., the individual solutions, encoded as real-valued vectors. The particles represent potential solutions within the problem search space. Visualizing the current state of a particle

[1] In PSO, each individual of the swarm is called a particle.

Table 1 Basic particle swarm optimization algorithm

Particle Swarm Optimization Algorithm

1. **Initialization**: Initialize the particle population of size μ by stochastically assigning locations and velocities:

$$\mathbf{P}(t) = (\mathbf{p}_1,\ldots,\mathbf{p}_i,\ldots,\mathbf{p}_\mu),\ \mathbf{V}(t) = (\mathbf{v}_1,\ldots,\mathbf{v}_i,\ldots,\mathbf{v}_\mu).$$

2. **Evaluation**: Evaluate the fitness of all particles:

$$f(\mathbf{P}(t)) = (f(\mathbf{p}_1),\ldots,f(\mathbf{p}_i),\ldots,f(\mathbf{p}_\mu)).$$

3. **Local best**: Keep track of the locations where each individual had its highest fitness so far:

$$\mathbf{p}^{best} = (\mathbf{p_1}^{best},\ldots,\mathbf{p_i}^{best},\ldots,\mathbf{p}_\mu{}^{best}).$$

4. **Global best**: Keep track of the position with the global best fitness:

$$\mathbf{p_g}^{best} = max_{\mathbf{p}\in P}\,(f(\mathbf{p}_i)).$$

5. **Velocity update**: Modify the particle velocities based on the previous best and global best positions for all $1 \leq \mathrm{i} \leq \mu$. φ_1 and φ_2 are constants that determine how much the particle is directed towards good positions.

$$\mathbf{v_i}^{new} = \mathbf{v}_i + \varphi_1 * Random[] * (\mathbf{p_i}^{best} - \mathbf{p}_i) + \varphi_2 * Random[] * (\mathbf{p_g}^{best} - \mathbf{p}_i).$$

6. **Location update**: $\mathbf{p_i}^{new} = \mathbf{p}_i + \mathbf{v_i}^{new}$ for $1 \leq i \leq \mu$.

7. **Termination**: Stop if the termination criterion is met.

8. **Update**: $t := t + 1$ and $\mathbf{P}(t+1) = (\mathbf{p_1}^{new},\ldots,\mathbf{p_i}^{new},\ldots,\mathbf{p_i}^{\mu})$. Go to step 2.

Table 2 Benchmark functions used in the experiments are denoted in column 1. The range column gives the intervals for each dimension. The last column shows the best reported optimal function value. Note that we are maximizing each of these functions

Function	Range	$f(\overrightarrow{x}^*)$
Sphere: $f_1 = -\sum_{j=1}^{d} x_j^2$	[-5.12, 5.12]	0
Foxholes: $f_2 = (0.002 + \sum_{j=1}^{Length[a]} \frac{1}{j+\sum_{i=1}^{Length[x]}(x[[i]]-a[[j]][[i]])^6})^{-1}$ a = 0, 0, 0, 16, 16, 0, 16, 16	[-50, 50]	0

along with its fitness value provides a baseline for understanding the actual state and the structure of a particle. In *VISPLORE*, different views of a particle, $\mathbf{p}_i(t)$, at a given iteration, t, are provided through a dialog window as illustrated in Figure 2. Through the use of sliders the user can select the iteration and individual number for which he/she wishes to view the details. Any slider can be animated, hence turned

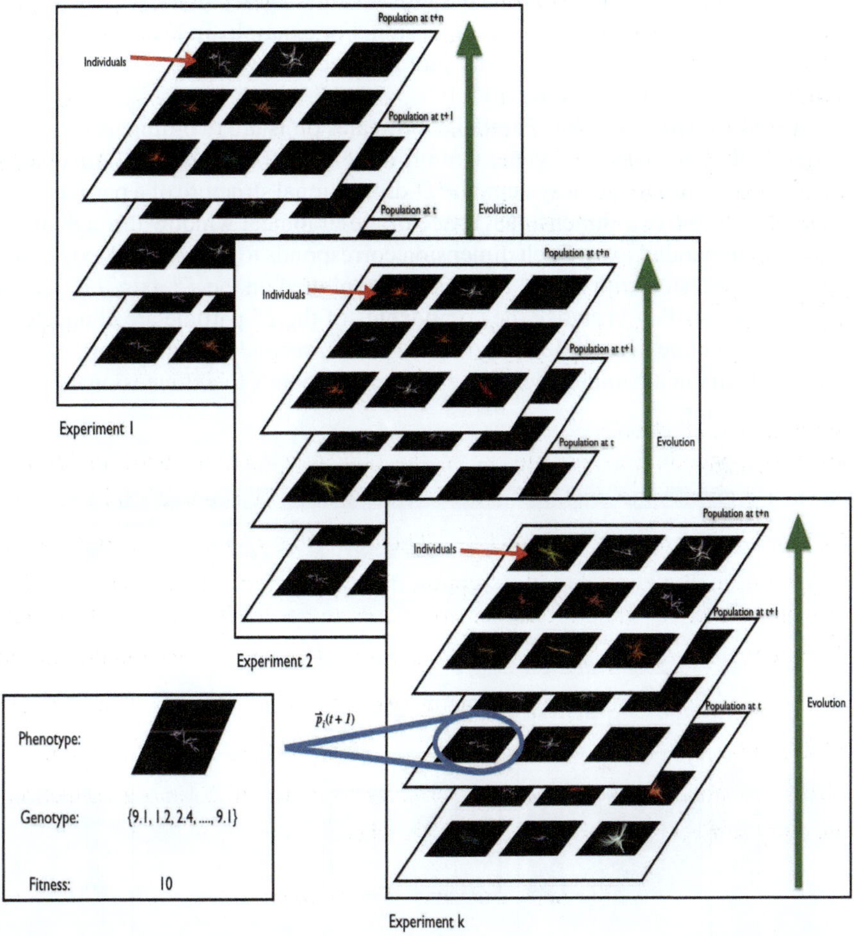

Fig. 1 Levels of data visualization in *VISPLORE*: Each particle can be visualized at either the genotype or the phenotype level. Here the genotype encodes parameters for a fractal generator [19]. A group of particles is visualized at a population level. Changes in the population over time are displayed at an experimental level. Finally, we can also view a set of experiments. Therefore, *VISPLORE* makes provisions for interactively visualizing a particle, a population of particles, an experiment, and a collection of experiments

into a movie so that one is able to go back and forth in time. Our slider control provides a movie-like-view of the individuals. This additional dynamic is obtained automatically in our system through *Mathematica*'s functionality. Figure 2 depicts the representations for individual 2 at timestep 4, $\mathbf{p}_2(4)$. Four different views of a particle, $\vec{p}_i(t)$, at a given iteration, t, are provided to the user:

1. **Density plot:** In density plots the parameter values are represented as gray scales. Each rectangle corresponds to a dimension. For example, in Figure 2 the rectangle at the left corner of the density plot is associated with the first dimension of individual 2 at timestep 4, $\mathbf{p}_{2,1}(4)$.
2. **Parallel coordinates plot:** Parallel coordinates plots are two-dimensional visual representations to express general multi-dimensional data sets [10]. In a parallel coordinates visualization system, the D-dimensional structure of a particle is projected onto the two-dimensional space of the graphical window through a set of D vertical parallel axes. Each dimension corresponds to one vertical axis. All the values associated with the jth dimension are plotted on the jth axis, $j \in 1,\ldots,D$. All the points that visualize the components of the ith particle are connected by a polygonal line, a so-called polyline (Figure 2, second column). Note that each vertical axis in a parallel coordinates plot corresponds to a dimensional block in a density plot.

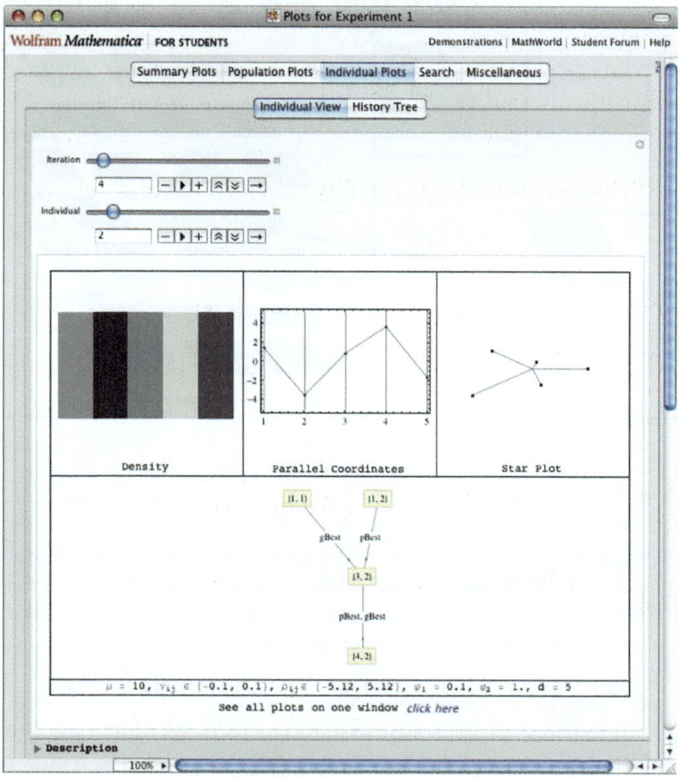

Fig. 2 Individual view in *VISPLORE* for Particle 2, timestep 4: $\mathbf{p}_2(4)$: The individual view presents the user with density, parallel coordinates, and star plots. In the lower part the family tree of an individual is displayed

3. **Star plot:** In a star plot, a D-dimensional vector is projected onto the two-dimensional space through a set of D lines originating at the origin and equally spaced over 360°. The length of each line reflects the value of the corresponding dimension (Figure 2, third column). Note that the line segment starting at the 3 o'clock position corresponds to dimension 1. The dimensions are plotted in a counter-clock wise manner.
4. **History plot:** Each PSO particle at a time, $\mathbf{p}_i(t)$, is dependent on the local best, $\mathbf{p}_i^{best}(t-1)$, and global best , $\mathbf{p}_g^{best}(t-1)$, determined from the previous timestep. We can think of the $\mathbf{p}_i^{best}(t-1)$ and $\mathbf{p}_g^{best}(t-1)$ as the parents of the current particle, $\mathbf{p}_i(t)$. Therefore, in order to view the origins of a particle we establish the relationships with its personal and global best of previous iterations. We visualize the pedigree tree of a particle in a graph structure, where the oldest generations appear at the top and the newer generations at the bottom (Figure 2, bottom section). In our toolkit, the users can select via sliders the iteration and individual number for when they wish to view the family tree (Figure 3). The "levels" slider determines the number of generations visualized for the selected particle. Each node of the graph is represented by the pair (iteration number, individual number). The links between the different nodes represent the relationship (*pBest*, *gBest*, or both), which the selected particle has with the previous timesteps. In Figure 3(a) we observe particle $\mathbf{p}_1(6)$; we display three levels of its history. It is clearly noticed that $\mathbf{p}_1(6)$ has particles $\mathbf{p}_2(5)$ and $\mathbf{p}_1(5)$ as its personal and global best, respectively. Furthermore, particle $\mathbf{p}_1(5)$ has $\mathbf{p}_2(1)$ as its global best and $\mathbf{p}_1(4)$ as its personal best parent. An interesting behaviour for particle $\mathbf{p}_2(27)$ is observed in Figure 3(b). Both the global best and local best of $\mathbf{p}_2(27)$ are propagated from timestep 20. This indicates that particle $\mathbf{p}_2$ has dominated the search area between timesteps 20 and 27. Note that clicking on any of the nodes of the tree provides the current state information of the selected particle (Figure 4).

3.2 *Visualization of a Population as a Collection of Particles*

A population at a given iteration, **P**(t), consists of many particles that represent potential solutions to the optimization problem. Therefore, it is important to understand how particles relate to other particles within a population. Visualizing parameter values of a collection of individuals provides insight into the distribution of the particles, detecting patterns or trends in the population, and identifying concentrations of particles in one or more areas of the search space. Hence, by visualizing a population of particles conveys a picture of the state of the population at a particular iteration. We use parallel coordinates plots, star plots, and density plots—as described above—but now apply these to a whole population. Note that. again, through the use of sliders the user can view a population at any iteration in the experiment.

1. **Density plot:** Figure 5 illustrates the density plots at three different time steps: $t=1$, $t=20$, and $t=60$, where the parameter (dimensional) values for each

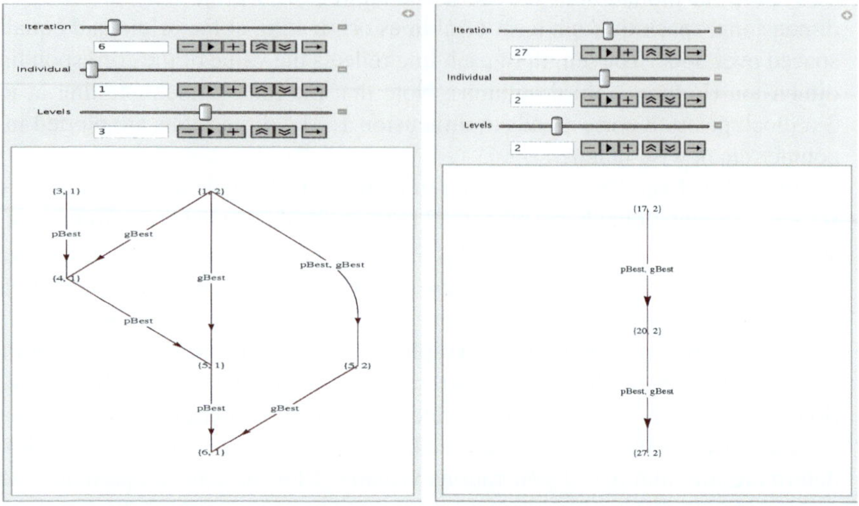

(a) Iteration 6, Individual 1: $\mathbf{p}_1(6)$ (b) Iteration 27, Individual 2: $\mathbf{p}_2(27)$

Fig. 3 Pedigree 'tree' of a particle in *VISPLORE*: Relationships between a particle and its previous best as well as its global best are represented in a graph structure. Each node is labelled with (iteration number, individual number)

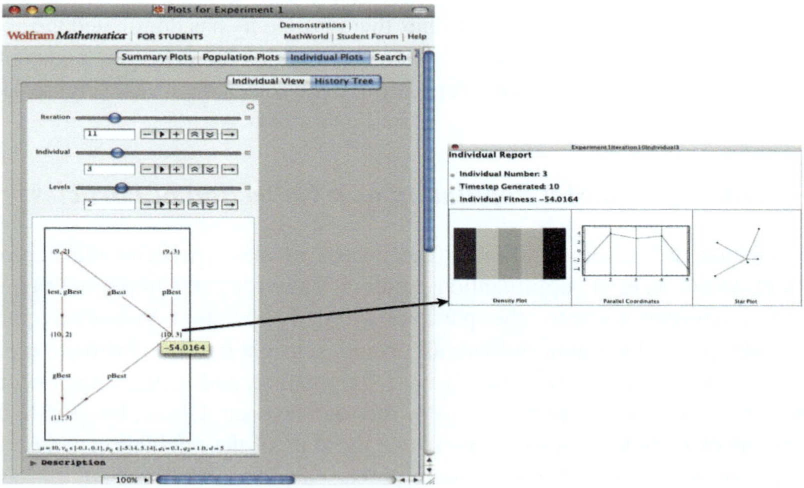

Fig. 4 User interactions in the family tree of a particle in *VISPLORE*

individual are plotted as a gray-scale rectangle. For example, the rectangle at the lower left corner of the plot (Figure 5(a)) is associated with the value of the first particle at the first dimension. We observe that at iteration 1 (Figure 5(a)) the values are uniformly random, indicating that we have diversity in our population. Over time all the parameter values have converged to a particular value indicated by the shared gray-scale value for each individual (Figure 5(c)).

2. **Parallel coordinates plot:** We use parallel coordinates plots as described earlier to visualize the population of particles. Column 2 in Figure 5 shows the parallel coordinates plot for a PSO run at three different timesteps: $t = 1$, $t = 20$, and $t = 60$. As seen in the figure, parallel coordinates can be very useful for visualizing the individuals of particle swarm algorithms. A structured overview of the individuals can be displayed, thus allowing us to recognize patterns, identify trends, and establish relationships among the dimensions of various individuals of a population. Using the parallel coordinates plot we can also examine areas of the search space that are covered at a particular iteration thus providing insights into the search space and into the algorithm progress at a particular moment in time. Note that moving the mouse over any of the polylines provides the fitness value of the selected particle. Clicking on a polyline invokes the particle information window.
3. **Star plot:** We use the star plots as described earlier to visualize the population of particles. Column 1 in Figure 6 displays the star plot for a population at three different timesteps: $t = 1$, $t = 20$, and $t = 60$. Star plots can be very useful for visualizing the structure of the particles, establishing relationships, and identifying trends. For instance, in Column 1 of Figure 6 individual 1, $\mathbf{p}_1$, on the top left keeps its original structure over time in comparison to individual 3, $\mathbf{p}_3$. The colour represents the fitness value (cooler colours represent better fitnesses in comparison to warm colours) of the particular particle relative to the best fitness found during the entire search experiment. Note that the position of each star plot in Figure 6 is constant throughout the experiment. Moving the mouse over any of the stars shows the fitness value of the particular particle and clicking on the star displays particle information.
4. **Concentric circles plot:** Concentric circle plots provide a dimensional view of the population. Each set of concentric circles represents one dimension (Column 2 in Figure 6). Each concentric circle represents one particle. The inner most concentric circle depicts individual 1, whereas the outermost portrays individual 10. The points on the concentric circle denote the parameter value of the particle at that dimension. It is observed in Figure 6(d) that all the particle values in different dimensions are uniformly distributed. However, by iteration 60 we observe clusters being formed for different particles indicating the convergence of the dimensional value of the particles (Figure 6(f)).

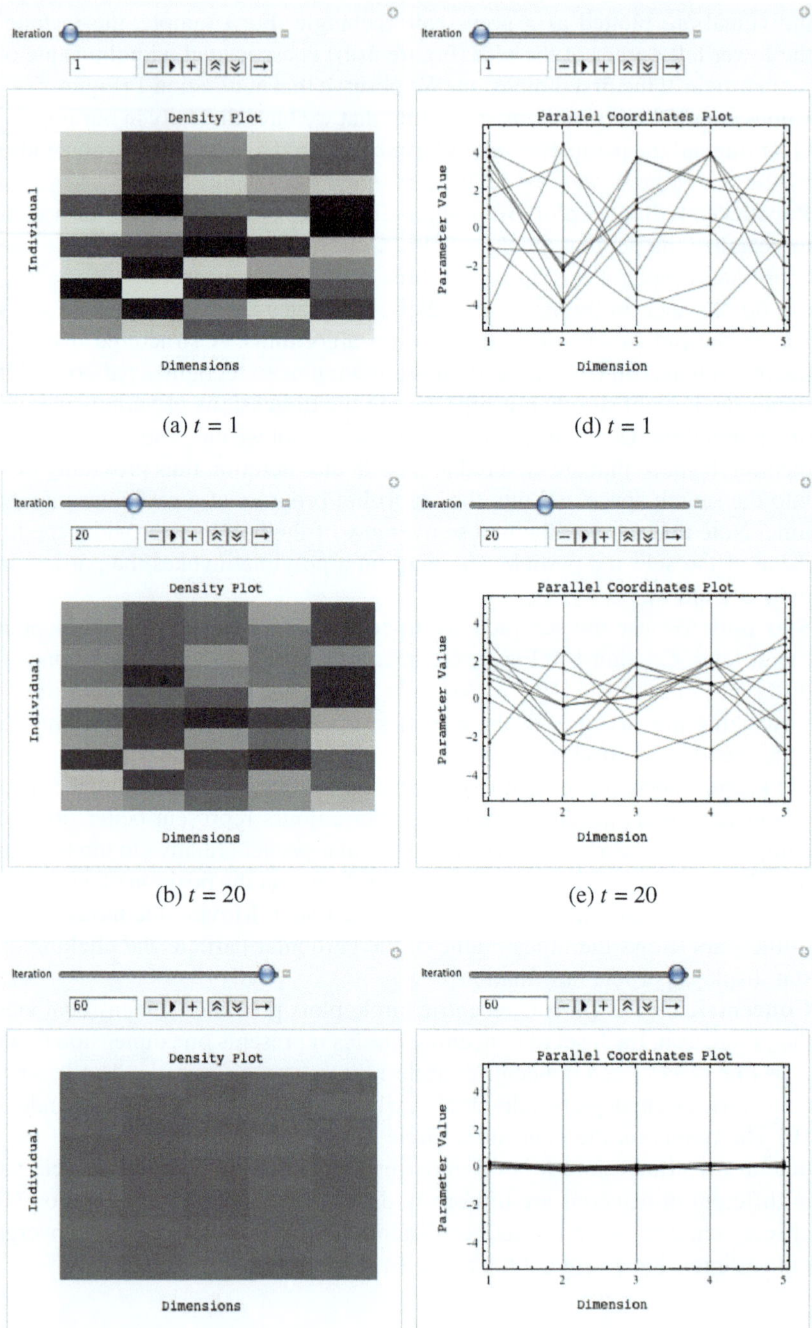

(a) $t = 1$ (d) $t = 1$

(b) $t = 20$ (e) $t = 20$

(c) $t = 60$ (f) $t = 60$

Fig. 5 Population Overview (1): Density and parallel coordinates plots in *VISPLORE*

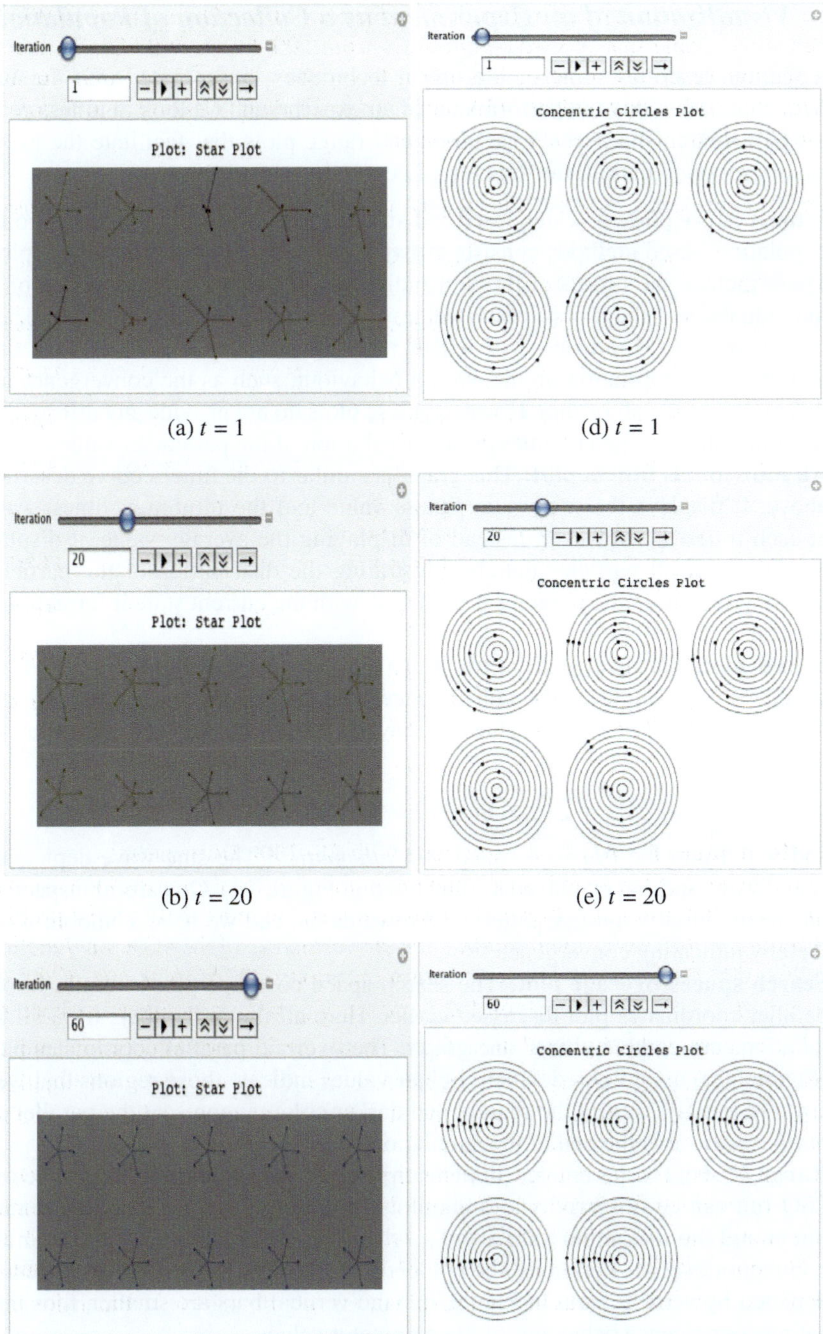

(a) $t = 1$ (d) $t = 1$

(b) $t = 20$ (e) $t = 20$

(c) $t = 60$ (f) $t = 60$

Fig. 6 Population Overview (2): Star and concentric circles plots in *VISPLORE*

3.3 Visualization of an Experiment as a Collection of Populations

This section describes some of the useful techniques incorporated into the *VISPLORE* toolkit to study the performance of an experiment. We look at fitness plots, sound plot, search space coverage plots, and range plots that facilitate the exploration of gross population statistics for all generations in a PSO run.

1. **Fitness curve plot:** A widely used and straight-forward form of visualization in population-based methods consists of two-dimensional population fitness plots. These include the fitness of the best individual, the average fitness value of all individuals, and the worst fitness values at each timestep (Figure 7(a)). These plots give a good indication of whether an algorithm is improving over time and provide information on the overall behaviour, such as the convergence and divergence of the algorithm. However, these plots do not provide any information regarding the parameter values or the distribution of the parameter values.
2. **All individuals fitness plot:** This graph is similar to the fitness curve described above. It displays the maximum fitness value and the minimum fitness value at each timestep. However, instead of displaying the average value, it displays the fitness of all particles thereby illustrating the distribution of the particles. Clicking on any of the particles provides us with the current state information of a particle.
3. **Sound plot:** We diverge from traditional visualization methods by utilizing a sound wave to indicate the convergence of a population over time. At each timestep the sum of fitness values is converted to a sine wave by the following formula:

   $$\Sigma_{i=1}^{\mu} \sin[(1440 + 40 f(\mathbf{p}_i) * 2\pi]],$$

   and is fed into the *Play* command of *Mathematica*. The sine wave depicts the sound wave and can be played as illustrated in Figure 7(f). On careful inspection, one hears initially a noisy pattern and towards the end we have a smooth wave, thereby indicating convergence.
4. **Search space coverage plot:** The search space coverage plot is similar to the parallel coordinates plot discussed earlier. Here all the individuals from all the iterations are combined into one graph. The overlaid parallel coordinates plots (Figure 7(c)) are coloured, where lighter values indicate those regions that were covered initially during the search, and darker colours represent the parallel coordinates towards the end of the simulation.
5. **Range Plots:** The parameter (dimension) range changes during the course of a PSO run can be observed via range plots. In Figure 7(d), we display the minimum and the maximum values (i.e., a range) for each parameter of the initial population over all iterations above a certain threshold value. These ranges are depicted by vertical bars. In Figure 7(e) the vertical bars are smaller, thus indicating convergence behaviour of the parameter values.

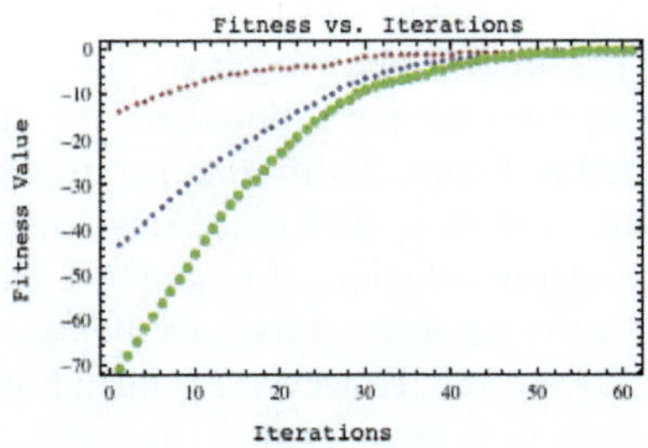

(a) Fitness plot

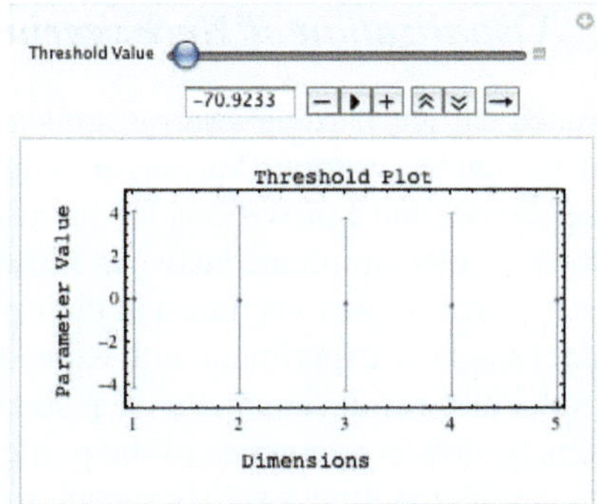

(d) Threshold plot for τ = -70.9233

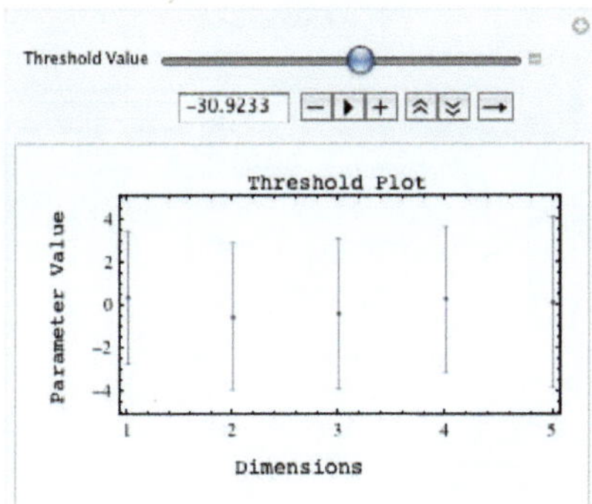

(b) Threshold plot for τ = -30.9233

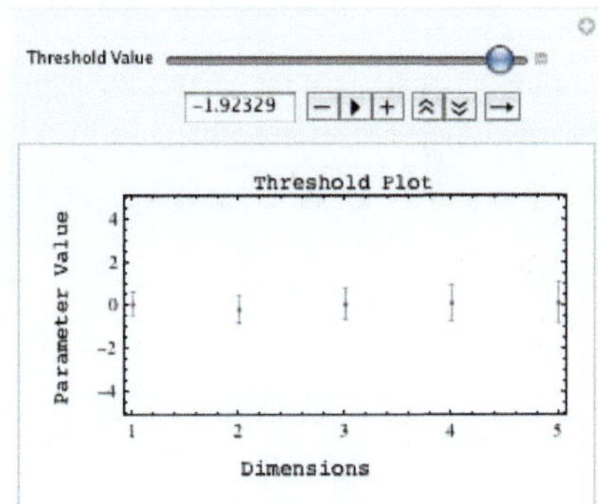

(e) Threshold plot for τ = -1.92329

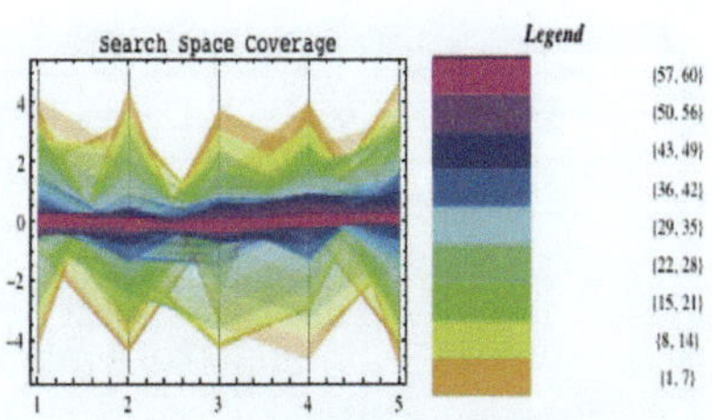

(c) Search space coverage plot

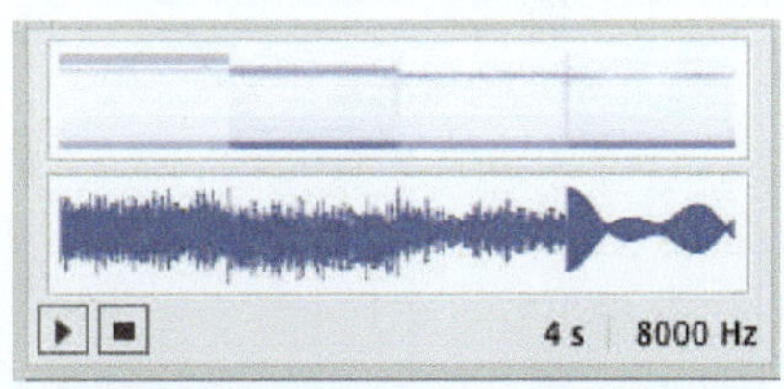

(f) Sound plot for every twentieth iteration

Fig. 7 Population level plots in *VISPLORE*

3.4 Visualization of Experiments as a Collection of Experiments

As stated earlier, particle swarms are stochastic in nature. As a result, no two runs yield the same solution. Various inherent parameters need to be set by the user. Therefore, we have provisions for the user to run multiple simulations and visualize their results simultaneously. As illustrated in Figure 8, different parameter settings are used by various runs and their effects can be studied at the same time. We conduct a sphere experiment with exceedingly low velocities (Figure 8(a)). Having low velocities results in smaller step sizes for the particles, consequently leading to extremely slow convergence of the particles over time. However, having a high velocity causes the fitness values at each timestep to fluctuate. Due to this artifact the

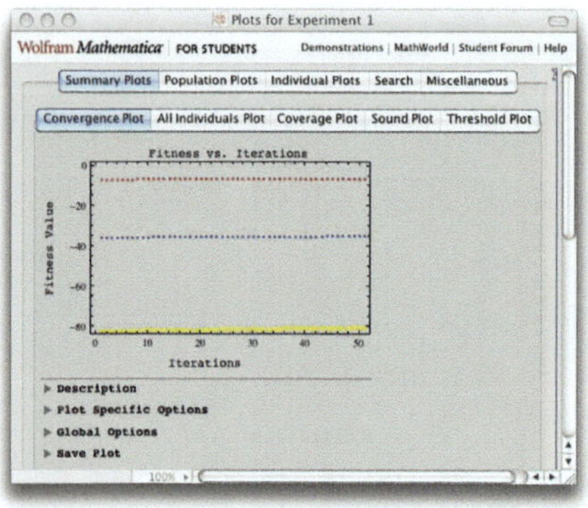

(a) Low velocity.

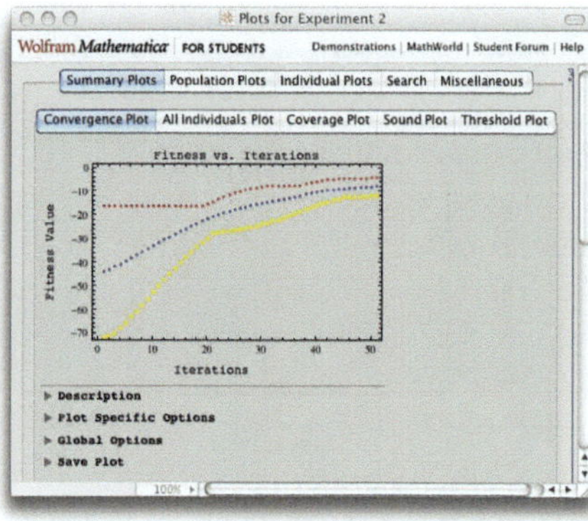

(b) Low exploration rate.

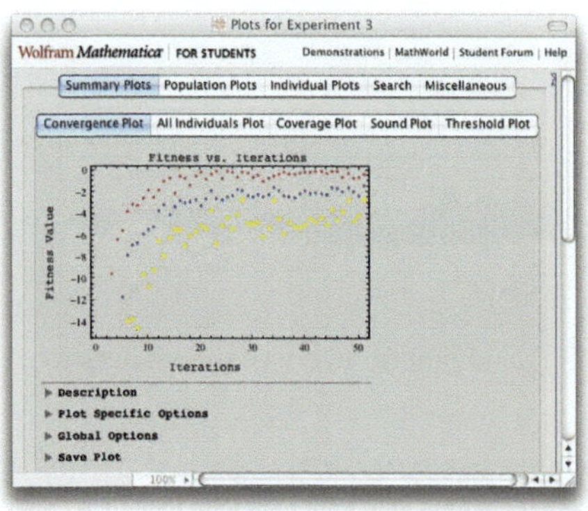

(c) High velocity.

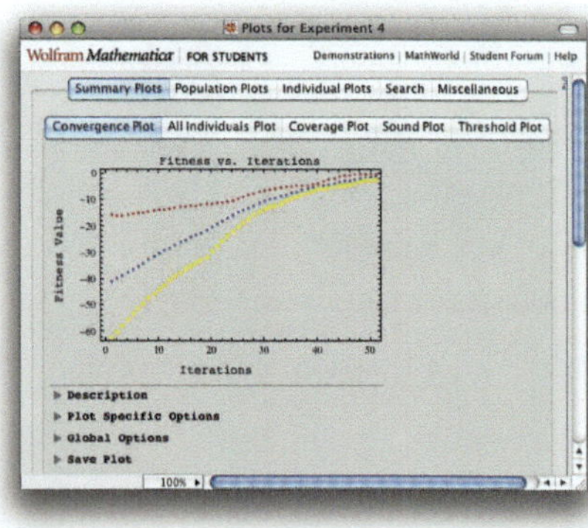

(d) Balance of exploration and exploitation.

Fig. 8 Many experiments can be visualized at the same time in our *VISPLORE* toolkit. The parameter settings used for the sphere function are: (a) $\mu = 10$, $v_{ij} \in [-0.001,0.001]$, $p_{ij} \in [-5.14,5.14]$, $\varphi_1 = 0.1$, $\varphi_2 = 1.0$, and $d = 5$. (b))$\mu = 10$, $v_{ij} \in [-0.1,0.1]$, $p_{ij} \in [-5.14,5.14]$, $\varphi_1 = 0.1$, $\varphi_2 = 0.01$, and $d = 5$. (c) $\mu = 10$, $v_{ij} \in [-1.0, 1.0]$, $p_{ij} \in [-5.14,5.14]$, $\varphi_1 = 0.1$, $\varphi_2 = 1.0$, and $d = 5$. (d) $\mu = 10$, $v_{ij} \in [-0.1,0.1]$, $p_{ij} \in [-5.14,5.14]$, $\varphi_1 = 0.1$, $\varphi_2 = 1.0$, and $d = 5$

population does not stabilize (Figure 8(c)). Figure 8(b) illustrates the convergence graph for really low exploration rates. Figure 8(d) demonstrates a balance between the exploration and exploitation rates.

The comparisons conducted point out various characteristics such as the convergence speed and the effects of the inherent parameters. The strengths and weaknesses of inherent parameters can also be exploited. Providing the user with the capability of visualizing many experiments at once provides an experimentation and inquiry platform for introducing particle swarms to those who wish to further their knowledge in the domain or require a streamlined platform to build prototypical strategies to solve their optimization tasks. Therefore, this interactive *VISPLORE* toolkit provides both a hands-on tutorial and experimental environment through simultaneous visualizations of multiple experiments.

4 Searching in *VISPLORE*

The basic search visualization technique is implemented as an interactive system that provides good visual feedback to help users locate particles with a certain fitness value. As illustrated in Figure 9, each particle is represented as a dot on a matrix. The search visualization generated is query-dependent. "Query-dependency" means that the particles matching the query exactly along with the particles that approximately satisfy the query are visualized [13]. Approximate results are determined by a relevance factor that is obtained by calculating distances between data and query values for each dimension. Examples of such distance functions can be numerical difference, lexicographical, character, or substring difference. In our toolkit, we compute the numerical difference with the fitness values of particles and the query values (minimum and maximum fitness values provided by the user). The particles are then sorted according to their fitness relevance with respect to the search query. Colours are then assigned depending on the relevance factor—the darker the hue values the further away a particle is from the search criteria. The particles in the middle of the window satisfy the query conditions (within the specified fitness range) and the less relevant particles are arranged in a square spiral shape towards the outside of the window.

Figure 9 illustrates this concept. By default, the minimum/maximum fitness search value is computed automatically by the system as the minimum/maximum fitness value of all the particles created by the search algorithm over all timesteps (Figure 9(a)). The user can then interactively change the search interval for the fitness values via two sliders (Figure 9(b)).

The dynamic query modification is a useful feature because it can provide immediate feedback about the effects on the change of data. From this change the user can get an impression about the amount of data retrieved, on the restrictiveness of the constraints, on the distribution of the distances and on special areas (e.g., hotspots, anomalies) the user may be interested in. For example, if the region in the center is growing larger, then we can draw a conclusion that more particles fulfill the search

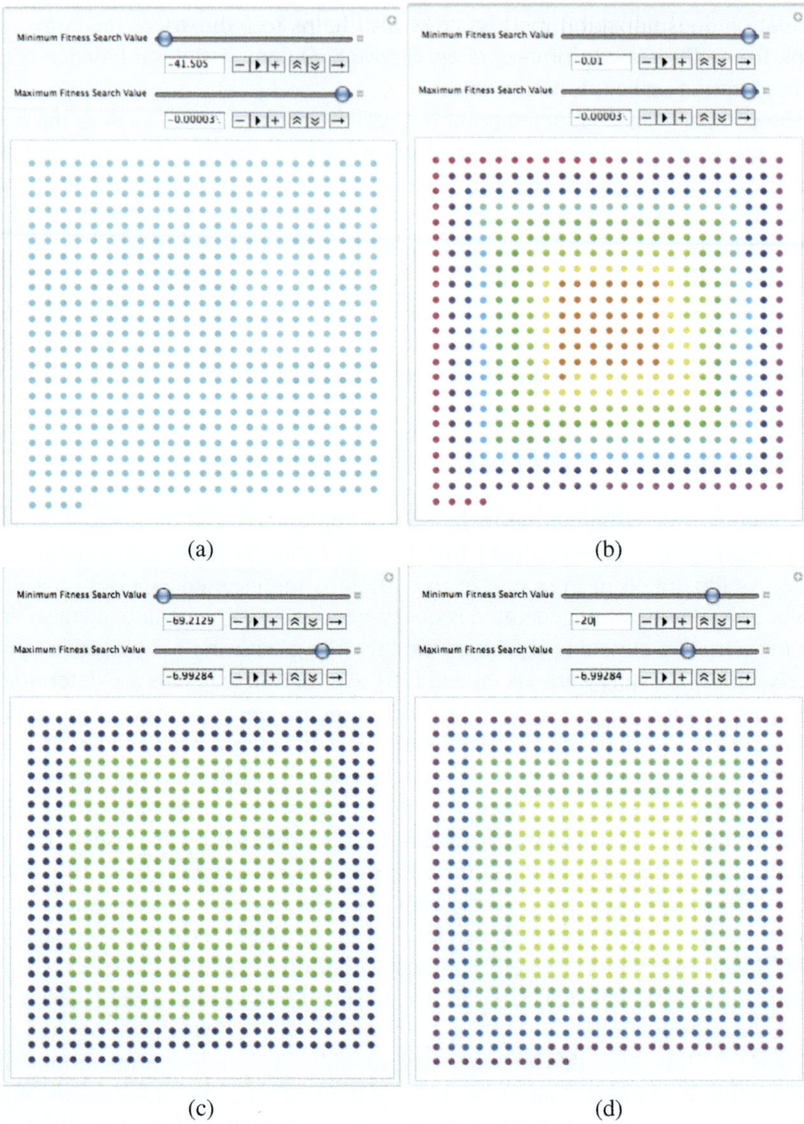

Fig. 9 Search mechanism: Moving the mouse over the dots displays the fitness value of the selected particle and clicking on a dot gives the particle's report. The parameter settings for the sphere function used here are: μ = 10, $v_{ij} \in$ [-0.1,0.1], $p_{ij} \in$ [-5.14,5.14], φ_1 = 0.1, φ_2 = 1.0, and d = 5. (a) By default all the particles created during the search are displayed. The minimum and maximum fitness values are computed automatically by the toolkit. (b) Each particle p_i in the middle of the window matches the search criteria exactly (i.e., $-0.00003 \leq f(p_i) \leq -0.01$). (c) $-69.2129 \leq f(p_i) \leq -6.99284$. (d) $-20 \leq f(p_i) \leq -6.99284$

interval. This visualization methodology also helps to focus on specific details of interest by keeping related data accessible.

5 Different Views in *VISPLORE*

There are five distinct tab views in our VISPLORE tool: Individual view, Population view, Search view, Summary view, and Miscellaneous view (Figure 10). The individual view (Figure 10(a)) provides mechanisms to visualize the particle biography. Density plots, parallel coordinates plots, star plots, and the history of a particle can be visualized. The population perspective (Figure 10(b)) provides population-level

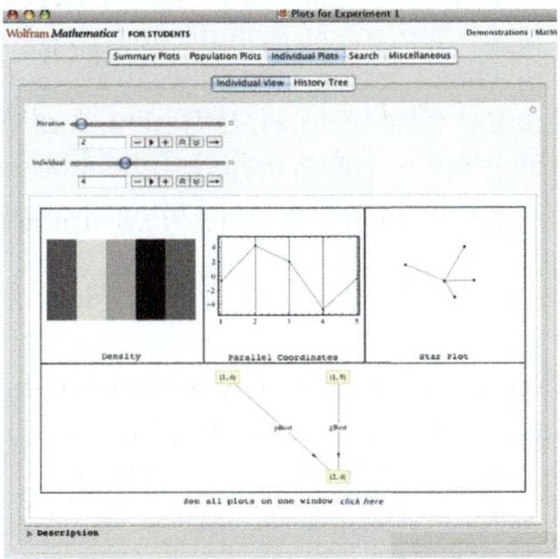

(a) Individual View

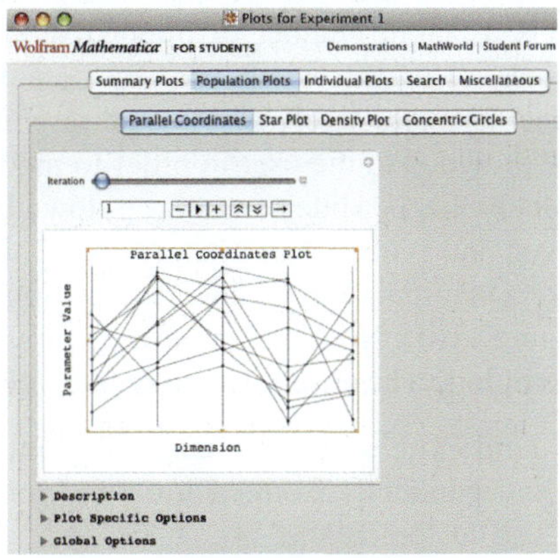

(b) Population View

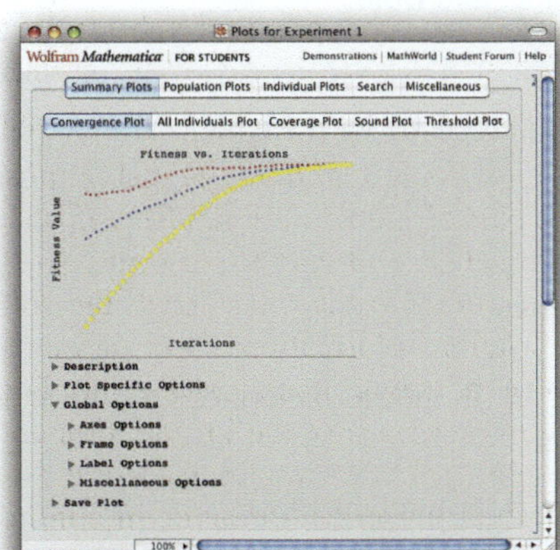

(c) Experiment View

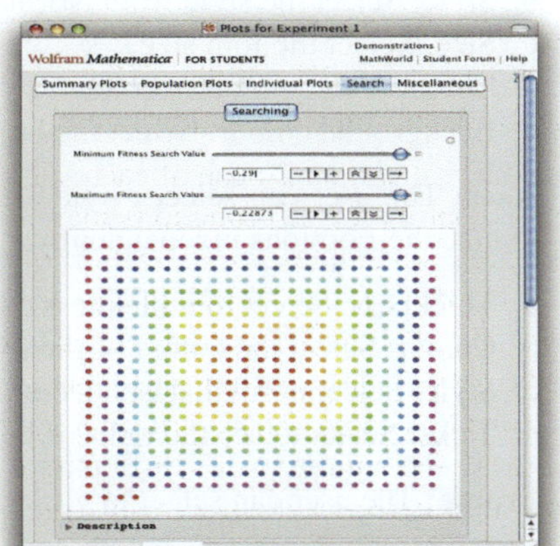

(d) Search View

Fig. 10 Different views of an experiment in *VISPLORE*

visualizations such as parallel coordinates, density plots, star plots, and concentric circles. The summary view (Figure 10(c)) provides summary level plots at the experimental level. Summary plots such as fitness curves, sound plots, range plots, and search space coverage plots can be conceptualized by the users. The search view (Figure 10(d)) enables users to perform a search depending on the fitness criteria. The Miscellaneous tab provides the user with inherent parameter settings (exploration rate, exploitation rate, velocities, etc.) for the conducted experiment and also offers a mechanism to save a particular experiment.

6 Customizing Plots in *VISPLORE*

We provide the user with a set of tools for creating their own picture compositions. A large number of interactive components are made generic and reusable across all the visualizations. By default each visualization comes with a control panel to interactively manipulate or configure the visualization. Different controls are provided to the user in order to modify the graphs. These include: checkboxes, colour wheels, sliders, and drop-down lists. These custom-designed images can be saved and retrieved at a later time. This way we provide mechanisms to perform direct manipulation on the visualizations. A sample options window is displayed in Figure 11(d). Note that the user is provided with 30 different options to customize their plots. These options are broadly classified into four sections (Figure 11(d)):

1. **Font:** The user can change the font for the "x-axis", "y-axis", and "plot title" independently from each other. From the "label options" opener view the user can customize the:
   a. Font: Select the font type from Courier, Times, and Helvetica.
   b. Style: Select the style from Normal, Italic, and Bold.
   c. Size: Select the font size.
   d. Colour: Select the font colour.
   e. Background colour: Select the background colour for the label.
2. **Axes:** The user can select to view either the "x-axis" or the "y-axis" and customize their options independently:
   a. Thickness: Select the thickness of the axes.
   b. Colour: Select the colour of the axes.
   c. Style: Select normal, dashed, dotted, or dot-dashed as the options for the axes.
   d. Ticks: Select to see the ticks of the frame.
3. **Frame:** The user can select to view the frame around the plots:
   a. Thickness: Select the thickness of the frame.
   b. Colour: Select the colour of the frame.
   c. Style: Select normal, dashed, dotted, or dot-dashed as the options for the frame.
   d. Ticks: Select to see the ticks of the frame.

4. **Miscellaneous:** The user can set the background colour for the plot or may select to view the gridlines.

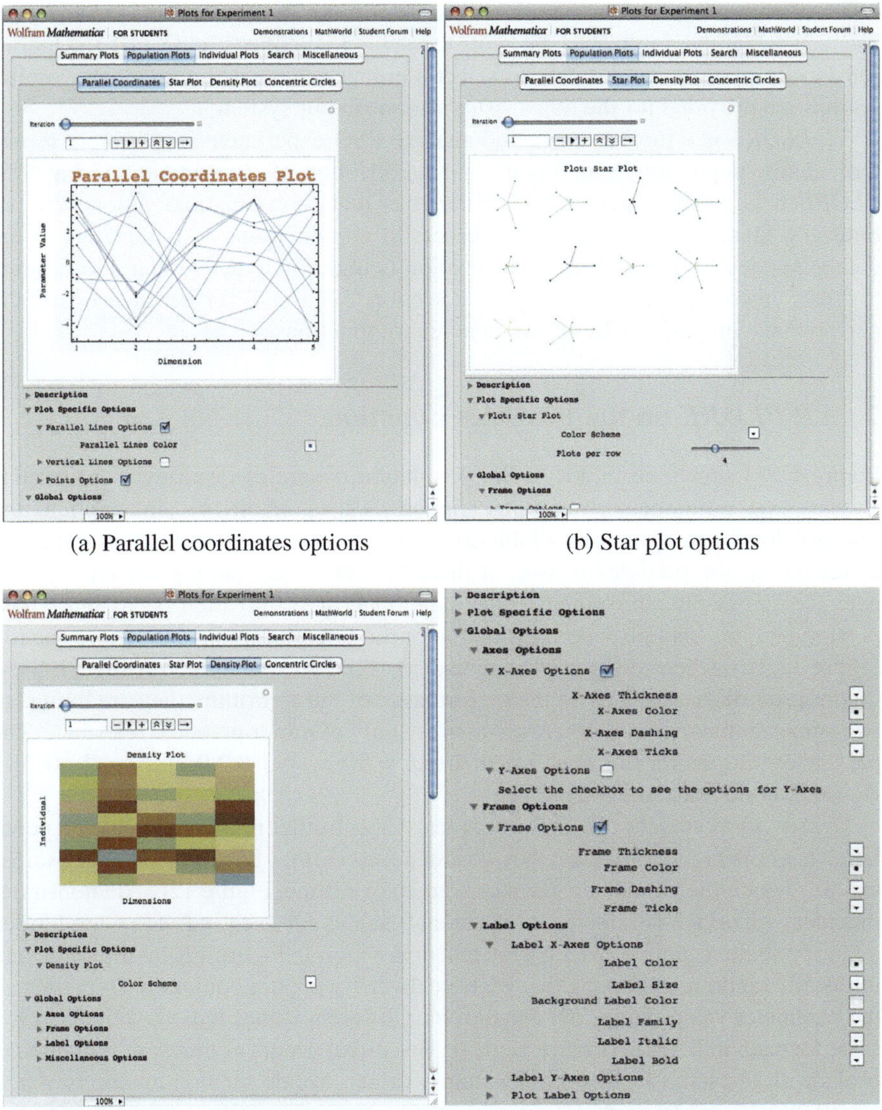

(a) Parallel coordinates options (b) Star plot options

(c) Density plot options (d) Global options

Fig. 11 Sample options available in *VISPLORE* for customizing its visualizations

We further supply the user with customization options valid for specific visualizations. We list a few of these options here. The user is provided with the option of viewing the vertical lines, polylines, and the dots. These options are selected via a checkbox as shown in Figure 11(a). The user can further select the colour for the different components of the parallel coordinates plot. Through the use of sliders the user can select the number of star plots to view on one line (Figure 11(b)). Options for density plot allows the user to select a colour scheme (Figure 11(c)). Similar options are provided for the other visualizations in our system.

VISPLORE is a toolkit to run and analyze PSO experiments at different levels: individuals, population experiment, and a collection of experiments. Using *VISPLORE* it is much easier to find visual cues that show relations among the parameters, clusters in the data, and successful combinations of parameters in the data generated by PSOs. Each visualization is highly interactive that allows users to create customizable graphs. The next section (Section 7) further investigates the performance of *VISPLORE* on two benchmark functions.

7 *VISPLORE* on the Foxholes Function

Figure 12(1) shows examples of the population dynamics resulting from particle swarms over a number of iterations applied to the Foxholes benchmark function for particle swarm optimizers (Table 2). The particles are represented as dots. The behaviour of the particles is seen at different iterations, making it easy to compare and contrast the movement of the individuals and study their behaviour. These phenotype plots also provide information on finding multiple solutions, as seen in Figure 12(1c), where the particles have discovered all four peaks. Although these graphs provide visual clues on the performance of the algorithm, they are limited to two-dimensional search problems, which are only of minor practical relevance. The parameter settings for this experiment are: $\mu = 50$, $v_{ij} \in [-0.2, 0.2]$, $p_{ij} \in [-50, 50]$, $\varphi_1 = 0.2$, $\varphi_2 = 2.0$, and $d = 2$.

Density plots such as the ones in Figures 12(2) further illustrate whether a population has converged. However, they go a step further; they can even tell us the values towards which the parameters have converged over time. Figure 12(2) demonstrates this idea, where the parameter (dimensional) values for each individual are plotted as a gray-scale rectangle. We observe that at iteration 1 (Figure 12(2a) the values are uniformly random, indicating that we have diversity in our population. Over time all the parameter values have converged to four different values indicated by the gray-scale for each individual (Figure 12(2c)). This effect occurs as there are four distinct peaks with the same or very similar fitness values. We do see combinations of values (0,0), (0,1), (1,0), and (1,1) represented by their respective gray-scale patterns. This is in sync with the phenotype plots for this particular experiment, where the individuals have converged to four different peaks.

Parallel coordinates can be very useful for visualizing the individuals of particle swarm algorithms. A structured overview of the individuals can be displayed, thus allowing us to recognize patterns, identify trends, and establish relationships among

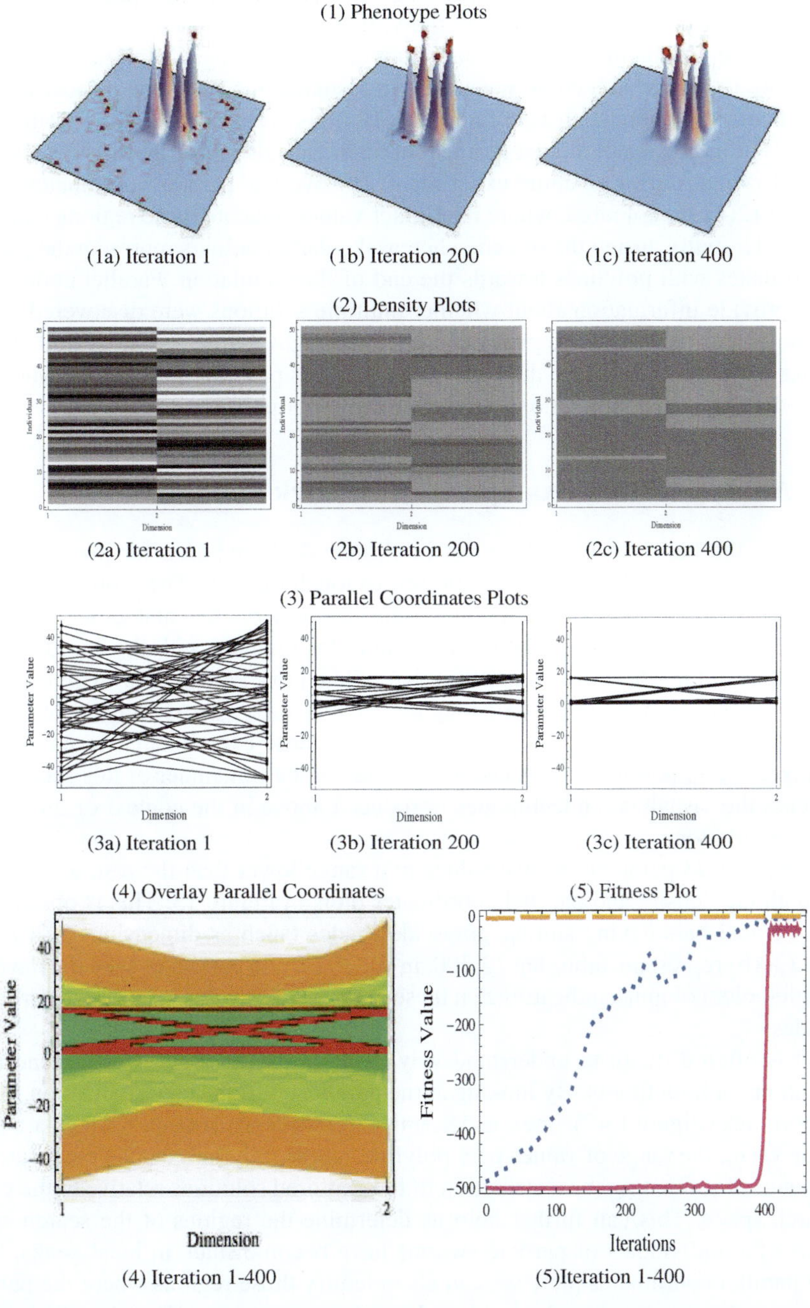

Fig. 12 Snapshots from a PSO evolution over the two-dimensional foxholes function. (1) Phenotype plots at iteration 1, 200, and 400. (2) Density plots at iteration 1, 200, and 400. (3) Parallel coordinates plots at iteration 1, 200, and 400. (4) Overlay Parallel Coordinates. (5) Fitness plot (solid line-worst, dotted line-average, long dashed line-best)

the dimensions of various individuals of a population (or experiment(s)). This is an important visualization aspect, as individuals represent potential solutions to the optimization problem and are the most important elements of a particle swarm algorithm. Using parallel coordinates we can visualize the structure of individuals at a particular timestep (such as in Figure 12(3)), thus providing insights into the algorithm's progress at a particular moment in time. One also gains an overall picture of all individuals during an entire experiment. The overlaid parallel coordinates plot in Figure 12(4) are coloured, where the lighter values indicate those regions that were covered initially during the search, whereas the darker colours represent the parallel coordinates with polylines towards the end of the simulation. Parallel coordinates also provide information about whether multiple solutions were discovered. In the foxholes example, we have peaks of the same height. Particle swarms find all four peaks, nicely illustrated by the parallel coordinates plot in Figure 12(4), where the four-dimensional combinations are depicted by four polyline groups.

8 An Application Example: Soccer Kick Simulation

Kicking the ball is one of the most important skills required for playing soccer since the arms and hands are not allowed to touch the ball. The motion of the leg kicking the ball involves 17 different muscle groups in the foot and toes, talus, thigh, shank, etc. In a kinematic model, the 17 muscles together with the coordinates of the ball result in a 56-dimensional search problem [15]. Particle swarm optimization was used to optimize the modeled leg movement, such that when the foot touches the ball, high ball velocity is obtained. During these experiments we particularly realized the importance of visualization tools. In the remainder of this section, we discuss the visualization techniques introduced above in the context of this soccer kick simulation.

The first 23 parameters have values in a range lower than the rest, as is visible in both the density and parallel coordinates plots in Figure 13. This is observed in the vertical bars having similar grayscale shades (such as dimensions 1-20). Figure 13(1b) represents time step 12,300, in which certain parameters are displayed as single-color columns, indicating that these parameters have been locked at particular values.

It is often difficult to understand why an algorithm is successful in generating solutions of high fitness. By looking at the parallel coordinates visualization for the soccer kick (Figure 13(2), one can determine the successful ranges for the parameter values, i.e., the range of values on a polyline. These polylines can be considered as a means of estimating the distribution of the explored solutions relative to the entire search space. This can further help us determine the regions of the search space where the individuals of particle swarms have become stuck in local peaks. With the parallel coordinates plots we can also identify those regions where the particle swarm optimizer has found a local peak on the search space. The identified ranges of values for each dimension can further narrow down our search space by adding constraints to the fitness function. Parallel coordinates can be relatively simple and

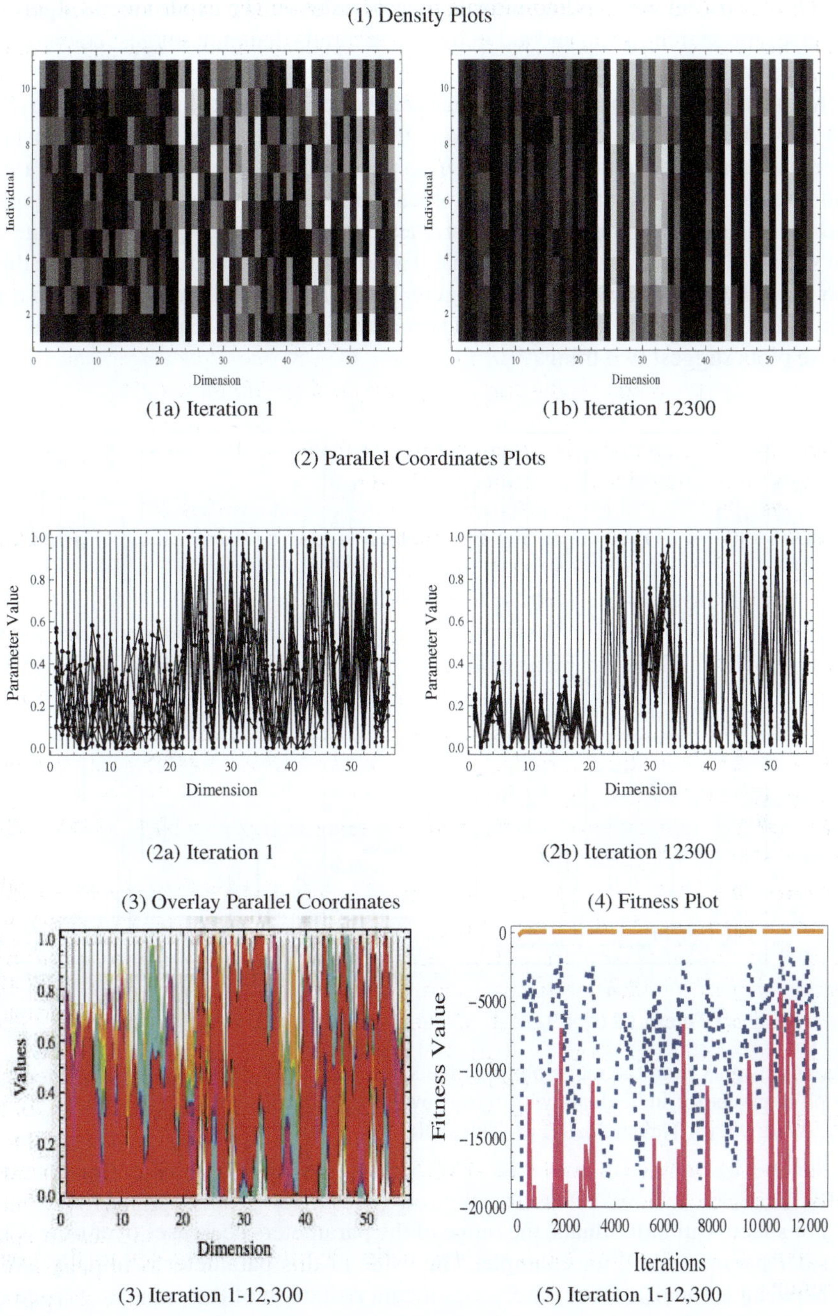

Fig. 13 Plots for the 56-dimensional soccer kick simulation. (1) Density plots at iteration 1 and 12,300. (2) Parallel coordinates plots at iteration 1 and 12,300. (3) Parallel coordinates for all generations. (4) Fitness plot (solid line-worst, dotted line-average, long dashed line-best)

can be very useful in providing information about how the particle swarm algorithm traverses the search space, and aid in detecting trends that may suggest convergence or divergence.

Another variant of the range plots was created for the soccer kick simulation (Figure 14), where the vertical bars represent the range for each of the 56-dimensions, over all iterations, limited to all those solutions that have a fitness above a certain threshold (τ). The range plots graphically reveal the subset of the search space and the successful combination of parameters that yields a high fitness value for the soccer kick simulation. For example, Figure 14(a) represents all the individuals that have their fitness over 80, and Figure 14(d) includes all individuals with a fitness value over 99 (the maximum ball speed obtained in our experiments is 99.39). These plots suggest two things:

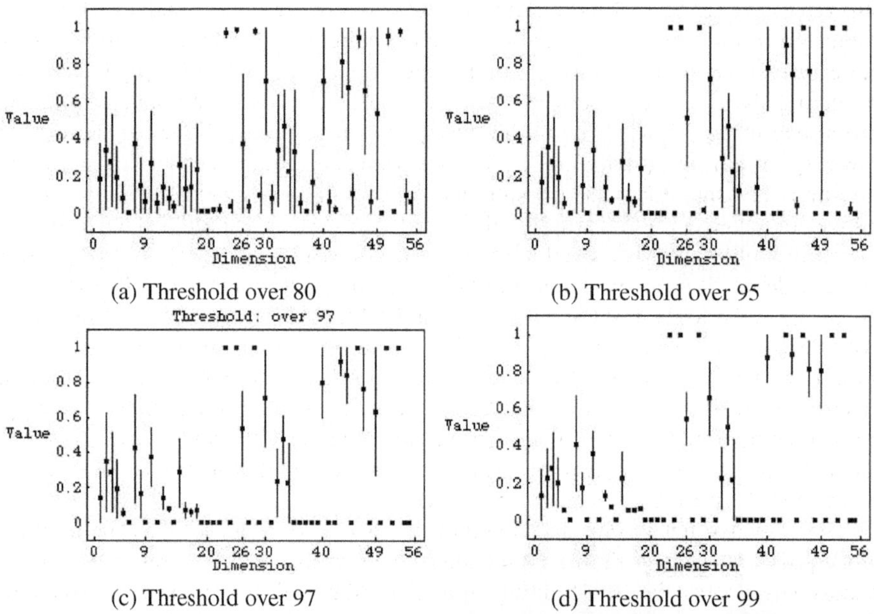

(a) Threshold over 80

(b) Threshold over 95

(c) Threshold over 97

(d) Threshold over 99

Fig. 14 Insights from the range plots for the soccer kick simulation The plots show the location intervals for individuals with fitness over a certain threshold. (a) Threshold: 80. (b) Threshold: 95. (c) Threshold: 97. (d) Threshold: 99

1. For successful individuals, the range of the parameters decreases over time. Consider parameter 49, for example. The value of this parameter is initially varied between 0.1 and 1. Over time, a significant reduction of the parameter space occurs and the interval length shrinks to at least a third (0.6 and 1) for individuals having a fitness over 99. A similar pattern is observed for parameter 26.
2. We also observe that some of the parameter values get completely locked in. As the fitness increases, more of these parameter values have a certain maintained

value. For example, parameter 9 was in the range of 0 and 0.123214. For individuals with a fitness value of over 95, the value of this parameter is 0. Also the majority of the parameters for individuals with a higher fitness are locked in at either 0 or 1.

9 Conclusion

VISPLORE is a toolkit to run and analyze data produced from population-based optimizers, in particular the particle swarm optimizer at different levels: individuals, population, experiment, and a collection of experiments (Figure 15). Using *VISPLORE*, it is much easier to find visual cues that show relations among the parameters, clusters in the data, and successful combinations of parameters in the data generated by PSOs. Each visualization is highly interactive that allows users to create customizable graphs.

Data visualization techniques such as parallel coordinates, range plots, glyphs, density plots, family trees, etc. can easily facilitate exploration of population-based algorithms. These visualization methods provide a good overview of the data and aid in identifying patterns and recognizing trends in multi-dimensional data sets. Our extensive library of visualization tools for population-based algorithms is implemented in *Mathematica*. These visualization components are part of our experimental platform that provides interactive capabilities, consists of an experiment browser and animations, and has sophisticated two- and three-dimensional graphics. The built-in symbolic, interactive, and visualization capabilities of *Mathematica* made the development of our system a lot easier. Customized dynamic interfaces were created with one- and two-dimensional sliders, buttons, menus, and checkboxes. *Mathematica* Version 6 also supported our development of two- and three-dimensional real-time graphics for different functions and data. These new graphical capabilities enabled us to further explore the dynamics of various algorithms. Our system is an example of integrating various *Mathematica* capabilities, such as symbolic computations, interactive components, convenient built-in functions such as *Dynamic[]*, *Manipuate[]*, and easy to navigate user interfaces. The *Mathematica* notebooks also provide us with insights regarding the relations between control parameters and performance of the optimizers. Consequently, we gained a better understanding of the algorithms on multi-dimensional real-world problems

VISPLORE is implemented as a *Mathematica* palette to provide a large experimental and inquiry platform for introducing evolutionary and swarm-based optimization techniques to those who wish to further their knowledge in the evolutionary computation domain. The best way to understand and utilize evolutionary- and swarm-based algorithm heuristics is through practical experience, which can be gained most efficiently on smaller scale problems. The database of *VISPLORE* notebooks, along with the evolution- and swarm-based algorithms, provide an experimental and inquiry platform for introducing evolutionary and swarm-based optimization techniques to those who also require a streamlined platform to build prototypical strategies to solve their optimization tasks.

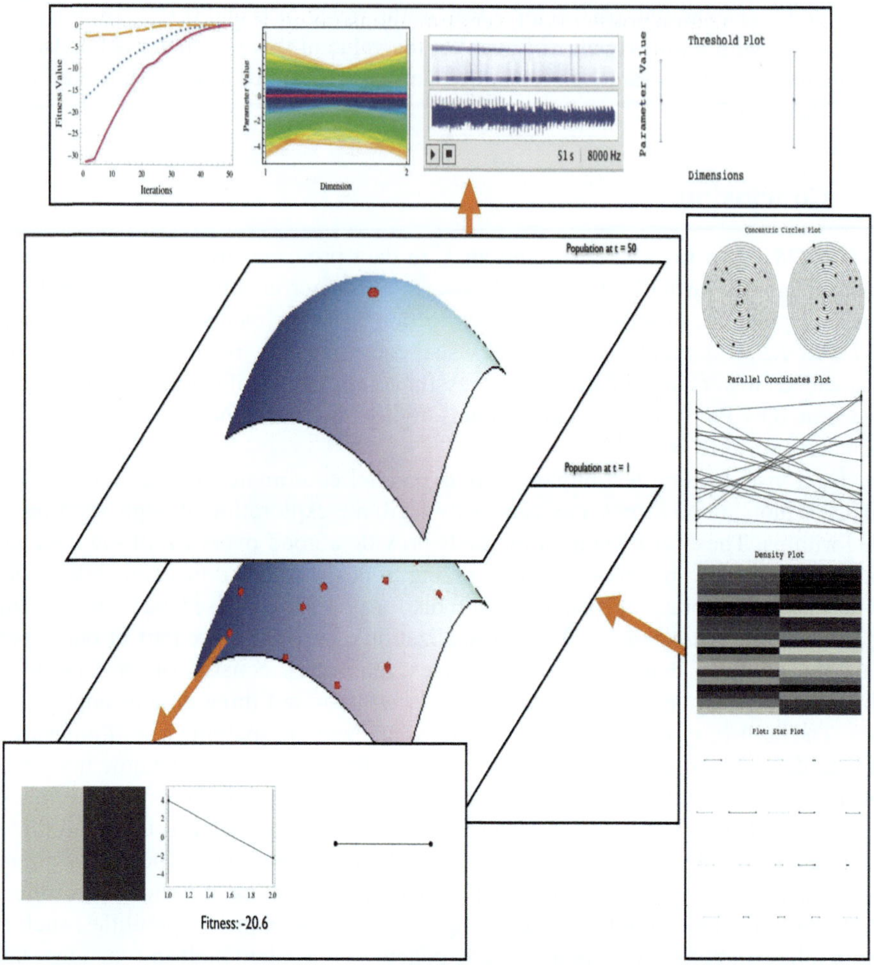

Fig. 15 Levels of data visualization in *VISPLORE*: Each particle can be visualized at either the genotype or the phenotype level. A group of particles is visualized at a population level in the form of concentric plots, parallel coordinates plots, density plots, and star plots. Changes in the population over time are displayed at an experimental level via fitness curves, search space coverage plots, sound plots, and threshold plots. Finally, we can also view a set of experiments. Therefore, *VISPLORE* makes provisions for interactively visualizing a particle, a population of particles, an experiment, and a collection of experiments

Data visualization techniques such as parallel coordinates, range plots, and density plots can easily facilitate exploration of multi-dimensional data such as the ones produced by the soccer kick simulation or any data in general that changes over time. These visualization methods provide a good overview of the data and aid in identifying patterns and recognizing trends. Note that any multi-dimensional data can be visualized with minor adjustments to the *VISPLORE* package.

Most optimization problems are very time-consuming as it can take several weeks (as was the case in the soccer kick simulation) to simulate the results of one experiment. It is not the processing of evolutionary operations that usually takes up a lot of time. The majority of the optimization process is spent evaluating the fitness (objective) function. Hence, we investigate the building of a component to conduct parallel execution of the fitness function evaluation by using the Parallel Computing Toolkit (PCT) provided by *Mathematica*. The PCT is an add-on package that allows for parallel computation on several independent *Mathematica* kernels running on a cluster of computers. Furthermore, making our notebooks available through a *web-Mathematica* site means that anyone with access to the newly built web pages will have instant access to a wide range of optimization and visualization algorithms. This interactive site will provide a hands-on tutorial and an experimental environment through a selection of benchmark functions along with visualization tools. The control parameters for the population-based algorithms are tuned manually, which basically leads to a trial and error approach. This is not only time consuming but does not guarantee optimal results at all. Consequently, automating the procedure of optimizing these control parameters is another objective of ours, which will lead to the development of a hierarchical meta-optimizer. Parallel coordinates are very powerful for visualizing multi-dimensional solutions. However, with higher dimensions they can become cluttered and difficult to read. The primary issue with this visual representation is to address the problem of overlapping polylines. This occurs even with a small number of solutions. In some situations it is hard to read the visualization to determine the values of the dimensions for a specific solution. Therefore, one would have to investigate methods for overcoming these issues as illustrated in [6] or investigate in new visualization techniques for higher dimensional solutions such as in [2].

References

1. Amor, H.B., Rettinger, A.: Intelligent exploration for genetic algorithms: using self-organizing maps in evolutionary computation. In: GECCO '05: Proceedings of the 2005 conference on Genetic and evolutionary computation, pp. 1531–1538. ACM, New York (2005), `http://doi.acm.org/10.1145/1068009.1068250`
2. Carpendale, S., Agarawala, A.: Phyllotrees: Harnessing nature's phyllotactic patterns for tree layout. In: INFOVIS 2004: Proceedings of the IEEE Symposium on Information Visualization, p. 215.3. IEEE Computer Society, Washington (2004), `http://dx.doi.org/10.1109/INFOVIS.2004.53`
3. Chipperfield, A., Fleming, P., Pohlheim, H., Fonseca, C.: Genetic algorithm toolbox for use with matlab, `http://citeseer.ist.psu.edu/502345.html`
4. Collins, T.D.: Understanding evolutionary computing: A hands on approach. In: 1998 IEEE Int. Conf. on Evolutionary Computation (ICEC), Piscataway, NY (1998), `http://citeseer.ist.psu.edu/collins97understanding.html`
5. Daida, J.M., Hilss, A.M., Ward, D.J., Long, S.L.: Visualizing tree structures in genetic programming. In: Cantú-Paz, E., Foster, J.A., Deb, K., Davis, L., Roy, R., O'Reilly, U.-M., Beyer, H.-G., Kendall, G., Wilson, S.W., Harman, M., Wegener, J., Dasgupta, D., Potter, M.A., Schultz, A., Dowsland, K.A., Jonoska, N., Miller, J., Standish, R.K. (eds.) GECCO 2003. LNCS, vol. 2723, Springer, Heidelberg (2003)

6. Fanea, E., Carpendale, S., Isenberg, T.: An interactive 3d integration of parallel coordinates and star glyphs. In: INFOVIS 2005: Proceedings of the Proceedings of the 2005 IEEE Symposium on Information Visualization, p. 20. IEEE Computer Society, Washington (2005), http://dx.doi.org/10.1109/INFOVIS.2005.5
7. Führer, T., Jacob, C.: Evolvision - an evolvica visualization tool. In: Proceedings of the Genetic and Evolutionary Computation Conference (2001)
8. Hart, E., Ross, P.: Gavel - a new tool for genetic algorithm visualization. Evolutionary Computation 5(4), 335–348 (2001), doi:10.1109/4235.942528
9. Mathematica website, http://www.wolfram.com
10. Inselberg, A.: Multidimensional detective. In: INFOVIS 1997: Proceedings of the 1997 IEEE Symposium on Information Visualization (InfoVis 1997), p. 100. IEEE Computer Society, Washington (1997)
11. Jacob, C.: Illustrating Evolutionary Computation with Mathematica. Morgan Publishers, San Francisco (2001)
12. Jacob, C., Khemka, N.: Particle swarm optimization in mathematica an exploration kit for evolutionary optimization. In: Proceedings of the Sixth International Mathematica Symposium (2004)
13. Keim, D.A., Kriegel, H.: VisDB: Database exploration using multidimensional visualization. Computer Graphics and Applications (1994), http://citeseer.ist.psu.edu/keim94visdb.html
14. Kennedy, J., Eberhart, R.: Particle swarm optimization. In: Proceedings of the Sixth International Symposium on Micromachine and Human Science (1995)
15. Khemka, N.: Comparing particle swarm optimization and evolution strategies: Benchmarks and application. Master's thesis. University of Calgary (2005)
16. Khemka, N., Jacob, C.: A comparative study between particle swarms and evolution strategies on benchmarks and soccer kick simulation. IEEE Transactions on Evolutionary Computation (2005)
17. Khemka, N., Jacob, C.: Visualization strategies for evolutionary algorithms. In: Proceedings of the Ninth International Mathematica Symposium (2008)
18. Khemka, N., Jacob, C.: What hides in dimension x? a quest for visualizing particle swarms. In: ANTS Conference, pp. 191–202 (2008)
19. Moniz, R.D., Jacob, C.: Interactively evolving fractals using grid computing (submitted). In: EvoWorkshops (2009)
20. Pohlheim, H.: Visualization of evolutionary algorithms - set of standard techniques and multidimensional visualization. In: Banzhaf, W., Daida, J., Eiben, A.E., Garzon, M.H., Honavar, V., Jakiela, M., Smith, R.E. (eds.) Proceedings of the Genetic and Evolutionary Computation Conference, vol. 1, pp. 533–540. Morgan Kaufmann, Orlando (1999), http://citeseer.ist.psu.edu/pohlheim99visualization.html
21. Shi, Y., Eberhart, R.: Parameter selection in particle swarm optimization. In: Evolutionary Programming VII: Proceedings of the Seventh Annual Conference on Evolutionary Programming, New York, pp. 591–600 (1998)
22. Shi, Y., Eberhart, R.: Empirical study of particle swarm optimization. In: Congress on Evolutionary Computation (1999)
23. Wu, A.S., Jong, K.A.D., Burke, D.S., Grefenstette, J.J., Ramsey, C.L.: Visual analysis of evolutionary algorithms. In: Angeline, P.J., Michalewicz, Z., Schoenauer, M., Yao, X., Zalzala, A. (eds.) Proceedings of the Congress on Evolutionary Computation, vol. 2, pp. 1419–1425. IEEE Press, Washington D.C (1999), http://citeseer.ist.psu.edu/wu99visual.html

Index

Adaptive partitioning 117
AES 98
Agent 3
Animat 143
Ant colony optimization 161
APAA 165
Artificial life 249

Bidding agent 208
Bidding strategy 207

CCEA 118
CCEA-AVP 127
Coevolution 117
Competitive behavior 101
Constraint handling 50
Continuous ant algorithm 163
Convergence 173
Correlations 125

Decomposable functions 34
Divide and conquer 117
Dynamic Optimization 97
Dynamic Test Problems 104

EMAS 182
Emergent Behaviour 139
Equality constraints 55
Evolutionary Algorithms 3
Evolutionary robotics 241

Feedback 87
Foxholes function 276

General equilibrium 244
Genetic algorithms 204

HMAGA 38

iEMAS 190

LSLP 51

MAGA 19
MacroAEM 34
Memetic algorithms 53
Mission capability planning 81
Multi-agent systems 2
Multiobjective optimization 233

Options Production Layer 85

Parallel ACO 161
Particle Swarm optimization 245, 278
Partitioning 122
Predator, prey 143
Risk analysis 78
Risk Tolerance Layer 86

SAMC 167
SANUM, SANUX 102
Scheduling 79
Soccer kick simulation 256
Statistics based learning 101
Stochastic modeling 179

VISPLORE 259
Visualization 259

6 Conclusion

References

GPSR Compliance

The European Union's (EU) General Product Safety Regulation (GPSR) is a set of rules that requires consumer products to be safe and our obligations to ensure this.

If you have any concerns about our products, you can contact us on ProductSafety@springernature.com

In case Publisher is established outside the EU, the EU authorized representative is:

Springer Nature Customer Service Center GmbH
Europaplatz 3
69115 Heidelberg, Germany

Batch number: 10584688

Printed by Printforce, the Netherlands